ROSS & WILSON

Anatomy and Physiology in Health and Illness

Kathleen J. W. Wilson

OBE BSc PhD RGN SCM RNT

Formerly: Senior Lecturer, Department of Nursing Studies,
University of Edinburgh; Nursing Research Liaison Officer,
West Midlands Regional Health Authority/University of
Birmingham; Principal Tutor, Royal
Infirmary, Edinburgh.

ROSS & WILSON

Anatomy and Physiology in Health and Illness

Sixth Edition

Kathleen J. W. Wilson OBE

CHURCHILL LIVINGSTONE
EDINBURGH LONDON MELBOURNE AND NEW YORK 1987

CHURCHILL LIVINGSTONE
Medical Division of Longman Group UK Limited

Distributed in the United States of America by Churchill Livingstone
Inc., 1560 Broadway, New York, N.Y. 10036, and by associated
companies, branches and representatives throughout the world.

First edition 1963
Second edition 1966
Third edition 1968
Fourth edition 1973
Fifth edition 1981
Sixth edition 1987
  Reprinted 1988

Previous editions published under the title
Foundations of Anatomy and Physiology.

ISBN 0-443-03530-X

British Library Cataloguing in Publication Data
Ross, Janet S.
  Anatomy and physiology in health and illness — 6th ed.
  1. Human physiology
  I. Title II. Wilson, Kathleen J. W. III. Ross, Janet S. Foundations of
anatomy & physiology.
  612′.0024612   RT69

Library of Congress Cataloging in Publication Data
Ross, Janet S.
  Ross and Wilson anatomy and physiology in health and illness.
  Rev. ed. of: Foundations of anatomy and physiology.
5th ed. 1981.
  Includes index.
  1. Anatomy, Human. 2. Human physiology. 3. Pathology. I. Wilson,
Kathleen J. W. II. Ross, Janet S, Foundations of anatomy and
physiology. III. Title. IV. Title: Anatomy and physiology in health and
illness.
[DNLM: 1. Anatomy—nurses' instruction. 2. Physiology—nurses'
instruction. QS 4 R824f]
RT69.R66   1987   611   86-23268

Produced by Longman Group (FE) Ltd
Printed in Hong Kong

Preface to the sixth edition

This book is an extended version of the publication previously entitled *Foundations of Anatomy and Physiology*. The purpose of the present edition is to provide nurses and other health workers with a knowledge of the structure and function of a healthy human body and of the changes which take place when disease interferes with normal processes.

The new material consists mainly of descriptions of commonly occurring diseases: their causes, the abnormal physiology, the likely progress and probable outcome of untreated disease. Diagnostic procedures and the treatment of disease are not included.

Chapter 1 remains unchanged from the previous edition, its purpose being to provide a brief introduction to the interrelationships between all body systems. To avoid unnecessary repetition Chapters 2 and 3 include, respectively, general descriptions of *tumours* and *inflammation* which can be applied to organs and tissues throughout the body. The *healing process* is dealt with in Chapter 11 (The Skin). Specific skin conditions are not included.

Although the overall systemic arrangement of material has been retained, some changes have been made in its order and presentation to show clearly the links between the normal and the abnormal. In some chapters the description of diseases is inserted immediately after the anatomy and physiology of the affected organs, in others it seemed more appropriate to place together all diseases associated with a particular system. The opening page of each chapter shows which order of presentation has been adopted.

I am most grateful to readers of previous editions for their constructive comments and suggestions.

I would like to express my special thanks to Dr R. D. Eastham, Honorary Consultant Pathologist, Frenchay Health District, Bristol, for his helpful comments on the new material and parts of the physiology; Mr I. Ramsden for drawing some new illustrations and amending others; and the staff of Churchill Livingstone for their practical assistance in the preparation of this new extended edition.

Peebles, 1987   Kathleen J. W. Wilson

Contents

Introduction to the Body as a Whole

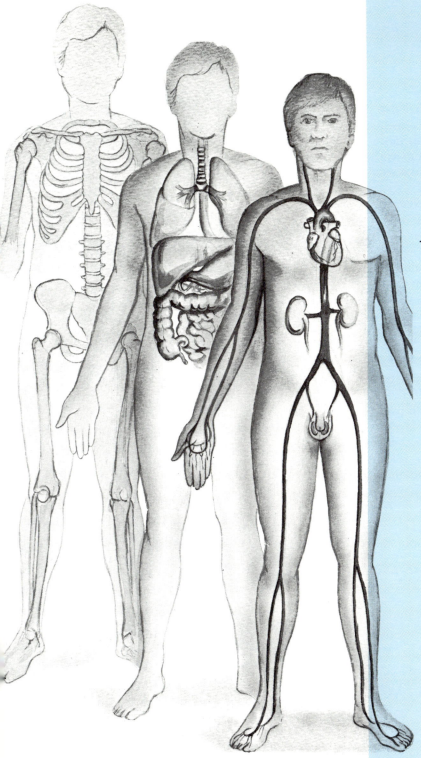

1. Introduction to the Body as a Whole

The human animal is a very complex multicellular organism in which the maintenance of life depends upon a vast number of physiological and biochemical activities. The sum of these activities enables the human being to live in and utilise his environment, and to maintain the species by reproducing.

There is considerable variation in the complexity of organisms. At one end of the continuum there is the *single cell organism* such as the amoeba, and at the other, the highly complex *multicellular human animal*. A cell is the smallest functional unit of an organism, thus a single cell organism is the simplest kind of organism that can exist independently.

In order to survive, each species, simple and complex, must be able to perform certain functions. A single cell organism can carry out all these functions because all its parts have easy access to its external environment. The inside of the single cell organism is separated from its environment by only one porous membrane, which is described as *semipermeable* because small particles can pass through it while large ones cannot (Fig. 1:1). It is not possible for all the cells of the multicellular human animal to be in close contact with the environment so, in order to survive, *specialisation of cells* has evolved. Functional specialisation has taken place in parallel with structural

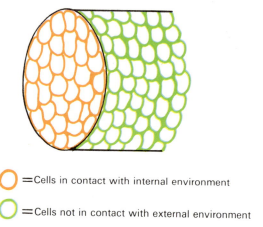

○ = Cells in contact with internal environment

○ = Cells not in contact with external environment

Figure 1:2 Diagram of a cylindrical multicellular structure.

specialisation. Figure 1:2 shows a multicellular structure.

The cells of the body are too small to be seen with the naked eye, but when magnified using a microscope, different types of cells can be distinguished by their size, shape and the types of dye they absorb when stained in the laboratory.

Groups of cells which have the same physical characteristics tend to have similar specialised functions, and a large number of cells that perform the same functions are described as *tissues*. *Organs* are made up of a number of different types of tissue, and *systems* consist of a number of organs and tissues. Each system contributes to one or more of the vital functions of the body. However, because of specialisation of cells, none of the systems can exist in isolation.

BASIC NEEDS OF THE BODY AND ASSOCIATED SYSTEMS

It will be noted from the list on page 3 that some of the systems contribute to a number of different basic needs. The contribution that each system makes to each need is described *briefly* in this chapter. *Internal transportation* is described first, because all the systems depend upon the circulation of blood. The description of the systems involved in meeting the other needs assumes that an effective circulatory system is operating.

A more detailed description of the structure, functions

• = Small particles able to pass through pores in the cell membrane

▲ = Large particles—outside—cannot pass into the cell

■ = Large particles—inside—cannot pass out of the cell

● = Large particles—inside and outside the cell which cannot pass through its membrane

Figure 1:1 Diagram of a single cell with a semipermeable membrane.

and the commonly occurring malfunctions of each system is provided in later chapters.

Needs	System(s)
1. Internal transportation	Circulatory (Chs 3,4 and 5) Lymphatic (Ch. 6)
2. Intake of raw materials A. Food B. Oxygen	Digestive (Ch. 9) Respiratory (Ch. 7)
3. Elimination of waste materials	Urinary (Ch. 10) Respiratory Digestive
4. Communication A. With the outside world	Nervous (Ch. 12) Special senses (Ch. 13) Respiratory — voice production Skeletal (Ch. 16) Muscular (Ch. 18)
B. Within the body	Nervous Endocrine (Ch. 14)
5. Protection against the external environment	Skin (Ch. 11) Membranes lining passages which open on to the surface of the body
6. Reproduction	Male and female reproductive (Ch. 15) Endocrine
7. Movement within the external environment	Skeletal Muscular Nervous Special senses

1. INTERNAL TRANSPORTATION

The need for functional specialisation among the cells of the body was mentioned earlier. Because of specialisation, a sophisticated transport system is required to ensure that all cells have access to the *external* and *internal environments* of the body.

1. The external environment surrounds the body and provides the oxygen and nutritional materials required by all the cells of the body. Waste products of cell activity are excreted into the external environment.

2. The internal environment provides chemical substances produced by specialised cells. All living cells in the body are bathed in fluid (interstitial fluid). Oxygen, nutritional materials, chemicals produced by the body, and waste products, pass through the interstitial fluid between the cells and the internal transportation systems.

The systems involved are the *circulatory* and *lymphatic*.

CIRCULATORY SYSTEM (Chs 3, 4 and 5. Figs 1:3 and 1:4)

This system consists of the *blood*, the *blood vessels* and the *heart*.

BLOOD

This consists of two parts — a sticky fluid called *plasma*, and *cells* which float in the plasma.

Plasma

This consists of water and chemical substances dissolved or suspended in it. These are:
1. Nutrient materials absorbed from the intestine
2. Oxygen absorbed from the lungs
3. Chemical substances synthesised by body cells
4. Waste materials produced by body cells to be eliminated from the body by excretion.

Blood cells (Fig. 1:3)
There are three distinct groups, classified according to their functions.

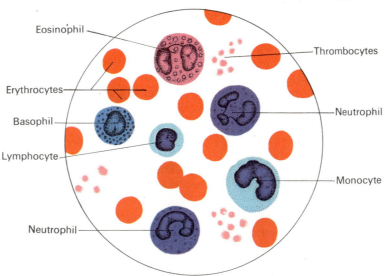

Figure 1:3   Blood cells after staining in the laboratory viewed through a microscope.

1. *Erythrocytes* (red blood cells) are concerned with the transport of oxygen and carbon dioxide between the lungs and all body cells. They contain *haemoglobin* which combines with oxygen and carries it from the lungs to the cells. After giving up oxygen it combines with carbon dioxide, carrying it from the cells to the lungs for excretion. Both the amount of oxygen needed and of carbon dioxide to be removed increase as cell activity increases, e.g., during hard physical exercise the blood supply to the muscles involved increases.

There are about 5×10^{12} erythrocytes in each litre of blood and the adult body contains between 5 and 6 litres of blood.

2. *Leukocytes* (white blood cells) are mainly concerned with the protection of the body against microbes and other potentially damaging substances that gain entry to the body. They are also involved in the removal of cells at the end of their normal life-span and those damaged by disease and injury. There are several different types of leukocytes which carry out their protective functions in different ways. These cells are larger than erythrocytes and are less numerous, the body containing about 5×10^9 to 9×10^9 per litre of blood.

3. *Thrombocytes* (platelets) are tiny cell fragments which play an essential part in the very complex process of blood clotting. A blood clot is a 'plug' consisting of blood cells and fibrous material which forms in the cut or torn ends of a blood vessel. It prevents excessive loss of blood.

There are 200×10^9 to 350×10^9 thrombocytes per litre of blood.

BLOOD VESSELS

There are three types:
1. *Arteries*, which convey blood away from the heart
2. *Veins*, which return blood to the heart
3. *Capillaries*, which link the arteries and veins. These are tiny blood vessels with very thin walls consisting of only one layer of cells. Between these cells there are very small openings or pores, which allow some of the constituents of blood to pass through, such as oxygen, nutritional materials, some chemical substances synthesised in the body and waste products from cells. Larger sized substances, such as erythrocytes and large molecule proteins, cannot pass through the *semipermeable* capillary walls (Fig. 1:1)

HEART

The heart is a muscular sac. It pumps the blood into the blood vessels which transport it:
1. To the lungs (*pulmonary circulation*) where oxygen is absorbed from the air in the lungs and at the same time carbon dioxide is excreted from the blood into the air
2. To the cells in all parts of the body (*general circulation*)

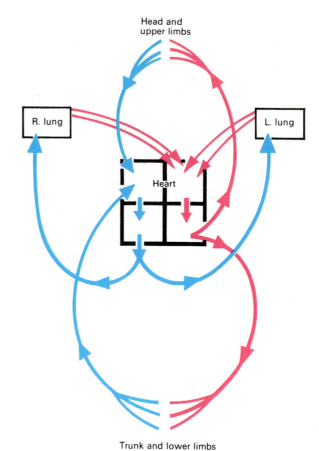

Figure 1:4   Diagram representing the circulatory system.

The muscle in the heart wall is not under the control of the will. At rest, the heart contracts between 65 and 75 times per minute. The rate may be greatly increased during physical exercise when the oxygen and nutritional needs of the muscles moving the limbs are increased and in some emotional states.

The rate at which the heart beats can be counted by taking the *pulse*. This is a wave of expansion of artery walls which occurs when the heart contracts and pushes blood into the *aorta*, the first artery of the general circulation. The pulse can be felt most easily where an artery lies close to the surface of the body and can be pressed gently against a bone. The wrist and the temple are the sites most commonly used for this purpose.

LYMPHATIC SYSTEM (Ch. 6. Fig. 1:5)

This system is a subsidiary of the circulatory system. It consists of a number of lymph vessels which begin as blind end tubes in the area containing tissue fluid between the blood capillaries and the cells (Fig. 1:5). Structurally they are similar to veins and blood capillaries but the pores in the walls of the lymph capillaries are larger than those of the blood capillaries. This means that water containing large molecule substances, fragments of damaged cells and foreign matter such as microbes drains away by passing

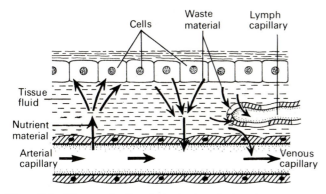

Figure 1:5 Diagram showing the beginning of a lymph capillary in the interstitial space.

into the lymph vessels. The two largest lymph vessels empty lymph into large veins near the heart.

There are collections of tissue called *lymph nodes* situated at various points along the length of the lymph vessels. Lymph from the interstitial spaces is filtered as it passes through the lymph nodes, and microbes, noxious substances and some waste materials are removed.

2. INTAKE OF RAW MATERIALS

A. FOOD (Chs 8 and 9)

Food is one of the sources of the raw materials that cells must obtain from the external environment, but it is not always in a form that cells can use. A specialised system has developed to modify or *digest* food to make it usable.

DIGESTIVE SYSTEM (Ch. 9. Figs 1:6 and 1:7)

This system consists of:
1. *The alimentary canal* which begins at the mouth and continues through the pharynx, oesophagus, stomach, small and large intestines, rectum and anus
2. *Glands*★ which produce *digestive enzymes*† and discharge them into the canal through ducts, or little tubes. Many glands are in the walls of the alimentary canal but those situated outside the canal with ducts leading into it are the *salivary glands*, the *pancreas* and the *liver*

Food, which is chemically complex, is taken in at the mouth and broken down by *physical* and *chemical* means into forms in which it can pass through the walls of the alimentary canal into the blood to be transported around the body (Fig. 1:7). The substances absorbed are:

★ *Glands* are aggregates of secretory cells enclosed within a sheet, or membrane, of connective tissue.
† *Enzymes* are chemical compounds which cause, or speed up, chemical changes in substances with which they are in contact. They are organic catalysts.

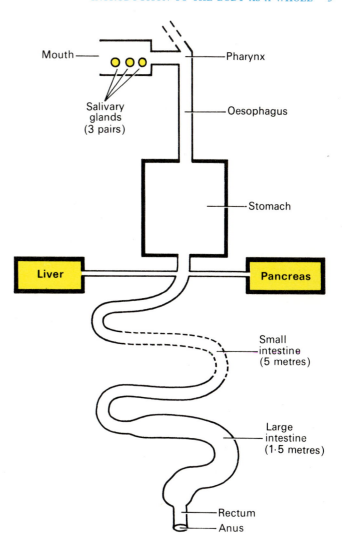

Figure 1:6 Diagram representing the digestive system.

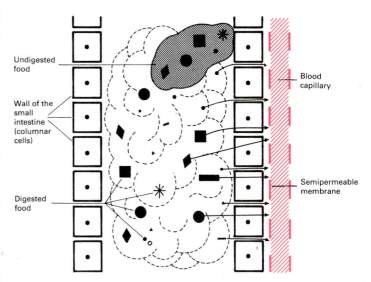

Figure 1:7 Diagram representing the digestion and absorption of food.

Water
Carbohydrates in the form of *monosaccharides*
Fats in the form of *fatty acids* and *glycerol*
Proteins in the form of *amino acids*
Minerals
Vitamins

These substances provide the materials required by the body for:

Energy production	—carbohydrates and fats
Cell building and repair	—proteins
Chemical synthesis	—minerals, proteins, fats, carbohydrates, vitamins
Medium for all chemical activity	—water

B. OXYGEN (Ch. 7)

Oxygen is a gas which makes up about 21% of the atmospheric air. It is essential for human life because most of the chemical activities which take place in the cells can only occur in its presence. Oxygen is involved in the series of chemical changes that result in the release of the energy from nutrient materials. This energy is essential for all cellular activities.

Metabolism

This is the sum total of the chemical activity in the body. It consists of two groups of processes:

1. *Anabolism*, building or synthesising new products
2. *Catabolism*, breaking down substances, either for excretion as waste or to provide raw materials for anabolism

The release of energy from nutritional materials, such as carbohydrates and fats, is a catabolic process.

RESPIRATORY SYSTEM (Ch. 7. Fig. 1:8)

This is the system through which oxygen is taken into the body from the external environment and carbon dioxide, a waste product of cell metabolism, is excreted. It consists of a series of tubes or *respiratory passages* which carry air from the nose to the *lungs* and a network of *blood capillaries* in the lungs. The respiratory passages, which are continuous with each other, are the *nose*, the *pharynx* (also part of the alimentary canal), the *larynx* (voice box), the *trachea*, two *bronchi* (one bronchus to each lung) and a large number of *bronchial tubes* which subdivide and lead to millions of tiny air sacs called *alveoli*. Air may also enter the pharynx through the mouth.

The *lungs* are two in number and are situated one on each side of the heart in the thoracic cavity. They consist of bronchial tubes, alveoli, blood and lymph vessels and nerves, all of which are supported by connective tissue. Each alveolus is surrounded by a dense network of blood capillaries.

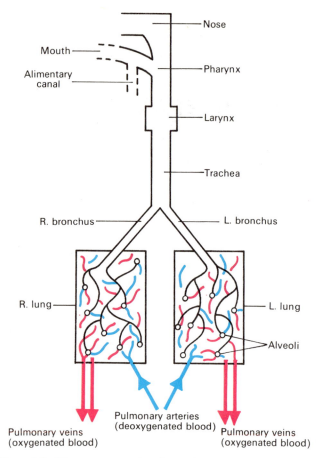

Figure 1:8 Diagram representing the respiratory system.

Respiration

Air containing oxygen and carbon dioxide is breathed into the lungs, filling the alveoli, and it is separated from the blood in the capillaries by two semipermeable membranes. These are the walls of the alveoli and the capillary walls, each of which is only one cell thick. Oxygen is in higher concentration in the alveoli so it passes from the alveoli to the blood; carbon dioxide is in higher concentration in blood so it passes in the opposite direction, from the blood to the alveoli (Fig. 1:9). Oxygen is carried in the blood in *solution in the blood water* and in *chemical combination with haemoglobin in the red blood cells*. The cells throughout the

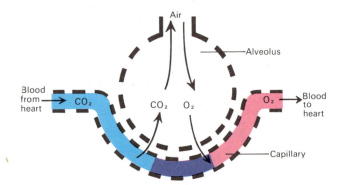

Figure 1:9 Diagram showing the interchange of gases between air in an alveolus and the blood in a capillary.

body obtain oxygen and get rid of carbon dioxide by the reverse process. Oxygen is in higher concentration in the blood than in the cells and the interstitial fluid around the cells, so it diffuses down the concentration gradient from the capillaries to the cells. Carbon dioxide, which is a waste product of cell metabolism, is in higher concentration in the cells and interstitial fluid and diffuses down the concentration gradient from the cells to the blood.

Breathing is the regular inflation and deflation of the lungs which maintains a steady concentration of atmospheric gases in the alveoli, i.e., the constant intake of oxygen and output of carbon dioxide.

Nitrogen, which is present in atmospheric air, is breathed in and out with oxygen and carbon dioxide but, in this gaseous form, it cannot be used by the body. The nitrogen needed by the body is present in protein foods, mainly meat and fish.

3. ELIMINATION OF WASTE MATERIALS

Substances in the body are regarded as waste materials if they cannot be used by the cells and if their accumulation will upset the fine balances which must be maintained between chemical substances in the internal environment. One of the most important of these is the acidity/alkalinity balance for which the *scale of measurement* is pH. The scale extends from 0 to 14 with 7 the *neutral point*. From 7 to 0 represents *increasing acidity* and from 7 to 14, *increasing alkalinity* (Fig. 1:10).

Varying quantities of acids and alkalis are produced by metabolism. It is an important function of the systems involved in elimination to control excretion of these substances in order to maintain the optimum blood pH, i.e., about 7.4. In addition, other waste materials which do not affect the pH are also excreted. The *respiratory*, *urinary* and *digestive systems* are those involved in elimination.

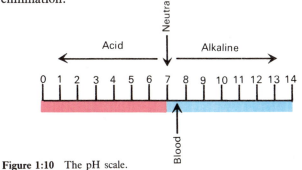

Figure 1:10 The pH scale.

RESPIRATORY SYSTEM

This system excretes carbon dioxide, as described above. Carbon dioxide dissolved in water forms an acid which must be excreted in appropriate amounts to maintain the optimum pH of the blood at about 7.4.

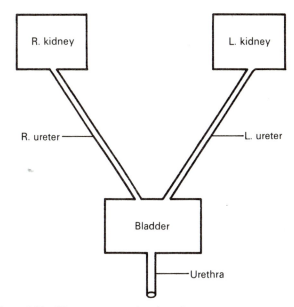

Figure 1:11 Diagram representing the urinary system.

URINARY SYSTEM (Ch. 10. Fig. 1:11)

This system is involved in:
1. Maintaining the appropriate balance between water and substances dissolved in it, and between acids and alkalis
2. The elimination of the waste products resulting from the catabolism of cell protein (urea and uric acid)

The system consists of:
2 kidneys, situated one on each side of the vertebral column on the posterior wall of the abdominal cavity
2 ureters, one of which extends from each kidney to the bladder
1 urinary bladder, situated in the pelvis
1 urethra, extending from the bladder to the exterior

DIGESTIVE SYSTEM

The large intestine excretes *faeces* which contain:
1. Food residue that remains in the alimentary canal because it cannot be digested and absorbed
2. Bile from the liver, which contains the waste products resulting from the catabolism of erythrocytes
3. Large numbers of microbes

4. COMMUNICATION

In order to live in and be able to adapt to the external environment, the individual must be able to communicate with it. Similarly, communications are necessary for the stimulation, regulation and co-ordination of activities within the body. In both cases communication involves a cycle of receiving, collating and giving information.

A. WITH THE OUTSIDE WORLD

NERVOUS SYSTEM (Ch. 12. Figs 1:12 and 1:13)

The brain receives communications from outside the body through the five senses (Fig. 1:12). These senses and the special organs involved are:

Sight   —eyes
Hearing —ears
Smell   —nose
Taste   —tongue
Touch   —skin

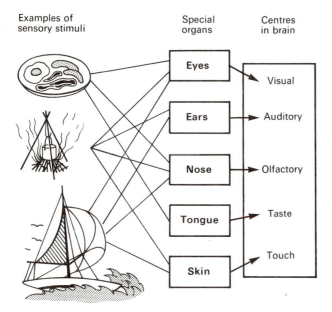

Figure 1:12   Diagram representing the special senses.

Although the senses are considered different and separate from each other, one sense is rarely used on its own. For example, when the smell of smoke is perceived, other senses such as sight and hearing are likely to be used to try to locate the source of the smoke. Similarly, taste and smell are closely associated in the enjoyment, or otherwise, of food.

Nerve endings are stimulated by phenomena outside the body and the resultant nerve impulses are transmitted to the brain by nerve fibres for 'interpretation' or *perception*. The brain collates this material with information obtained from the memory, and the result is co-ordinated and regulated communication with the outside world. The nervous system consists of:

1. The *brain*, situated inside the skull
2. The *spinal cord*, which extends from the base of the brain to the lumbar region and is protected from injury by the bones of the vertebral column
3. *Nerve fibres*, of two types: *sensory* or *afferent nerves*, which provide the brain with 'input' from organs and tissues, and *motor* or *efferent nerves*, which convey nerve impulses from the brain to organs and tissues

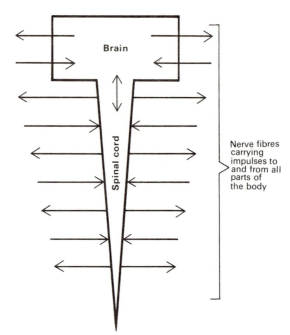

Figure 1:13   Diagram representing the nervous system.

VOICE PRODUCTION

This mechanism provides a means of communication with the outside world. Sound is produced in the larynx as a result of blowing air through the space between the *vocal cords* during expiration. What is known as speech is the *manipulation of sound* by the contraction of the muscles of the throat and the cheeks and the movements of the tongue and lower jaw.

SKELETAL AND MUSCULAR SYSTEMS (Chs 16, 17 and 18)

These systems combined produce the posture and movements associated with non-verbal communication with the outside world. The skeletal system provides the bony framework of the body, and movement takes place at joints between bones. The muscles which move the bones lie between them and the skin. They are stimulated by the part of the nervous system under the control of the will. Some non-verbal communication, e.g., changes in facial expression, may not involve the movement of bones.

B. WITHIN THE BODY

NERVOUS SYSTEM

The brain receives information from inside the body by means of:

1. Nerve impulses transmitted by sensory nerve fibres
2. Chemical substances circulating in the blood supplying the brain

The brain collates this information and responds by initiating efferent (motor) nerve impulses and secreting chemical substances. These pass from the brain to other parts of the body to stimulate, regulate and co-ordinate the activities of other organs and systems. Some of this activity is under the control of the will but much of it is involuntary, e.g., the individual cannot control the heart rate, emptying of the stomach or dilatation of the pupils of the eyes.

ENDOCRINE SYSTEM (Ch. 14)

This system consists of a number of *endocrine glands*, situated in different parts of the body. They communicate with each other, and with other organs and tissues of the body, by means of chemical substances, called *hormones*, which they synthesise. The glands are 'informed' about conditions within the body by variations in the concentration of chemical substances in the blood that supplies them and by nerve impulses. Changes in the amount of a hormone secreted result in stimulation or depression of activity in the organs and tissues they affect.

5. PROTECTION AGAINST THE EXTERNAL ENVIRONMENT

All the living cells in the body are in a watery environment and the substances which enter and leave the cells are either dissolved or suspended in water.

The *skin* and the *mucous membrane* lining the passages which open to the surface of the body provide a barrier between the 'dry' external environment of the body and the watery environment of the body cells.

SKIN (Ch. 11. Fig. 1:14)

The skin is described in two parts, the *epidermis* and the *dermis*.

The epidermis lies superficially and is composed of several layers of cells that grow towards the surface from its deepest layer. The surface layer consists of *dead cells* that are constantly being rubbed off by clothes, etc., and replaced from below. The epidermis constitutes the barrier between the moist environment of the living cells of the body and the dry atmosphere of the external environment.

The dermis is composed of collagen and elastic fibres. It contains tiny *sweat glands* that have little canals or ducts, leading to the surface. When sweat from these glands reaches the surface it evaporates, cooling the body and thus playing a major part in the maintenance of a temperature level inside the body consistent with the life of the cells.

Hair is not living material. Hairs grow from *follicles* in

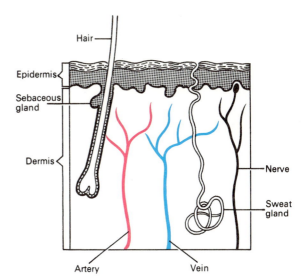

Figure 1:14 Diagram of the skin.

the dermis, consisting of invaginated epidermal cells, which have groups of specialised cells called *sebaceous glands*. These glands secrete *sebum* into the follicles and so on to the skin surface. The oily materials present in sebum keep the skin soft, pliable and to a large extent waterproof.

Sebum and tears contain chemical substances which kill microbes. In this way sebum enhances the protective function of the skin and tears protect the eyes.

Sensory nerve endings present in the dermis are stimulated by pain, temperature and touch. If the finger touches a very hot plate it is removed immediately. This cycle of events is called a *reflex action*, i.e., a very rapid motor response (contraction of muscles) to a sensory stimulus (stimulation of sensory nerve endings in the skin). This type of reflex action is an important protective mechanism.

MUCOUS MEMBRANE (Ch. 2)

This type of membrane consists of living cells. Their free surface is kept moist by the thick sticky fluid they secrete, called *mucus*. Mucus protects the mouth and oesophagus from mechanical injury by food and the lungs and respiratory passages from inhaled dust and microbes.

6. REPRODUCTION (Ch. 15. Fig. 1:15)

Successful reproduction is essential in order to ensure the continuation of a species from one generation to the next. *Bisexual reproduction* results from the fertilisation of the female egg cell or *ovum* by the male sperm cell or *spermatozoon*. Ova are produced by two *ovaries* situated in the female pelvis. Usually only one ovum is released at a time and it travels towards the *uterus* in the *uterine tube*. The spermatozoa are produced in large numbers by the two

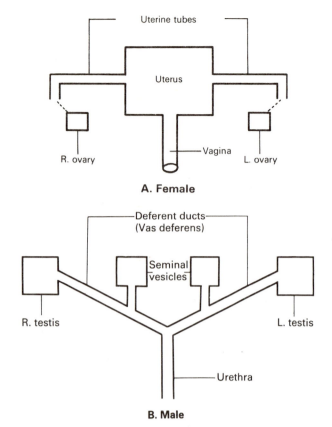

A. Female

B. Male

Figure 1:15   Diagram representing the female and male reproductive systems.

testes, situated in the *scrotum*. From each testis spermatazoa pass through a duct called the *deferent duct (vas deferens)* to the *seminal vesicle* then to the *urethra*. During coitus the spermatozoa are deposited in the female *vagina*. They then pass upwards through the uterus and fertilise the ovum in the uterine tube. The fertilised ovum (*zygote*) then passes into the uterus, embeds itself in the uterine wall and grows to maturity in about 40 weeks. When a baby is born it is entirely dependent on other people for the food and protection provided before birth by its mother's body.

One ovum is produced about every 28 days during the child-bearing years, beginning at *puberty*, at about 13 years of age, and ending at the *menopause*, about 35 years later. When the ovum is not fertilised it passes out of the uterus accompanied by bleeding, called *menstruation*. The cycle in the female, called the *menstrual cycle*, has recognisable phases associated with changes in the concentration of hormones. Although there is no similar cycle in the male, the same hormones are involved in the production and development to maturity of the spermatozoa.

7. MOVEMENT WITHIN THE EXTERNAL ENVIRONMENT

The essential purposes of the physical movement of the whole, or parts, of the body within the environment are:

To obtain food and water
To avoid injury
To reproduce

Most of the body movement is under the control of the will and is initiated at the level of consciousness in the brain. The exceptions are some protective movements which are carried out before the individual is aware of them, e.g., the reflex action of removing the finger from a very hot surface.

The *nervous system* and *joint movement*, which is dependent on the *muscular* and *skeletal systems*, are involved.

Before going on to discuss the individual bones which make up the skeletal framework it is necessary to be familiar with certain anatomical terms.

The anatomical position. This is the position assumed in all anatomical descriptions. The body is in the upright position with the head facing forward, the arms at the sides with the palms of the hands facing forward and the feet together.

Median plane. When the body, in the anatomical position, is divided *longitudinally* into right and left halves it has been divided in the median plane. Any structure which is *medial* to another is nearer the midline and any structure *lateral* to another is farther from the midline or at the side of the body.

Proximal and distal. These terms are used when describing the bones of the limbs. The·proximal end of a bone is the one nearest the point of attachment of the limb, and the distal end is farthest away from the point of attachment of the limb.

Anterior or ventral. This indicates that the part being described is nearer the front of the body.

Posterior or dorsal. This means that the part being described is nearer the back of the body.

Superior. This indicates a structure nearer the head.

Inferior. This indicates a structure farther away from the head.

Border. This is a ridge of bone which separates two surfaces.

Spine, spinous process or crest. This is a sharp ridge of bone.

Trochanter, tuberosity or tubercle. These are roughened bony projections, usually for the attachment of muscles or ligaments. The different names are used according to the size of the projection. Trochanters are the largest and tubercles the smallest.

Styloid process. This is a sharp downward projection of bone which gives attachment to muscles and ligaments.

Fossa (plural fossae). This is a hollow or depression.

Foramen (plural foramina). This is a hole in a structure.

Bony sinus. This is a hollow cavity within a bone.

Meatus. This indicates a tube-shaped cavity within a bone.

Articulation. This is a joint between two or more bones.

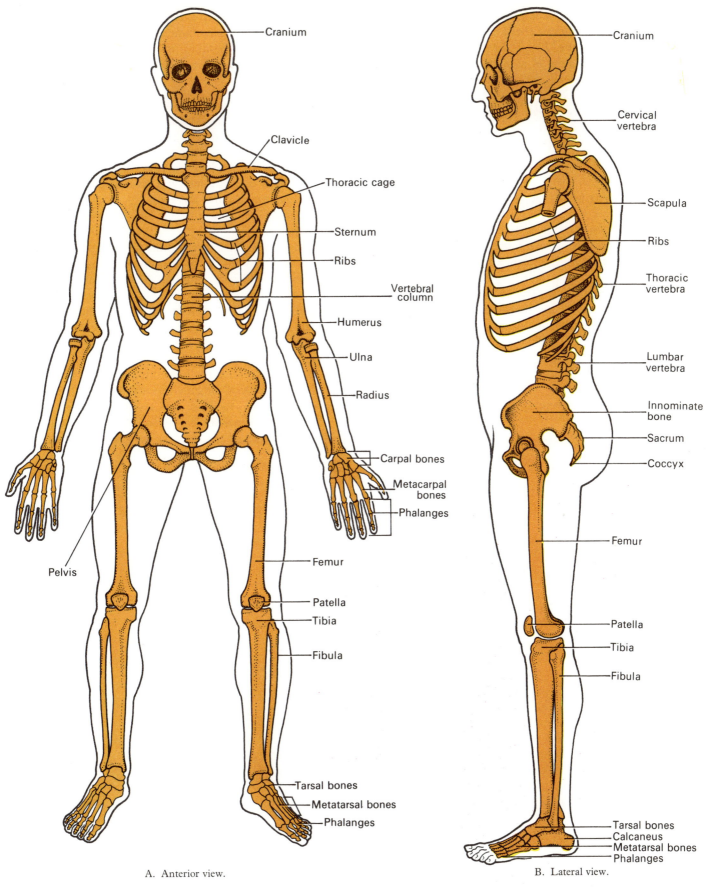

A. Anterior view.

B. Lateral view.

Figure 1:16   The bony skeleton.

Suture. This is the name given to an immovable joint, e.g., between the bones of the skull.

Articulating surface. This is the part of the bone which enters into the formation of a joint.

Facet. This is a small, generally rather flat, articulating surface.

Condyle. This is a smooth rounded projection of bone which takes part in a joint.

Septum. This is a partition separating two cavities.

Fissure or cleft. This indicates a narrow slit.

SKELETAL AND MUSCULAR SYSTEMS

The bones of the skeleton provide the framework for the body. Movement is achieved when the muscles, lying between the skin and the bones, contract and move them at the joints.

The central part of the skeleton is described as the *axial skeleton*. The bones of the limbs and those which attach them to the axis make up the *appendicular skeleton* (see Fig. 1:16A and B).

The axial skeleton consists of:
Skull
Vertebral column
Sternum or breast bone
Ribs

The appendicular skeleton consists of:
The bones of the upper limbs, the two clavicles and the two scapulae
The bones of the lower limbs and the two innominate bones of the pelvis

AXIAL SKELETON (Figs 1:17, 1:18 and 1:19)

SKULL

The skull is described in two parts, the *cranium* which contains the brain, and the *face*. It consists of a number of bones which develop separately but fuse together as they mature. The only movable bone is the mandible or lower jaw. The names and positions of the individual bones of the skull can be seen in Figure 1:17.

Functions

The various parts of the skull have specific and different functions:

1. The cranium protects the delicate tissues of the brain
2. The bony eye sockets provide the eyes with some protection against injury and give attachment to the muscles which move the eyes
3. The temporal bone protects the delicate structures of the ear
4. Some bones of the face and the base of the skull give resonance to the voice because they have cavities called *sinuses*, containing air. The sinuses have tiny openings into the nose
5. The bones of the face form the walls of the posterior part of the nasal cavities. They keep the air passage open, facilitating breathing
6. The maxilla and the mandible provide alveolar ridges in which the teeth are embedded
7. The mandible is the only movable bone of the skull and chewing food is the result of raising and lowering the mandible by contracting and relaxing some muscles of the face, the muscles of mastication

VERTEBRAL COLUMN

This consists of 24 separate movable bones (vertebrae) plus the sacrum and coccyx. The vertebral column is described in five parts and the bones of each part are numbered from above downwards (Fig. 1:18):

7 cervical
12 thoracic
5 lumbar
1 sacrum (5 fused bones)
1 coccyx (4 fused bones)

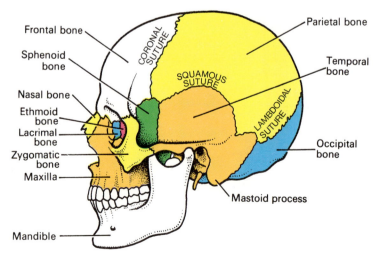

Figure 1:17   The bones of the skull and face — lateral view.

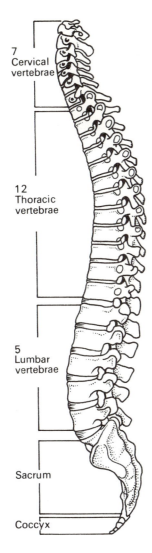

Figure 1:18 The vertebral column — lateral view.

The first cervical vertebra, called *the atlas*, articulates with the skull. Thereafter each vertebra forms a joint with the vertebrae immediately above and below. In the cervical and lumbar regions more movement is possible than in the thoracic region.

The sacrum consists of five vertebrae fused into one bone which articulates with the 5th lumbar vertebra above and an innominate (pelvic or hip) bone at each side.

The coccyx consists of the four terminal vertebrae fused into a small triangular bone which articulates with the sacrum above.

Functions

The vertebral column has several important functions:

1. It protects the spinal cord. In each bone there is a hole or *foramen* and when the vertebrae are arranged one above the other, as shown in Figure 1:18, the foramina form a canal. The spinal cord, which is an extension of nerve tissue from the brain, lies in this canal (Fig. 1:19).

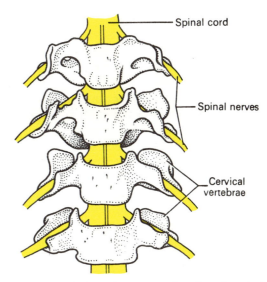

Figure 1:19 The cervical vertebrae separated to show the spinal cord and spinal nerves in yellow.

2. Adjacent vertebrae form openings through which spinal nerves pass from the spinal cord to all parts of the body (Fig. 1:19). There are 31 pairs of spinal nerves.

3. In the thoracic region the ribs articulate with the vertebrae forming joints which move during respiration.

THORACIC CAGE

The thoracic cage is formed by:

    12 thoracic vertebrae
    12 pairs of ribs
    1 sternum or breast bone

The arrangement of the bones can be seen in Figure 1:20.

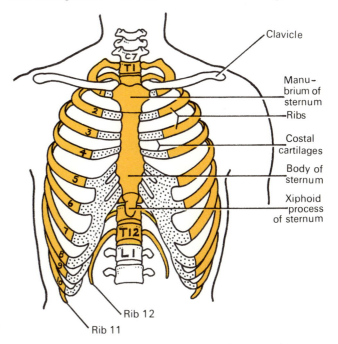

Figure 1:20 The structures forming the walls of the thoracic cage.

Functions

The functions of the thoracic cage are:

1. To protect the thoracic organs. The bony framework protects the heart, lungs, large blood vessels and other structures.

2. To form joints between the upper limbs and the axial skeleton. The upper part of the sternum, the *manubrium*, articulates with the clavicles forming the only joints between the upper limbs and the axial skeleton.

3. To give attachment to the muscles of respiration:

a. Intercostal muscles occupy the spaces between the ribs and when they contract the ribs move upwards and outwards, increasing the capacity of the thoracic cage, and inspiration (breathing in) occurs.

b. The diaphragm is a dome-shaped muscle which separates the thoracic and abdominal cavities. It is attached to the bones of the thorax and when it contracts it assists with inspiration. Structures which extend from one cavity to the other pass through the diaphragm.

APPENDICULAR SKELETON

The appendages are:

1. The upper limbs and the shoulder girdles
2. The lower limbs and the innominate bones of the pelvis

The names of the bones involved, their position and their relationship to other bones are shown in Figure 1:21.

Functions

The appendicular skeleton has two functions:

1. Voluntary movement. The bones, muscles and joints of the limbs are involved in voluntary movement. This may range from the very fine movements of the fingers associated with writing, to the co-ordinated movement of all the limbs associated with running and jumping.

2. Protection of delicate structures. Structures such as blood vessels and nerves lie along the length of bones of the limbs and are protected from injury by the muscles and skin. These structures are most vulnerable where they cross joints and where bones can be felt near the skin.

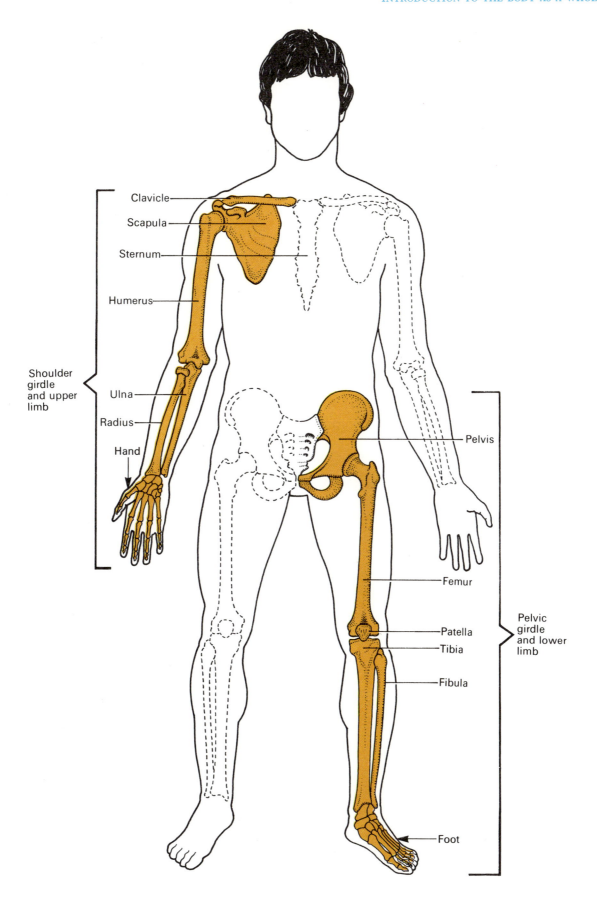

Figure 1:21   The bones of the upper and lower limbs and their relationships to the axial skeleton.

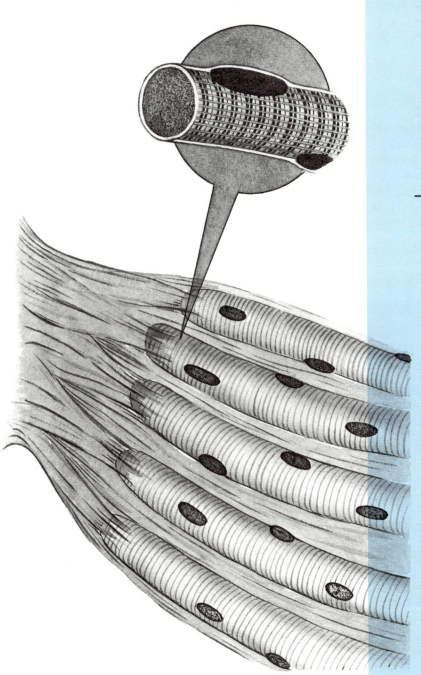

2

The Cells, Tissues, Systems and Cavities of the Body

2. The Cells, Tissues, Systems and Cavities of the Body

CELL

The human body develops from a single cell called the *zygote* which results from the fusion of the ovum (female egg cell) and the spermatazoon (male germ cell). Cell multiplication follows and, as the fetus grows, cells with different structural and functional specialisation develop, all with·the same genetic characteristics as the zygote. Individual cells are too small to be seen with the naked eye but *tissues*, consisting of millions of specialised cells can be, e.g., skin and the lining membrane of the mouth. When very thin slices of tissue are stained in the laboratory, individual cells can be seen when viewed through a microscope. Different types of cells are identified by the dye they absorb and their size and shape.

STRUCTURE (Fig. 2:1)

All living cells are made up of *protoplasm* — a slightly opaque, colourless, soft jelly-like substance consisting of water and organic and inorganic compounds in solution or suspension. The materials from which these compounds develop are:

Carbohydrates
Fats (lipids)
Amino acids obtained from proteins
Minerals

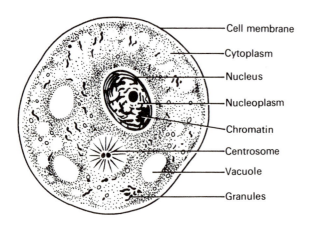

Cell membrane
Cytoplasm
Nucleus
Nucleoplasm
Chromatin
Centrosome
Vacuole
Granules

Figure 2:1   The simple cell.

CELL MEMBRANE

The protoplasm of the cell is contained within a very fine membrane consisting of protein threads and lipids. Substances pass into and out of the cell across this membrane in a variety of ways.

1. *By diffusion.* The cell membrane has tiny pores between the protein threads and lipids through which substances with small enough molecules can pass. Diffusion always takes place *down* the concentration gradient, i.e., from the high concentration side of the membrane to the low. Substances diffuse into and out of cells, e.g., oxygen diffuses in and carbon dioxide diffuses out.

2. *By dissolving in the lipids in the cell membrane.* Fat-soluble substances which cannot diffuse through the pores of the membrane pass through by dissolving in the lipid part.

3. *By active transport.* This system applies to substances which are too large to diffuse through the cell membrane, are not soluble in fat or must be transported *up* the concentration gradient. The substance transported is bound to a carrier substance in the cell membrane. It is then 'carried' across and released on the other side. Carrier substances are specific, e.g., glucose and amino acids are transported in this way but the glucose carrier cannot transport amino acids and vice versa.

Some proteins on the surface of cell membranes are associated with the individual's immunological identity. They are the *antigens* responsible for the blood groups and the compatibility (or incompatibility) of transplanted tissues, including blood used for transfusion.

CYTOPLASM

The cytoplasm is the protoplasm inside the cell but outside the nucleus (Fig. 2:1). It contains molecules of *ribonucleic acids* (RNA) and granular structures called *mitochondria*. Mitochondria are involved in oxidative reactions which result in the controlled release of energy from the nutrient materials and the formation of *adenosine triphosphate* (ATP), the main *energy carrier* within the cell. Within the cytoplasm there are clear circular spaces called *vacuoles* which may contain waste materials or secretions formed by the cell.

NUCLEUS

With the exception of erythrocytes, all cells contain a dark-staining mass of deoxyribose nucleic acid (DNA) called the *nucleus* enclosed in a *nuclear membrane*. In addition the nucleus contains RNA which is responsible for the synthesis of protein. Near the nucleus there is a small spherical body called the *centrosome* surrounded by a radiating thread-like structure. The centrosome contains two minute, dark-staining, circular bodies called *centrioles* which participate in the early stages of cell division.

CELL DIVISION (MULTIPLICATION)

Within the nucleus there are *chromatin* threads which carry the *genes*, composed of *deoxyribonucleic acid* (DNA). The genes are the inherited material: they ensure that when cells divide, the 'daughter cells' are identical to the 'parent cell'. This type of cell division or multiplication, called *mitosis*, goes on throughout the life of the individual. A number of genes are linked together to form thread-like structures called *chromosomes*. Each cell has 46 chromosomes arranged in 23 pairs.

The ovum and the spermatazoon each have 23 chromosomes. When they combine and form the zygote it has the full complement of 46 chromosomes. Thus the new individual has a mixture of DNA, half obtained from the mother and half from the father. This type of cell division is called *meiosis*.

Mitosis

The multiplication of cells by mitosis occurs throughout the life of the individual. It occurs at a more rapid rate until growth is complete and thereafter new cells are formed to replace those which have died. Nerve cells are the exception: when they die they are not replaced.

There are several fairly well-defined stages in the process of cell division by mitosis, of which four are described here (Fig. 2:2). The period during which cells are not dividing is called the *interphase*.

1. *Prophase*. The centrosome divides into two *centrioles* which migrate to opposite poles of the cell, but remain attached by fine thread-like *spindles*. The chromatin becomes concentrated and forms dark rod-shaped structures called *chromosomes*. Each chromosome divides longitudinally into two identical daughter chromosomes, called *chromatids*, which are held together at the middle by a *centromere*.

2. *Metaphase*. The nuclear membrane disappears and the chromosomes arrange themselves at the centre of the cell and are attached to the spindles of the centrosomes, at opposite poles of the cell.

3. *Anaphase*. The centromere divides and the two identical chromosomes move apart, still attached to the spindles. They arrange themselves at opposite poles of the cell. At the end of this stage the spindles break.

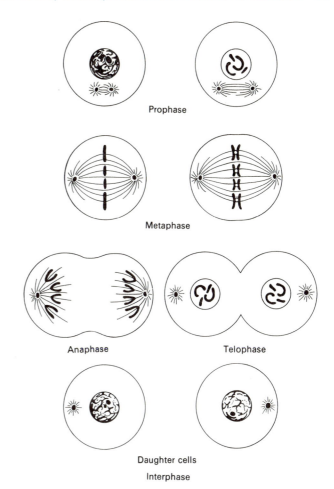

Figure 2:2 Some stages in cell multiplication by mitosis.

4. *Telophase*. Nuclear membranes enclose the two groups of chromosomes, the spindles disappear and a constriction develops round the middle of the cell body.

5. *Interphase*. The constriction of the cytoplasm increases until the cell divides, the chromosomes disappear and the two daughter cells enter the interphase stage.

CHROMOSOMES

As indicated above, the nucleus of every human cell contains 46 chromosomes. The chromosomes are arranged in pairs, one member of each pair being inherited from the male parent and the other from the female parent.

The chromosomes contain the basic hereditary substances which determine the individual's characteristics and traits, including colour of hair and eyes, structure of bones and teeth, height and the ability to produce enzymes necessary for the metabolic processes.

Determination of sex depends on *one pair of sex chromosomes*. In the female the two sex chromosomes are the same in size and shape and are called X chromosomes. In the male they are slightly different, one is an X chromosome and the other is a slightly smaller Y chromosome.

Therefore the female sex chromosomes are XX and the male XY.

In the ovum and in the spermatozoon the number of chromosomes is reduced to 23 single chromosomes; there is therefore only one sex chromosome in each. This means that *all ova* have an X chromosome and *one half of the spermatozoa* have an X chromosome and *the other half* have a Y chromosome.

In conception, when an X-bearing spermatozoon fertilises an ovum the offspring is female and when a Y-bearing spermatozoon fertilises an ovum the offspring is male.

Sperm X + ovum X = offspring XX = female
Sperm Y + ovum X = offspring XY = male

Within a pair of genes for a particular characteristic, one may exert a stronger influence than the other. The gene exerting the stronger influence is called *dominant* and the gene which is less effective is described as *recessive*. The characteristics of the offspring such as height, colour of eyes and hair and other familial traits depend upon the dominance of the parents' genes.

TISSUES

The tissues of the body consist of large numbers of cells and they are classified according to the size, shape and functions of these cells. There are four main types of tissue, each of which has subdivisions. They are:

Epithelial tissue or epithelium
Connective tissue
Muscle tissue
Nervous tissue

EPITHELIAL TISSUE

The cells are very closely packed together and the intercellular substance, called the *matrix*, is minimal. The cells usually lie on a *basement membrane* which is an inert connective tissue.

Epithelial tissue may be *simple* or *compound*.

SIMPLE EPITHELIUM

Simple epithelium consists of a single layer of cells and is divided into four types. They are named according to the shape of the cells, which differ according to their functions.

Squamous (pavement) epithelium

This is composed of a single layer of flattened cells (Fig. 2:3). The cells fit closely together like flat stones and they form a very smooth membrane.

Figure 2:3 Squamous epithelium.

The function of squamous epithelium is to provide a thin, smooth, inactive lining for the following structures:
Heart
Blood vessels
Alveoli of the lungs
Lymph vessels

Cuboidal (cubical) epithelium

This consists of cube-shaped cells fitting closely together lying on a basement membrane (Fig. 2:4). It forms the tubules of the kidneys and is found in some glands. Cuboidal eithelium is actively involved in secretion and absorption.

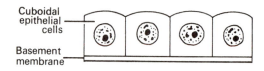
Figure 2:4 Cuboidal epithelium

Columnar epithelium

This is formed by a single layer of cells, rectangular in shape, on a basement membrane (Fig. 2:5). It is found lining the organs of the alimentary tract and consists of a mixture of cells, some absorb the products of digestion and others secrete *mucus*. Mucus is a thick sticky substance secreted by special columnar cells called *goblet cells*.

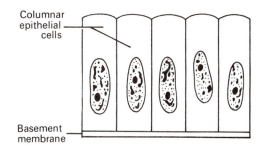
Figure 2:5 Columnar epithelium.

Ciliated epithelium (Fig. 2:6)

This is formed by columnar cells which have fine, hair-like processes on their free surface, called *cilia*. The wave-like movement of many cilia propels the contents of the tubes that they line in one direction only.

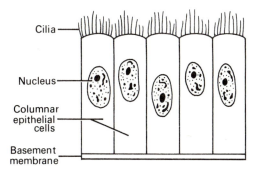

Figure 2:6   Ciliated columnar epithelium.

Ciliated epithelium is found lining the uterine tubes and most of the respiratory passages. In the uterine tubes the cilia propel ova towards the uterus (Ch. 15) and in the respiratory passages they propel mucus towards the throat (Ch. 7).

COMPOUND EPITHELIUM

Compound epithelium consists of several layers of cells. The superficial layers grow up from below. Basement membranes are usually absent. The main function of compound epithelium is to protect underlying structures. They are two main types: stratified and transitional.

Stratified epithelium (Fig. 2:7)
This is composed of a number of layers of cells of different shapes. In the deepest layers the cells are mainly columnar in shape and, as they grow towards the surface, they become flattened.

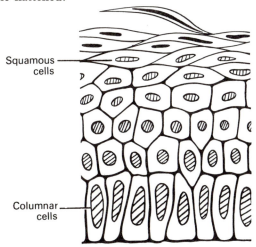

Figure 2:7   Stratified epithelium.

Non-keratinised stratified epithelium is found on wet surfaces that may be subjected to wear and tear, such as, the conjunctiva of the eyes and the lining of the mouth, the pharynx and the oesophagus.

Keratinised stratified epithelium is found on dry surfaces, that is, skin, hair and nails. The surface layers of keratin-

ised cells are dead cells. They give protection to, and prevent drying of, the cells in the deeper layers from which they develop. The surface layer of skin is rubbed off and is replaced from below.

Transitional epithelium (Fig. 2:8)
This is composed of several layers of pear-shaped cells and is found lining the ureters and the urinary bladder.

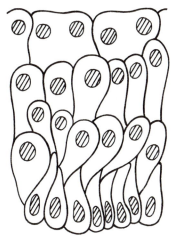

Figure 2:8   Transitional epithelium.

CONNECTIVE TISSUE

The cells forming the connective tissues are more widely separated from each other than those forming the epithelium, and the intercellular substance is present in considerably larger amounts. There may or may not be fibres present in the matrix, which may be of a semisolid jelly-like consistency or dense and rigid, depending upon the position and function of the tissue.

CELLS OF CONNECTIVE TISSUE

Connective tissue, excluding blood (see Ch. 3), is found in all organs supporting the specialised tissue. The different types of cell involved include:
    Fibroblasts
    Macrophages
    Plasma cells
    Mast cells
    Fat cells

Fibroblasts
These are large flat cells with irregular processes. They produce *white collagen fibres*, the basic material of *fibrous tissue*. They may also produce elastic fibres. Fibroblasts are particularly active in tissue repair where they may bind together the cut surfaces of wounds or form *granulation tissue*, following tissue destruction (see p. 205). The collagen fibres formed during healing shrink as they grow

old, sometimes interfering with the functions of the organ involved.

In some glands a type of fibroblast called *reticular cells* produce fine fibrous strands called *reticulin* and are closely associated with phagocytosis.*

Macrophages

These are irregular-shaped cells with granules in the cytoplasm. Some are *fixed*, i.e., attached to connective tissue fibres, and others are *motile*. They are actively phagocytic. Although the relationship is not clear, their activities are similar to those of cells of the lymphoreticular system, e.g., monocytes in blood, phagocytes in the alveoli of the lungs, Kupffer cells in liver sinusoids, reticular cells in lymph nodes and spleen and microglial cells in the brain.

Plasma cells

Plasma cells are derived from B-lymphocytes (see p. 40). They secrete specific antibodies into the blood in response to the presence of foreign material, such as microbes.

Mast cells

These cells are similar to basophil leukocytes (see p. 39). They are found under the fibrous capsule of some organs, e.g., liver and spleen, and in considerable numbers round blood vessels. They produce *heparin*, *serotonin* (5-hydroxytryptamine) and *histamine* which are released when the cells are damaged by disease or injury. Histamine and serotonin are involved in local and general inflammatory reactions, and may be associated with the development of allergies and hypersensitivity (see p. 41). Heparin prevents coagulation of plasma which may aid the passage of protective substances from blood to affected tissues.

Fat cells

These cells occur singly or in groups, especially in adipose tissue under the skin. They vary in size and shape according to the amount of fat they contain.

AREOLAR TISSUE (Fig. 2:9)

This is the most generalised of all connective tissue. The matrix is described as semisolid with *fibroblasts* widely separated by *yellow elastic* and *white collagen fibres*. It is found in almost every part of the body connecting and supporting other tissues, for example:

Under the skin
Between muscles
Supporting blood vessels and nerves
In the alimentary canal
In glands supporting secretory cells

* Phagocytosis is the process by which some specialised cells ingest and remove dead and damaged cells, microbes and other foreign matter from organs and tissues.

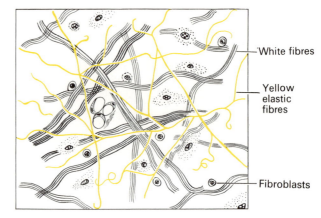

Figure 2:9   Areolar tissue.

ADIPOSE TISSUE (Fig. 2:10)

Adipose tissue consists of a collection of *fat cells* containing fat globules in a matrix of areolar tissue. It is found supporting the kidneys and the eyes, between bundles of muscle fibres and under the skin.

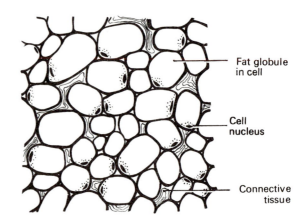

Figure 2:10   Adipose tissue.

WHITE FIBROUS TISSUE (Fig. 2:11)

This is a strong connecting tissue made up mainly of closely packed bundles of *white collagen fibres* with very little matrix. There are very few cells, lying in rows between the bundles of fibres. Fibrous tissue is found:

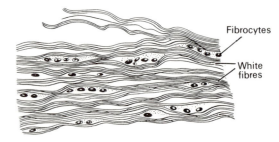

Figure 2:11   White fibrous tissue.

1. Forming the ligaments which bind bones together
2. As an outer protective covering for bone, called *periosteum*
3. As an outer protective covering of some organs, e.g., the kidneys, lymph nodes and the brain
4. Forming muscle sheaths, called *muscle fascia*, which extend beyond the muscle to become the tendon that attaches the muscle to bone

YELLOW ELASTIC TISSUE (Fig. 2:12)

Yellow elastic tissue is capable of considerable extension and recoil. There are few cells and the matrix consists mainly of masses of *elastic fibres* believed to be secreted by fibroblasts. It is found in organs where alteration of shape is required, e.g., in blood vessel walls and the elastic cartilage of the epiglottis and lobes of the ears.

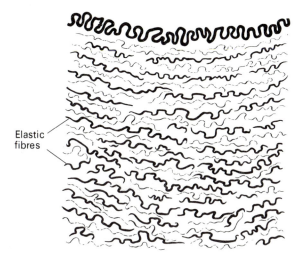

Figure 2:12   Yellow elastic tissue.

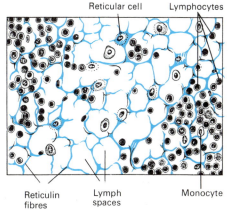

Figure 2:13   Lymphoid tissue.

LYMPHOID TISSUE (Fig. 2:13)

This tissue has a semisolid matrix with fine branching reticulin fibres. The highly specialised cells are called *lymphocytes* (see p. 392). They are found in blood and in lymphoid tissue in the:

Lymph nodes
Spleen
Palatine and pharyngeal tonsils
Vermiform appendix
Solitary and aggregated glands in the small intestine
Wall of the large intestine

CARTILAGE

Cartilage is a much firmer tissue than any of the other connective tissues; the matrix is quite solid. There are three types:

Hyaline
White fibrocartilage
Yellow or elastic fibrocartilage

Hyaline cartilage (Fig. 2:14)
Hyaline cartilage appears as a smooth bluish-white tissue.

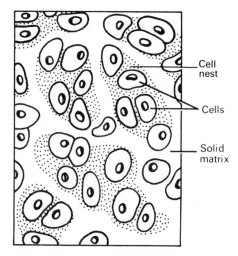

Figure 2:14   Hyaline cartilage.

The cells are in small groups and the matrix is solid and smooth. Hyaline cartilage is found:
1. On the surface of the parts of the bones which form joints
2. Forming the costal cartilages which attach the ribs to the sternum
3. Forming part of the larynx, trachea and bronchi

White fibrocartilage (Fig. 2:15)
This consists of dense masses of white fibres in a solid matrix with the cells widely dispersed. It is a tough, slightly flexible tissue found:
1. As pads between the bodies of the vertebrae, called the intervertebral discs
2. Between the articulating surfaces of the bones of the knee joint, called semilunar cartilages

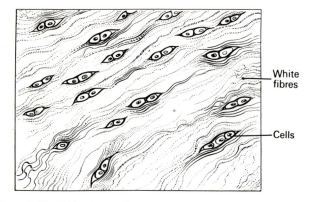

Figure 2:15   White fibrocartilage.

3. On the rim of the bony sockets of the hip and shoulder joints, deepening the cavities without restricting movement.

Yellow elastic fibrocartilage (Fig. 2:16)

This consists of yellow elastic fibres lying in a solid matrix.

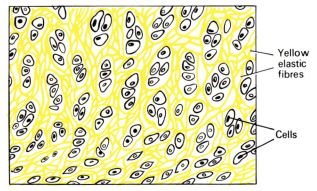

Figure 2:16   Yellow elastic fibrocartilage.

The cells lie between the fibres. It forms the pinna or lobe of the ear, the epiglottis and part of the tunica media of blood vessel walls.

BLOOD

This is a fluid connective tissue and is described in detail in Chapter 3.

BONE

The different types of bone are described in Chapter 16.

MUSCLE TISSUE

There are three main types of muscle tissue:
    Striated or voluntary muscle
    Smooth or involuntary muscle
    Cardiac muscle

Striated muscle tissue (Fig. 2:17)

This may be described as *skeletal*, *striated*, *striped* or *voluntary* muscle. It is called voluntary because contraction is under control of the will.

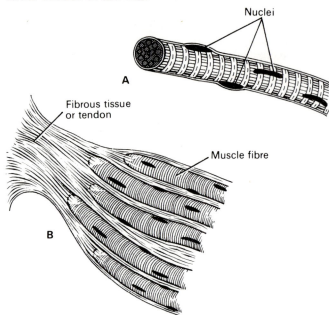

Figure 2:17   A. A striated muscle fibre. B. A bundle of striated muscle fibres and their connective tissue.

When voluntary muscle is examined microscopically the cells are found to be roughly cylindrical in shape and their length varies from 1 to 40 millimetres. Each fibre has several nuclei situated just under the *sarcolemma*, which is a fine sheath surrounding each muscle fibre. The muscle fibres lie parallel to one another and, when viewed under the microscope, they show well-marked transverse dark and light bands, hence the name striated or striped muscle.

A muscle consists of a large number of muscle fibres. In addition to the sarcolemma mentioned previously, each fibre is enclosed in and attached to fibrous tissue called *endomycium*. Small bundles of fibres are enclosed in *perimycium*, and the whole muscle in *epimycium*. The fibrous tissue enclosing the fibres, the bundles and the whole muscle extends beyond the muscle fibres to become the *tendon* which attaches the muscle to bone or skin.

Smooth muscle tissue (Fig. 2:18)

Smooth muscle may be described as *involuntary*, *plain* or *visceral* muscle. It is not under the control of the will. It is found in the walls of blood and lymph vessels, the alimentary tract, the respiratory tract, the urinary bladder and the uterus.

When examined under a microscope, the cells are seen to be spindle-shaped with only one central nucleus. There is no distinct sarcolemma but a very fine membrane surrounds each fibre. Bundles of fibres form sheets of muscle which form parts of the walls of the above structures.

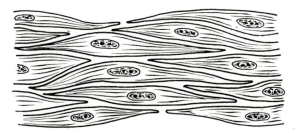

Figure 2:18 Smooth muscle fibres.

Cardiac muscle (Fig. 2:19)

This type of muscle tissue is found exclusively in the wall of the heart. It is not under the control of the will but, when viewed under a microscope, cross stripes characteristic of voluntary muscle, can be seen. Each fibre (cell) has a nucleus and one or more branches. The ends of the cells and their branches are in very close contact with the ends and branches of adjacent cells. Microscopically these 'joints', or *intercalated discs*, can be seen as lines which are thicker and darker than the ordinary cross stripes. This arrangement gives cardiac muscle the appearance of a sheet of muscle rather than a very large number of individual fibres. The end to end contiguity of cardiac muscle cells has significance in relation to the way the heart contracts.

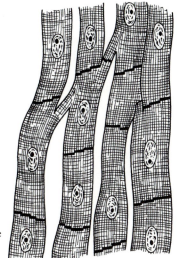

Figure 2:19 Cardiac muscle fibres.

Each fibre does not need to be stimulated as the impulse of contraction spreads from cell to cell across the intercalated discs (see p. 68).

FUNCTION OF MUSCLE

Muscle functions by alternate phases of contraction and relaxation. When the fibres contract they become thicker and shorter. Skeletal muscles are stimulated by nerve impulses originating in the brain or spinal cord. Smooth muscle and cardiac muscle have intrinsic ability to initiate contraction and, in addition, may be stimulated by nerves. When muscle fibres contract they follow the *all or none law*, i.e., each fibre contracts to its full capacity. The *strength* of contraction, e.g., lifting a weight, depends on the *number* of fibres contracting at the same time. When effort is sustained groups of fibres contract in series.

In order to contract when it is stimulated, a muscle fibre must have an adequate blood supply to provide sufficient oxygen and nutritional materials and to remove waste products.

Muscle tone

This is a state of partial contraction of muscles. It is achieved by the contraction of a few muscle fibres at a time. Muscle tone in relation to striated muscle is associated with the maintenance of posture in the sitting and standing positions. The muscle is stimulated to contract through a system of spinal reflexes. Stretching a muscle or its tendon stimulates the reflex action (Ch. 12). A degree of muscle tone is also maintained by smooth and cardiac muscle.

Muscle fatigue

If a muscle is stimulated to contract at very frequent intervals, it gradually becomes depressed and will cease to respond. This is called muscle fatigue and is usually due to an inadequate blood supply. Fatigue is prevented during sustained muscular effort because the fibres usually contract in series. All the fibres of a muscle rarely contract at the same time but if maximum effort is made it can be sustained if for only a short time.

Energy source for muscle contraction

The energy which muscles require is derived from the breakdown (catabolism) of carbohydrate and fat. Protein molecules inside the fibres are used to provide energy when supplies of carbohydrate and fat are deficient. Each molecule undergoes a series of changes and, with each change, small quantities of energy are released. For the complete breakdown of these molecules and the release of all the available energy, an adequate supply of oxygen is required. If the individual undertakes excessive exercise, the oxygen supply may be insufficient to meet the metabolic needs of the muscle fibres. This may result in the accumulation of intermediate metabolic products, such as lactic acid. Where the breakdown process and the release of energy are complete, the waste products are carbon dioxide and water (see Ch. 9).

Further features of striated muscle

The voluntary muscles are those which produce body movements. Each muscle consists of a *fleshy* part made up of striped fibres and *tendinous* parts consisting of white fibrous tissue usually at both ends of the fleshy part. The muscle is attached to bone or skin by these tendons. When the tendinous attachment of a muscle is broad and flat it is called an *aponeurosis*.

To be able to produce movement at a joint, a muscle or its tendon *must stretch across the joint*. When a muscle contracts, its fibres *shorten and it pulls* one bone towards another, e.g., bending the elbow.

The muscles of the skeleton are arranged in groups, some of which are *antagonistic* to each other. To produce movement at a joint, one muscle or group of muscles contracts while the antagonists relax, e.g., to bend the knee the muscles on the back of the thigh contract and those on the front relax. The constant adjustment of the contraction and relaxation of antagonistic groups of muscles is well demonstrated in the maintenance of balance and posture when sitting and standing. These adjustments usually occur without conscious effort.

Individual muscles and groups of muscles have been given names that reflect certain characteristics, e.g.:

The shape of the muscle — the trapezius is shaped like a trapezium

The direction in which the fibres run — the oblique muscles of the abdominal wall

The position of the muscle — the tibialis is associated with the tibia

The movement produced by contraction of the muscle — flexors, extensors, adductors.

The number of points of attachment of a muscle — the biceps muscle has two tendons at one end.

The names of the bones to which the muscle is attached — the carpi radialis muscles are attached to the carpal bones in the wrist and to the radius in the forearm.

A more detailed description of voluntary muscles is given in Chapters 17 and 18.

NERVOUS TISSUE

Nervous tissue is described in Chapter 12.

FIBROSIS

Fibrous tissue is formed during healing when the cells destroyed do not regenerate, e.g., following chronic inflammation, ischaemia, suppuration, large scale trauma. The process begins with the formation of granulation tissue then, over a period of time, the new capillaries and inflammatory material are removed leaving only the collagen fibres secreted by the fibroblasts. Fibrous tissue may have long-lasting damaging effects.

Adhesions consisting of fibrous tissue may limit movement, e.g., between the layers of pleura, preventing inflation of the lungs; between loops of bowel, interfering with peristalsis.

Fibrosis of infarcts. Blockage of an end-vessel by a thrombus or an embolus causes an infarct (area of dead tissue). Fibrosis of one large infarct or of numerous small infarcts may follow, leading to varying degrees of organ dysfunction, e.g., in heart, brain, kidneys, liver.

Fibrous tissue shrinkage occurs as it ages. The effects depend on the site and extent of the fibrosis, e.g.,

1. Small tubes, such as blood vessels, air passage, ureters, the urethra and ducts of glands may become narrow or obstructed and lose their elasticity

2. Contractures (bands of shrunken fibrous tissue) extending across joints, e.g., in a limb or digit there may be limitation of movement and following burns of the neck the head may be pulled to one side

CELL REGENERATION

When cell regeneration occurs it is essential that some of the original cells are available to replicate by mitosis. The extent to which regeneration is possible depends on the normal rate of physiological turnover of particular types of cell. Those with a rapid turnover regenerate most effectively. There are three types.

Labile cells are those in which replication is normally a continuous process. They include cells in:

1. Epithelium of, e.g., skin, mucous membrane, secretory glands, ducts, endometrium
2. Bone marrow
3. Blood
4. Spleen and lymphoid tissue

Stable cells have retained the ability to replicate but do so infrequently. They include:

1. Liver, kidney and pancreatic cells
2. Fibroblasts
3. Smooth muscle
4. Osteoblasts and osteoclasts in bone

Permanent cells are unable to replicate after normal growth is complete. They include:

1. Nerve cells
2. Skeletal and cardiac muscle

MEMBRANES

Some of the tissues which line or cover organs are described as *membranes*. These are composed of types of cells which have already been described. When they are called membranes, the cells form *sheets* which cover or line structures. The more important membranes are classified as:

Mucous

Serous

Synovial

Mucous membrane

This is the name given to the lining of the alimentary tract, respiratory tract and genito-urinary tracts. The membrane consists of epithelial cells, some of which produce a secretion called *mucus*. This is a slimy tenacious fluid formed within the cytoplasm of the cells. As it accumulates the cells become distended and finally burst, discharging the mucus on to the free surface. As the cells fill up with mucus they have the appearance of a goblet or flask and are known as *goblet cells* (Fig. 2:20). Organs lined by mucous membrane have a moist slippery surface. Mucus protects the lining membrane from mechanical and

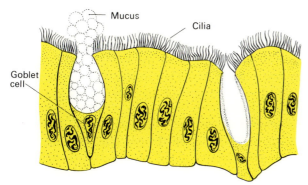

Figure 2:20 Ciliated columnar epithelium with goblet cells.

chemical injury and in the respiratory tract it traps inhaled foreign particles, preventing them from entering the alveoli of the lungs.

Serous membrane

Serous membranes consist of a double layer of epithelial tissue. The *visceral* layer closely invests organs and the *parietal* layer lines the space that they occupy. The visceral layer is reflected off the organ to become the parietal layer and the two layers are separated by a watery or *serous fluid* secreted by the membrane.

There are three sites in which serous membranes are found. In the thoracic cavity the *pleura* surround the lungs and the *pericardium* surrounds the heart, and in the abdominal cavity the *peritoneum* surrounds the abdominal organs.

The serous fluid between the visceral and parietal layers enables an organ to move without being damaged by friction between it and adjacent organs. For example, the heart changes its shape and size during each beat and friction damage is prevented by the arrangement of pericardium and its serous fluid.

Synovial membrane

This membrane is found lining the joint cavities and surrounding tendons which could be injured by rubbing against bones, e.g., over the wrist joint. It is made up of a layer of fine, flattened epithelial cells on a layer of delicate connective tissue.

Synovial membrane secretes clear, sticky, oily *synovial fluid* which acts as a lubricant to the joints and helps to maintain their stability. Synovial fluid also helps to nourish the hyaline cartilage which covers the articular surfaces of the bones, and synovial cells remove damaging material from within the joint cavity.

GLANDS

Some groups of epithelial cells are called *glands* and they produce specialised secretions. Glands that discharge their secretion on to the epithelial surface of an organ, either directly or through a *duct*, are called *exocrine glands*. Other groups of cells that have become isolated from epithelial

surfaces discharge their secretions into blood and lymph. These are called *endocrine glands* (ductless glands) and their secretions are *hormones* (see Ch. 14).

Figure 2:21 shows the different lines of development of exocrine and endocrine glands. Exocrine glands vary considerably in size, shape and complexity as shown in Figures 2:22 and 2:23.

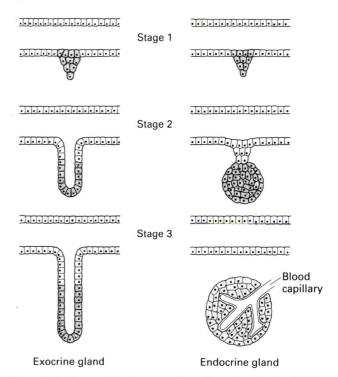

Exocrine gland   Endocrine gland

Figure 2:21 Stages of development of exocrine and endocrine glands.

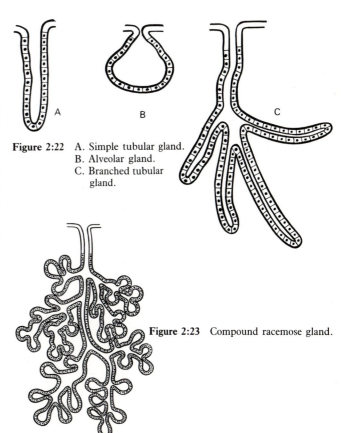

Figure 2:22 A. Simple tubular gland.
B. Alveolar gland.
C. Branched tubular gland.

Figure 2:23 Compound racemose gland.

DISORDERS OF TISSUES

NEOPLASMS OR TUMOURS

A neoplasm is a mass of tissue that grows in excess of normal in an unco-ordinated manner, and continues to grow after the initial stimulus has ceased. Tumours are classified as benign or malignant although a clear distinction is not always possible (see Table 2:1). Benign tumours only rarely change their character and become malignant.

Table 2:1   Differences between benign and malignant tumours

Benign	Malignant
Slow growth	Rapid growth
Cells well differentiated	Cells poorly differentiated
Usually encapsulated	Not encapsulated
Does not spread	Spreads:
	by local infiltration
	via lymph and
	blood

RATE OF GROWTH

A tumour forms when the rate of cell multiplication is greater than that of cell death. Blood vessels grow with the proliferating cells, but in some malignant tumours the blood supply does not keep pace with growth and *ischaemia* (lack of blood supply) leads to tumour cell death, called *necrosis*. If the tumour is near the surface this may result in skin ulceration and infection. In deeper tissues there is fibrosis, e.g., retraction of the nipple in breast cancer is due to the shrinkage of fibrous tissue in a necrotic tumour. The mechanisms controlling the life span of tumour cells are not known.

CELL DIFFERENTIATION

Differentiation of cells into types with particular structural and functional characteristics occurs at an early stage in fetal development, e.g., epithelial cells develop different characteristics from lymphocytes. Later, when cell replacement occurs, daughter cells have the same distinctive features as the parent. In benign tumours the cells from which they originate are easily recognised, i.e., well differentiated. In malignant tumours there is a wide range of differentiation. In some cases most identifying features are retained, causing only a slight degree of *dysplasia*. At the other extreme, when the parent cells cannot be identified, the tumour is described as *anaplastic*. Tumours with well-differentiated cells are usually benign but some may be malignant.

ENCAPSULATION AND SPREAD OF TUMOURS

Most benign tumours are contained within a fibrous capsule derived partly from the surrounding tissues and partly from the tumour. They neither infiltrate local tissues nor spread to other parts of the body, even when they are not encapsulated.

Malignant tumours are not encapsulated. They spread locally by infiltration, and tumour fragments may spread to other parts of the body in blood or lymph. Some spreading cells may be phagocytosed but others grow into *secondary (metastatic)* tumours.

Local spread

As tumours grow into surrounding tissues they may:
1. Erode blood and lymph vessel walls, causing spread of tumour cells to other parts of the body
2. Damage nerves, causing pain and loss of nerve control of other tissues and organs
3. Compress adjacent structures causing, e.g., ischaemia, necrosis, blockage of ducts, organ dysfunction or displacement, pain due to pressure on nerves

Lymphatic spread

This occurs when tumours grow into lymph channels. Groups of tumour cells break off and are carried to lymph nodes where they lodge and may grow into secondary tumours. There may be further spread through the lymphatic system, and eventually to blood via the subclavian veins.

Blood spread

This occurs when the walls of a blood vessel are eroded by a tumour. A *thrombus* (blood clot) may form at the site and *emboli*, consisting of fragments of tumour and blood clot, enter the blood stream. These emboli block small blood vessels, cause infarcts, and metastatic tumours develop. Phagocytosis of tumour cells in the emboli is unlikely to occur because they are protected by blood clot. The sites of blood-spread metastases depend on the site of the original tumour and the anatomy of the circulatory system, although the reasons why some organs are more frequently affected than others is not always clear.

Body cavities spread

This occurs when a tumour penetrates the wall of a cavity. The peritoneal cavity is most frequently involved. If, for example, a malignant tumour in an abdominal organ penetrates the visceral peritoneum, tumour cells may metastasise folds of peritoneum or any abdominal or pelvic organ. Where there is less scope for the movement of fragments within a cavity the tumour tends to bind layers of tissue together, e.g., a pleural tumour binds the visceral and parietal layers together, limiting expansion of the lung.

Table 2:2 shows common sites of primary tumours and their metastases.

Table 2:2 Common sites of primary tumours and their metastases

Primary tumour	Metastatic tumours
Bronchi	Adrenal glands, brain
Alimentary tract	Abdominal and pelvic structures, especially liver
Prostate gland	Pelvic bones, vertebrae
Thyroid gland	Pelvic bones, vertebrae
Breast	Vertebrae, brain
Many organs	Lungs

EFFECTS OF TUMOURS

PRESSURE EFFECTS

Both benign and malignant tumours may cause pressure damage to adjacent structures, especially if in a confined space. The effects depend on the site of the tumour but are most marked in areas where there is little space for expansion, e.g., inside the skull, under the periosteum of bones, in bony sinuses and respiratory passages. Compression of adjacent structures may cause ischaemia, necrosis, blockage of ducts, organ dysfunction or displacement, pain due to invasion of or pressure on nerves.

HORMONAL EFFECTS

Tumours of endocrine glands may secrete hormones, producing the effects of hypersecretion. The extent of cell dysplasia is an important factor. Well-differentiated benign tumours are more likely to secrete hormones than markedly dysplastic malignant tumours. Some malignant tumours produce *uncharacteristic hormones*. The cells of the tumour do not appear to originate from the appropriate endocrine gland. There is evidence of this phenomenon but the reasons for it are unclear. Endocrine glands may be destroyed by invading tumours, causing hormone deficiency.

CACHEXIA

This is the progressive weakness, loss of appetite, wasting and anaemia associated with cancer. The severity is usually indicative of the stage of development of the disease. The causes are not clear.

CAUSES OF DEATH IN MALIGNANT DISEASE

INFECTION

Acute infection is a common cause of death when superimposed on advanced malignancy. Predisposition to infection is increased by prolonged bedrest, and by depression of the immune system by cytotoxic drugs and irradiation. The most commonly occurring infections are pneumonia, septicaemia, peritonitis and pyelonephritis.

ORGAN FAILURE

A tumour may destroy so much tissue that an organ cannot function. Severe damage to vital organs, such as lungs, brain, liver and kidneys are common causes of death.

HAEMORRHAGE

This may occur when a tumour grows into and ruptures the wall of a vein or artery. The most common sites are the gastrointestinal tract, brain, lungs and the peritoneal cavity.

CAUSES OF NEOPLASMS

Normally cells divide in an orderly manner. Neoplastic cells have escaped from the normal controls and multiply in a disorderly manner. In malignancy they grow beyond their normal boundaries and develop varying degrees of dysplasia. Some factors are known to precipitate the changes found in tumour cells but the reasons for the uncontrolled cell multiplication are not known. The process of change is *carcinogenesis* and the agents precipitating the change are *carginogens*. Carcinogenesis may be of genetic and/or environmental origin and a clearcut distinction is not always possible.

CARCINOGENS

Environmental agents known to cause malignant changes in cells do so by progressive irreversible disorganisation and modification of the chromosomes and genes. It is impossible to specify a maximum 'safe dose' of a carcinogen. A small dose may initiate change but not be enough to cause malignancy unless repeated doses within a limited period of time have a cumulative effect. In addition there are widely varying latent periods between exposure and evidence of malignancy. There may also be other unknown factors. Environmental carcinogens include chemicals, irradiations and oncogenic viruses.

Chemical carcinogens
Some chemicals are carcinogens when absorbed, others are modified after absorption and become carcinogenic. Some known chemical carcinogens are:

Aniline dyes	Cigarette smoke
Arsenic compounds	Nickel compounds
Asbestos	Some fuel oils
Benzene derivatives	Vinyl chloride

Radiation carcinogens
Exposure to X-rays, radioactive isotopes, environmental

radiations and ultraviolet rays in sunlight may cause malignant changes in some cells and kill others. The cells are affected during mitosis so those normally undergoing continuous controlled division are most susceptible. These labile tissues include skin, mucous membrane, bone marrow, lymphoid tissue and germ cells in the ovaries and testes.

Oncogenic viruses

Viruses, some consisting of DNA and some of RNA, are known to cause malignant changes in animals and there are indications of similar involvement in man. Viruses enter cells and the addition of DNA or RNA to the cell's nucleus causes mutation. The mutant cells may be malignant.

Tumours of individual structures are described in the appropriate chapters.

SYSTEMS AND CAVITIES OF THE BODY

A *system* is a group of structures or organs which together carry out essential and related functions. The main systems of the body are:

Circulatory
Respiratory
Digestive
Urinary
Nervous
Endocrine
Reproductive
Osseus or skeletal
Muscular

The body as a whole is built round the bony framework or skeleton and consists of a number of different parts:
The *head* and *neck*
The *trunk*, which can be divided into the *thorax (chest)*, *abdomen* and *pelvis*
The *limbs*, both upper and lower

The four *cavities* that contain the main organs are:
Cranial
Thoracic
Abdominal
Pelvic

CRANIAL CAVITY

The cranial cavity contains the *brain*, and its *boundaries* are formed by the bones of the skull (Fig. 2:24).

Anteriorly	— 1 frontal bone
Laterally	— 2 temporal bones
Posteriorly	— 1 occipital bone
Superiorly	— 2 parietal bones
Inferiorly	— 1 sphenoid and 1 ethmoid bone and parts of the frontal, temporal and occipital bones

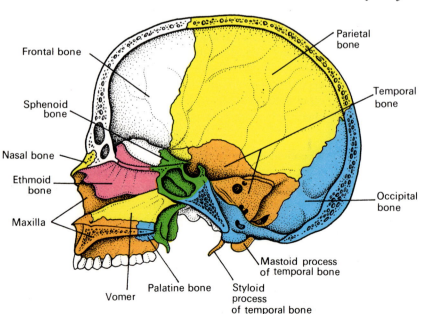

Frontal bone
Sphenoid bone
Nasal bone
Ethmoid bone
Maxilla
Vomer
Palatine bone
Styloid process of temporal bone
Mastoid process of temporal bone
Occipital bone
Temporal bone
Parietal bone

Figure 2:24 Bones forming the right half of the cranial cavity and the face — viewed from the left.

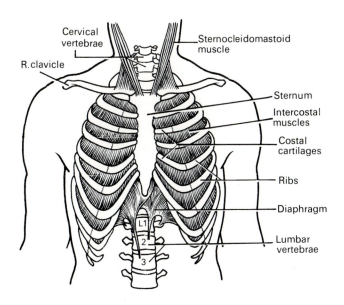

Figure 2:25   Structures forming the walls of the thoracic cavity and associated structures.

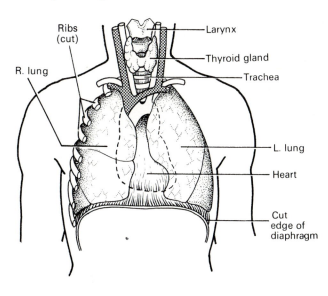

Figure 2:26   Main structures in the thoracic cavity and the root of the neck. Dotted line shows the position of the heart.

THORACIC CAVITY

This cavity is situated in the upper part of the trunk. Its *boundaries* are formed by a bony framework and supporting muscles (Fig. 2:25).

Anteriorly	— the sternum and costal cartilages of the ribs
Laterally	— 12 pairs of ribs and the intercostal muscles
Posteriorly	— the thoracic vertebrae and the intervertebral discs between the bodies of the vertebrae
Superiorly	— the structures forming the root of the neck
Inferiorly	— the diaphragm, a dome-shaped muscle

Contents
The main organs and structures contained in the thoracic cavity are (Fig. 2:26):

2 lungs
1 heart
1 trachea
2 bronchi
1 oesophagus
1 aorta
1 superior and 1 inferior vena cava
Numerous other blood vessels
Lymph vessels
Lymph nodes
Nerves
Glands

The *mediastinum* is the name given to the space between the lungs.

ABDOMINAL CAVITY

This is the largest cavity in the body and is oval in shape (Figs 2:27 and 2:28). It is situated in the main part of the trunk and its *boundaries* are:

Superiorly	— the diaphragm, which separates it from the thoracic cavity
Anteriorly	— the muscles forming the anterior abdominal wall
Posteriorly	— the lumbar vertebrae and muscles forming the posterior abdominal wall
Laterally	— the lower ribs and parts of the muscles of the abdominal wall
Inferiorly	— the pelvic cavity with which it is continuous

The abdominal cavity is divided into nine regions for the purpose of identifying the positions of the abdominal organs. These are named in Figure 2:29.

Contents
Most of the space in the abdominal cavity is occupied by the organs and glands involved in the digestion and absorption of food.

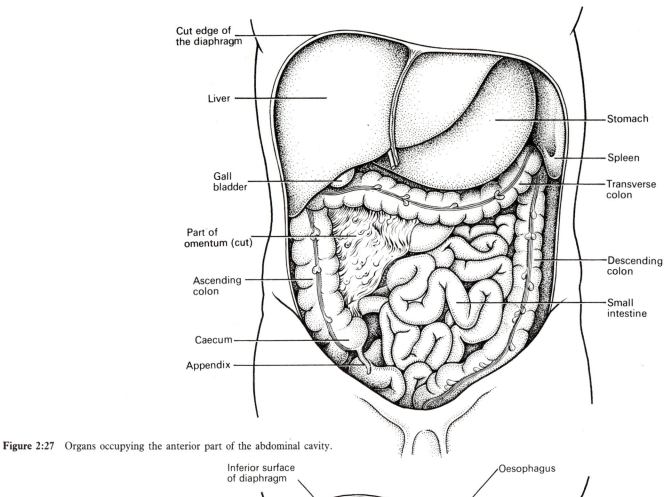

Figure 2:27   Organs occupying the anterior part of the abdominal cavity.

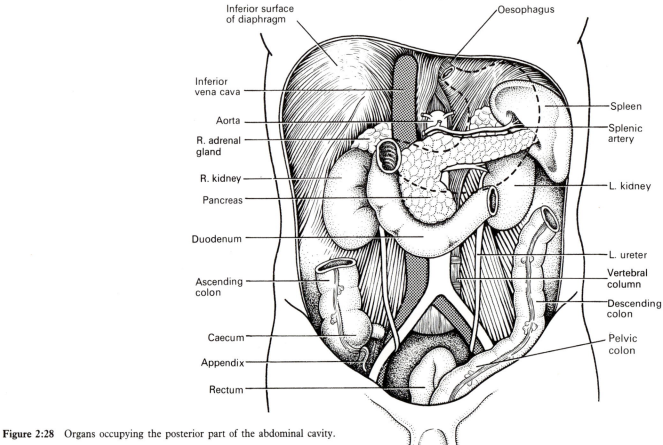

Figure 2:28   Organs occupying the posterior part of the abdominal cavity.

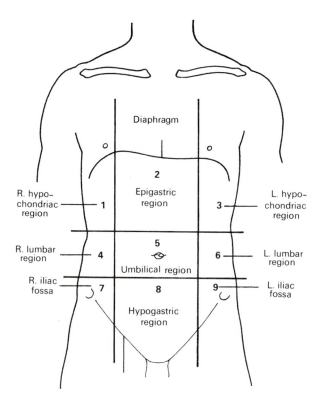

Figure 2:29   Regions of the abdominal cavity.

These are:
Stomach
Small intestine
Most of the large intestine
Liver
Gall bladder
Bile ducts
Pancreas

Other structures include:
Spleen
Kidneys and the upper part of the ureters
Adrenal (suprarenal) glands
Numerous blood vessels, lymph vessels, nerves
Lymph nodes

PELVIC CAVITY

The pelvic cavity is roughly funnel-shaped and extends from the lower end of the abdominal cavity (Figs 2:30 and 2:31). The *boundaries* are:

Superiorly — it is continuous with the abdominal cavity
Anteriorly — the pubic bones
Posteriorly — the sacrum and coccyx
Laterally — the innominate bones
Inferiorly — the muscles of the pelvic floor

Contents

The pelvic cavity contains the following structures:
Pelvic colon, rectum and anus
Some loops of the small intestine
Urinary bladder, lower parts of the ureters and the urethra
In the female, the organs of the reproductive system — the uterus, uterine tubes, ovaries and vagina (Fig. 2:30)
In the male, some of the organs of the reproductive system — the prostate gland, seminal vesicles, spermatic cords, deferent ducts (vas deferens) and ejaculatory ducts (Fig. 2:31)

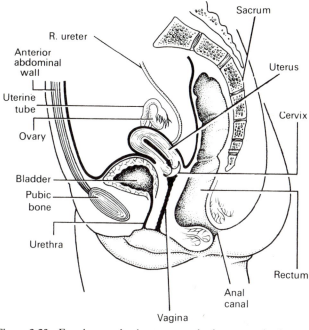

Figure 2:30   Female reproductive organs and other organs in the pelvic cavity.

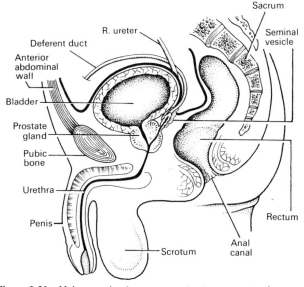

Figure 2:31   Male reproductive organs and other organs in the pelvic cavity.

3

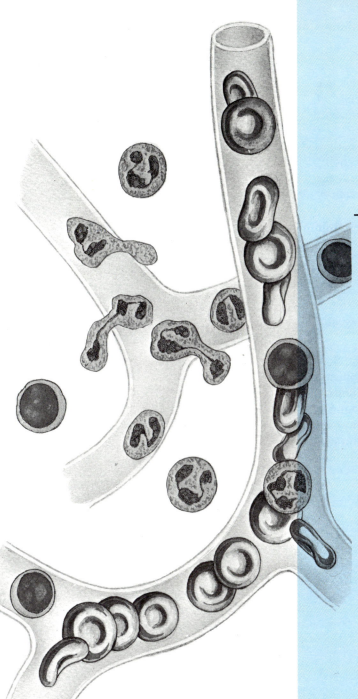

The Blood

3. The Blood

Blood is described as a connective tissue. It provides one of the means of communication between the cells of different parts of the body and the external environment, e.g., it carries:

1. Oxygen from the lungs to the tissues and carbon dioxide from the tissues to the lungs for excretion
2. Nutrient materials from the alimentary tract to the tissues and cell waste materials to the excretory organs, e.g., the kidneys
3. Hormones secreted by endocrine glands to their target glands and tissues
4. Heat produced in active tissues to other less active tissues
5. Protective substances, e.g., antibodies, to areas of infection
6. Materials that clot blood, preventing its loss from a ruptured blood vessel

Blood in the blood vessels is always in motion. The flow is such that body cells have a fairly constant environment. Changes in its composition remain within narrow limits.

Blood constitutes about 7% of body weight (about 5.6 litres in a 70 kg man). This proportion is less in women and considerably greater in children, gradually decreasing until the adult level is reached.

COMPOSITION OF BLOOD

Blood is composed of a straw-coloured transparent fluid, *plasma*, in which different types of *cells* are suspended. Plasma constitutes about 55% and cells about 45% of blood volume (see Fig. 3:1).

PLASMA

The constituents of plasma are water (90 to 92%) and dissolved substances, including:

Plasma proteins:
    albumin, globulin, fibrinogen, clotting factors
Inorganic salts (mineral salts):
    sodium chloride, sodium bicarbonate, potassium, magnesium, phosphorus, iron, calcium, copper, iodine, cobalt

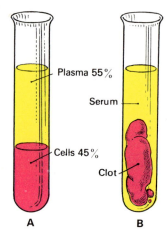

Figure 3:1 Blood in a test tube. A. Shows the percentage composition of plasma and cells. B. Shows the presence of a blood clot in serum.

Nutrient materials (from digested foods):
    monosaccharides from carbohydrates, amino acids from proteins, fatty acids and glycerol from fats, vitamins from most foods
Organic waste materials:
    urea, uric acid, creatinine
Hormones
Enzymes, e.g., various clotting factors
Antibodies
Gases:
    oxygen, carbon dioxide, nitrogen

PLASMA PROTEINS

Albumin. This is formed in the liver. It is the most abundant plasma protein and its main function is to maintain the plasma osmotic pressure at its normal level of about 25 mmHg (3.3 kPa).★

Globulins. Some are formed in the liver and some in lymphoid tissue. They are associated with a variety of activities:

1. The immune response to the presence of antigens (see p. 40).

★ 1 kilopascal (kPa) = 7.5 millimetres of mercury (mmHg)
    1 mmHg = 133.3 Pa = 1.3333 kPa

2. Transportation of some hormones and mineral salts, e.g., thyroid hormone, iodine, iron, copper
3. Inhibition of some proteolytic enzymes, e.g., trypsin, chymotrypsin

Clotting factors. These are substances essential for coagulation of blood (see p. 50).

Fibrinogen. This is synthesised in the liver and is essential for blood coagulation. *Serum* is plasma from which clotting factors have been removed.

MINERAL SALTS

These are involved in a wide variety of activities, including cell formation, contraction of muscles, transmission of nerve impulses, formation of secretions and maintenance of the balance between acids and alkalis. In health the blood is *slightly alkaline* in reaction. Alkalinity and acidity are expressed in terms of pH which is a measure of *hydrogen ion concentration* (see p. 57 and Fig. 3:2). The pH of blood is maintained at about 7.4 by an ongoing complicated series of chemical activities, involving buffering systems.

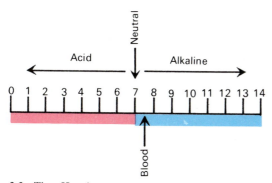

Figure 3:2   The pH scale.

NUTRIENT MATERIALS

Food is digested in the alimentary tract and the resultant nutrient materials are absorbed, i.e., monosaccharides, amino acids, fatty acids, glycerol and vitamins. Together with mineral salts they are required by all body cells to provide energy, heat, materials for repair and replacement, and for the synthesis of other blood components and body secretions.

ORGANIC WASTE PRODUCTS

Urea and uric acid are the waste products of protein metabolism. They are formed in the liver and conveyed in blood to the kidneys for excretion. Carbon dioxide, excreted by all cells, is conveyed to the lungs for excretion. It is carried bound to haemoglobin molecules and as part of bicarbonate ions.

HORMONES

These are chemical compounds synthesised by endocrine glands. Hormones pass directly from the cells of the glands into the blood which transports them to their target tissues and organs elsewhere in the body, where they influence activity.

ANTIBODIES

These are protective substances, consisting of complex protein molecules, produced by lymphoid tissue mainly in lymph nodes and in the spleen. Foreign material, e.g., microbes, act as *antigens*, stimulating lymphoid cells to produce protective antibodies (see p. 39).

GASES

Oxygen, carbon dioxide and nitrogen are transported round the body in solution in plasma. Oxygen and carbon dioxide are also transported in combination with haemoglobin in red blood cells (see p. 47). Atmospheric nitrogen enters the body in the same way as other gases and is present in plasma but it has no physiological function.

CELLULAR CONTENT OF BLOOD

There are three varieties of blood cells (Fig. 3:3).
Leukocytes or white cells
Erythrocytes or red cells
Thrombocytes or platelets

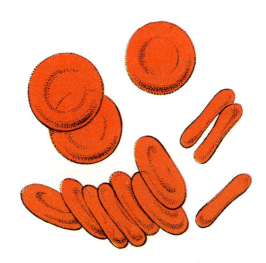

Figure 3:3   Normal blood cells after staining in the laboratory. Viewed through the microscope.

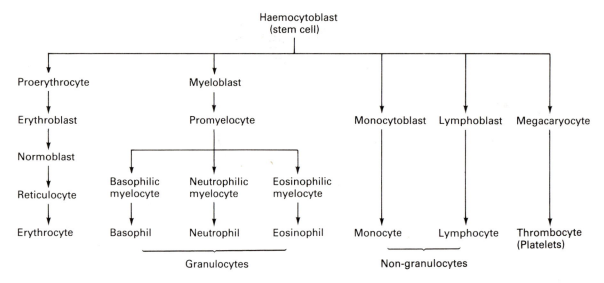

Figure 3:4 Stages in the development of blood cells in the red bone marrow.

LEUKOCYTES OR WHITE BLOOD CELLS

These are the largest blood cells. They contain nuclei and some have granules in their cytoplasm. There are three main types (see Table 3:1):

Granulocytes (polymorphonuclear leukocytes): neutrophils, eosinophils and basophils

Monocytes
Lymphocytes } non-granular leukocytes

Table 3:1 Numbers of different types of leukocyte in adult blood

Type of cell	Number × 10⁹/l	Percentage of total
Granulocytes		
Neutrophils	2.5 to 7.5	40 to 75
Eosinophils	0.04 to 0.44	1 to 6
Basophils	0.015 to 0.1	<1
Monocytes	0.2 to 0.8	2 to 10
Lymphocytes	1.5 to 3.7	20 to 50
TOTAL	5 to 9	100

GRANULOCYTES (Polymorphonuclear leukocytes)

These cells originate from stem cells (haemocytoblasts) in red bone marrow and go through several developmental stages before entering the blood. They follow a common line of development through *myeloblast* to *myelocyte* before differentiating into the three types. Their names represent the dyes they take up when stained in the laboratory. Eosinophils take up the red acid dye, eosin; basophils take up alkaline methylene blue; and basophils are purple because they take up both dyes.

Functions

Neutrophils. Their main function is to protect against foreign material that gains entry to the body, mainly microbes*, and to remove waste materials, e.g., cell debris. They are attracted in large numbers to any area of infec-

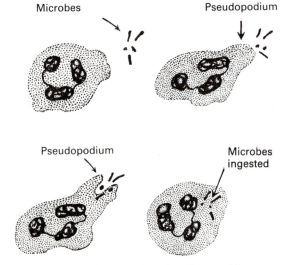

Figure 3:6 Diagram of phagocytic action of neutrophils.

* The term **microbe**, used throughout the text, includes all types of organisms that can only be seen by using a microscope. Specific microbes are named where appropriate.

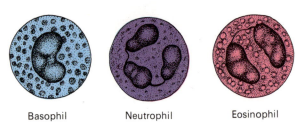

Figure 3:5 The granular leukocytes.

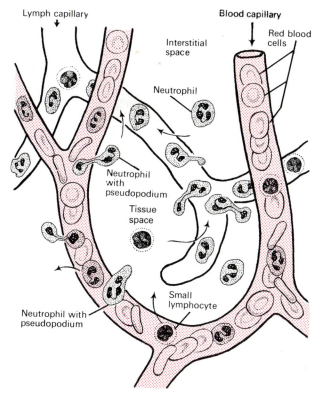

Figure 3:7 Diagram showing amoeboid movement of leukocytes.

tion by chemical substances, released by damaged cells, called *chemotaxins*. Neutrophils are capable of amoeboid movement and pass through the capillary walls in the infected area. Thereafter they ingest and kill the microbes by a process called *phagocytosis* (Figs 3:6 and 3:7). The pus that may form in the affected area consists of dead tissue cells, dead and live microbes, and phagocytes killed by microbes.

Eosinophils and basophils. They are less motile than neutrophils and their functions are not fully understood.

The number of eosinophils is increased in allergic conditions and parasitic infections. The cells contain *plasminogen*, the precursor of *plasmin* which breaks down *fibrin* in, e.g., blood clots. They also contain *histamine* which causes vasodilatation and increases the permeability of capillary walls, assisting the movement of phagocytes and protective substances into the tissue spaces.

Basophils contain the anticoagulant, *heparin*. Basophils and similar cells found in the *tissues* are called *mast cells*.

Figure 3:8 The non-granular leukocytes.

NON-GRANULAR LEUKOCYTES

The types of leukocyte with no granules in their cytoplasm are monocytes and lymphocytes and they make up 20% to 30% of all leukocytes (Fig. 3:8).

MONOCYTES

There are large mononuclear cells, believed to originate from haemocytoblasts in red bone marrow. Some circulate in the blood and are actively motile and phagocytic, while others migrate into the tissues where they develop into *macrophages*.

The macrophage system, sometimes called the *lymphoreticular system* consists of fixed phagocytic cells which multiply in situ. These cells are found in a wide variety of tissues, e.g., as:

Histiocytes in connective tissues
Microglia in the brain
Kupffer cells in liver sinusoids
Alveolar macrophages in the lungs
Sinus-lining macrophages (reticulum cells) in the spleen, lymph nodes and thymus gland

Macrophages function in close association with monocytes in the blood and with other cells in lymphoid tissue. They are actively phagocytic and if they encounter large amounts of foreign or waste material, they tend to multiply at the site and 'wall off' the area, isolating the material, e.g., in the lungs when foreign matter has been inhaled.

LYMPHOCYTES

Lymphocytes are associated with the protection of the body against foreign material. They develop from haemocytoblasts (stem cells) in red bone marrow then spread in the blood stream to lymphoid tissue elsewhere in the body where they are *activated*, i.e., they become able to respond to *antigens* (foreign material) (Fig. 3:9). There are two types of lymphocyte that sometimes function independently but usually in collaboration. They are *T-lymphocytes*, activated in the thymus gland, and *B-lymphocytes*, activated in lymphoid tissue elsewhere in the body, possibly in the walls of the intestine. Thereafter some cells of both types circulate in the blood and some settle in lymphoid tissue, mainly in lymph nodes and the spleen.

When *activated lymphocytes* encounter antigens they develop specific protective capabilities and each type divides into two groups; *effector cells* that promote destruction of their specific antigen, and *memory cells* that remain in lymphoid tissue and multiply, passing on their specific protective characteristics to subsequent generations of cells.

T-lymphocytes are *sensitised* when they encounter an antigen for the first time. Effector cells act directly against antigens in conjunction with phagocytes. The memory cells confer *cell-mediated* immunity and subsequent encounters with the same antigen lead to the proliferation of sensitised lymphocytes. Examples of such antigens include:

1. Cells regarded by lymphocytes as abnormal, e.g., those that have been invaded by viruses, cancer cells, tissue transplant cells
2. Pollen from flowers and plants
3. Fungi
4. Bacteria
5. Some large molecule drugs, e.g., penicillin

The fact that the body does not normally develop immunity to its own cells is due to the fine balance that exists between the immune reaction and its suppression. *Auto-immune diseases* are due to the disturbance of this balance.

B-lymphocytes are activated by microbes and their toxins. Then they grow into *plasma cells* which secrete *antibodies* (immunoglobulins). The antibodies promote phagocytosis of foreign particles and neutralise toxins. The memory cells confer *humoral immunity*, i.e., they react to subsequent encounters with the same antigen by stimulating the production of effector cells and antibodies. Small clusters of plasma cells secreting the same antibodies are called *clones*.

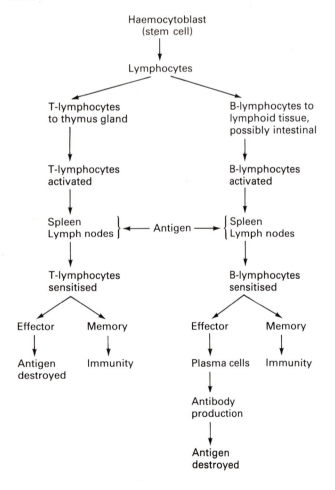

Figure 3:9 Summary of the lines of development of lymphocytes.

SPECIFIC DEFENCE MECHANISMS

IMMUNE RESPONSE (Antibody production)

When antigens, e.g., microbes, are encountered for the first time there is a *primary response* in which a low level of antibodies can be detected in the blood after about 2 weeks. This only persists if there is another encounter with the antigen within a short period of time (2 to 4 weeks). The second encounter produces a *secondary response* in which there is a marked increase in antibody production (see Fig. 3:10). Further increases can be achieved by later encounters but eventually a maximum is reached. This principle is used in active immunisation against infectious diseases.

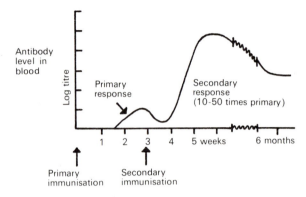

Figure 3:10 The antibody response.

ACQUIRED IMMUNITY

Immunity may be acquired *naturally* or *artificially* and both forms may be *active* or *passive*. Active immunity means that the individual has responded to an antigen and produced his own antibodies. In passive immunity he has been given antibodies produced by someone else (Fig. 3:11).

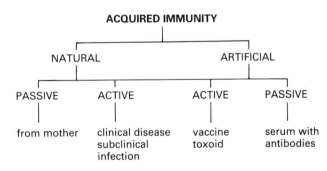

Figure 3:11 Summary of types of immunity.

Active naturally acquired immunity
The body may be stimulated to produce its own antibodies in two ways.

1. By *having the disease*. During the course of the illness, B-lymphocytes grow into plasma cells which produce antibodies (immunoglobulins) in sufficient quan-

tities to overcome the infecting microbes. After the individual recovers, the lymphocytes retain the ability to produce the specific antibodies, conferring immunity to future infection by the same microbes.

2. By *having a subclinical (subliminal) infection*. In this case the microbial infection is not sufficiently severe to cause clinical disease but is able to stimulate B-lymphocytes and establish immunity.

Active artificially acquired immunity
This type of immunity develops in response to the administration of dead or artificially weakened microbes (*vaccines*) or detoxicated toxins (*toxoids*). The vaccines and toxoids retain the antigenic properties that stimulate the development of immunity but they cannot cause the disease. Many microbial diseases can be prevented by immunisation, e.g.,

Anthrax	Poliomyelitis
Cholera	Smallpox
Diphtheria	Tetanus
Measles	Tuberculosis
Mumps	Whooping cough

Active immunisation against some infections confers lifelong immunity, e.g., diphtheria, whooping cough, mumps. In other infections the immunity may last for a number of years or for only a few weeks before revaccination is necessary. Apparent loss of immunity may be due to infection with a different strain of the same microbe which has different antigenic properties but causes the same clinical illness, e.g., viruses that cause the common cold and influenza. Age and nutrition are believed to be important factors in establishing and maintaining immunity. In the elderly and when nutrition is poor the production of lymphocytes, especially B-lymphocytes, is reduced and the primary and secondary response may be inadequate.

Passive naturally acquired immunity
This type of immunity is acquired before birth by the passage of maternal antibodies across the placenta to the fetus. The variety of different antibodies provided depends on the mother's active immunity. The child's lymphocytes are not stimulated and the immunity is shortlived.

Passive artificially acquired immunity
In this type, ready-made antibodies, in human or animal serum, are injected into the recipient. The source of the antibodies may be an individual who has recovered from the infection, or animals, commonly horses, that have been artificially actively immunised. Antiserum is usually administered *prophylactically* to prevent the development of disease in people who have been exposed to infection, or *therapeutically* after the disease has developed. Proteins in serum sometimes cause sensitisation of lymphocytes that may be damaging if encountered a second time.

ABNORMAL IMMUNE REACTIONS
ALLERGIC REACTIONS (Hypersensitivity)
These are harmful forms of immune reaction that occur following second or subsequent encounters with antigens. They may be:
1. Antibody-mediated, i.e., immediate
2. Cell-mediated, i.e., delayed
3. Mixed antibody- and cell-mediated

ANTIBODY-MEDIATED REACTIONS
The antibodies formed by plasma cells combine with their antigens to form *antigen/antibody complexes* (Ag/Ab or immune complexes).

Anaphylactic reactions
The Ag/Ab complexes stimulate mast cells (connective tissue cells) to release *histamine* and similar substances. When in excess these substances cause *anaphylactic shock* characterised by difficulty in breathing, spasm of bronchi, laryngeal oedema, dilatation and increased permeability of capillaries. The allergen is usually injected into the body, e.g., protein in antiserum used to treat infection, insect bite, large molecule drugs such as penicillin. Other forms of this type of allergy include food allergies, hay fever (see p. 112), asthma (see p. 115). The severity of anaphylaxis varies considerably. In extreme cases it may cause death.

CELL-MEDIATED REACTIONS
These reactions develop gradually over a period of 24 to 48 hours after the second encounter with the antigen. They are due to the activities of specifically activated T-lymphocytes and phagocytes. The lymphocytes release non-antibody substances called *lymphokines* that are believed to amplify the immune response, retain macrophages in the affected area and promote phagocytosis. The antigens include:
1. Intracellular microbes, e.g., those causing tuberculosis, measles, mumps
2. Some vaccines, e.g., against smallpox
3. Some metals and chemical compounds that combine with protein in the skin and cause allergic contact dermatitis

MIXED REACTIONS
These include the reaction to transfusion of incompatible blood (see p. 48) and many autoimmune diseases.

NON-SPECIFIC DEFENCE MECHANISMS

SKIN AND MUCOUS MEMBRANE

When skin and mucous membrane are intact and healthy they provide a physical barrier to invading microbes. The outer horny layer of skin can be penetrated by only a few microbes and the mucus secreted by mucous membrane traps microbes and other foreign material on its sticky surface. Sebum and sweat secreted on to the skin surface contain antibacterial and antifungal substances.

ANTIMICROBIAL SUBSTANCES IN BODY SECRETIONS

Hydrochloric acid, present in high concentrations in gastric juice, kills the majority of ingested microbes.

Lysozyme is a small molecule protein with antibacterial properties present in granulocytes, tears, and other body secretions. It is not present in sweat, urine and cerebrospinal fluid.

Immunoglobulins in nasal secretions and saliva are able to inactivate some viruses.

Saliva is secreted into the mouth and washes away food debris that may serve as culture media for microbes. Its slightly acid reaction inhibits the growth of some microbes.

Phagocytosis is stimulated by the presence of foreign material. Neutrophils and macrophages are attracted to areas invaded by microbes which they ingest and destroy.

INFLAMMATION

This consists of a series of changes that take place in the tissues in response to injury. The causes may be classified as follows:
1. Physical agents, e.g., heat, cold, mechanical injury, ultraviolet and ionising radiations
2. Chemical agents – organic, e.g., bacterial toxins
   – inorganic, e.g., acids, alkalis
3. Microbes, e.g., bacteria, viruses, protozoa, fungi
4. Immunological, i.e., the body's response to antigens

ACUTE INFLAMMATION

Episodes of acute inflammation are usually of short duration and may range from mild to very severe. The changes that take place in the tissues are active hyperaemia, exudation and migration of leukocytes.

ACTIVE HYPERAEMIA

This is due to increased arterial blood flow, as distinct from passive hyperaemia due to increased venous pressure.

Following injury there is arteriolar dilatation and increased blood flow to the capillaries. This may be influenced by modifications in the sympathetic nerve control of the arterioles but a number of chemical mediators are believed to have a more powerful effect, e.g., histamine and serotonin produced by damaged cells.

EXUDATION

In an area of inflammation the amount of interstitial fluid is increased and its composition changed. The amount depends on the net filtration pressure (hydrostatic and osmotic) across capillary walls, which is raised when arterioles dilate. It consists of molecules of water and other substances small enough to pass through pores in the walls. The composition is changed because the permeability of capillary walls is increased by histamine and similar substances released by damaged cells. This enables large molecules, including antibodies and fibrinogen, to enter the tissue spaces. Some interstitial fluid returns to the capillaries at the venous end but most of the inflammatory exudate, phagocytes and cell debris are removed in lymph vessels because the pores of lymph vessels are larger, and the pressure inside is lower, than in blood capillaries.

MIGRATION OF LEUKOCYTES

The migrating cells are the phagocytic granulocytes and monocytes. The number involved and the duration of the migration depend on the severity of the inflammation and, to some extent, on the cause. Pyogenic bacteria, such as *Streptococcus pyogenes*, *Staphylococcus aureus* and *Streptococcus pneumoniae* are associated with large scale migration. Exudation aids migration of phagocytes because it reduces the rate of blood flow and increases its viscosity. The leukocytes stick to the capillary walls then pass between the endothelial cells into the interstitial spaces by diapedesis. This process is assisted by the increased permeability of capillary walls.

CHEMOTAXIS

This is the chemical attraction of leukocytes to an area of inflammation. The role of chemotaxins and the way in which they work is not fully understood. They increase the permeability of capillary walls but their action also appears to be more specific. Large numbers of phagocytes accumulate in the inflamed area but substances of smaller size remain in the capillaries. The known chemotaxins include bacterial toxins, prostaglandins, antigen/antibody complexes, substances released by damaged cells.

EFFECTS OF ACUTE INFLAMMATION

Most of these assist the body to overcome the cause of inflammation.

Phagocytosis

Neutrophils and macrophages in the tissue spaces engulf particles of biological and non-biological origin. Biological material includes dead and damaged cells, microbes, damaged connective tissue fibres. Most biological material is digested by enzymes inside phagocytes. Some microbes resist digestion and provide a possible source of future infection, e.g., *Mycobacterium tuberculosis*. Non-biological materials which cannot be digested include inhaled dust particles, chemical substances and material contaminating wounds. Many phagocytes may die in an inflamed area if the material they ingest resists digestion, or if the number of particles is excessive. When this happens the phagocytes disintegrate and release material that may become fibrosed or cause further damage.

Immune response (see also p. 40)

When inflammation is caused by microbes, antibodies formed by lymphocytes following a previous encounter with a particular microbe or its toxin filter through the capillary walls in the inflammatory exudate. The antibodies promote phagocytosis of the microbes and neutralise their toxins. Harmful immune responses include rejection of transplanted tissues, and allergic reactions to normally innocuous substances, e.g., foods, pollen, talcum powder, clothing materials, feathers, dander from animal hair.

Toxin dilution

The water in inflammatory exudate dilutes damaging and waste materials in the area, and assists their removal from the site. This is of particular importance when injurious chemicals and bacterial toxins are involved.

Fibrin formation

Fibrinogen present in inflammatory exudate is acted upon by thromboplastin released from damaged cells and forms an insoluble fibrin network. This may:

1. Wall off the inflamed area, preventing the spread of the cause
2. Bind together the cut edges of a wound during primary healing

Tissue swelling

This is the result of the increased blood flow and exudation. The effects depend on the site:

1. In a joint — limitation of movement
2. In the larynx — interference with breathing
3. In a confined space, such as inside the skull or under the periosteum of bone — severe pain due to pressure on nerves

OUTCOMES OF ACUTE INFLAMMATION

Resolution

This occurs when the cause has been successfully overcome. The inflammatory process is reversed and:

1. Damaged cells are phagocytosed
2. Fibrin strands are broken down by enzymes
3. Waste material is removed in lymph and blood vessels

Suppuration (pus formation)

Pus consists of dead phagocytes, dead cells, cell debris, fibrin, inflammatory exudate and microbes. It is contained within a membrane of new blood capillaries, phagocytes and fibroblasts. The most common causative microbes are *Staphylococcus aureus* and *Streptococcus pyogenes*. Small amounts of pus form *boils* and larger amounts form *abscesses*. *Staphylococcus aureus* produces the enzyme coagulase which converts fibrinogin to fibrin, localising the pus. *Streptococcus pyogenes* produces streptolysins that promote the breakdown of connective tissue, causing spreading infection. Healing, following pus formation, is by granulation and fibrosis.

Superficial abscesses tend to rupture through the skin and discharge pus. Healing is usually complete unless there is extensive tissue damage.

Deep-seated abscesses may have a variety of outcomes. There may be:

1. Rupture and complete discharge of pus on to the surface, followed by healing
2. Rupture and limited discharge of pus on to the surface, then the development of a chronic abscess with an infected open channel (sinus)
3. Rupture and discharge of pus into an adjacent organ or cavity, forming an infected channel open at both ends (fistula)
4. Eventual removal of pus by phagocytes, followed by healing
5. Enclosure of pus by fibrous tissue that may become calcified, harbouring live organisms which may become a source of future infection

CHRONIC INFLAMMATION

The processes involved are the same as in acute inflammation. The main differences are:

1. The inflammation is less severe and lasts longer
2. There is destruction of considerably more tissue
3. More fibrous tissue is formed

Chronic inflammation may follow acute inflammation or be of slow onset.

FOLLOWING ACUTE INFLAMMATION

Any form of acute inflammation may develop into the chronic form if resolution is not complete, e.g., if live microbes remain at the site, as in some deep-seated abscesses, wound infections and bone infections.

SLOW ONSET OF INFLAMMATION

This may be caused by:

1. Infection by low virulence organisms in an area with a poor blood supply, e.g., endocarditis caused by non-haemolytic streptococci
2. Inorganic materials when:
   a. an internal stitch is not dissolved
   b. toxic silicic acid is formed when silicon compounds, inhaled in dust, dissolve
3. Hypersensitivity developed following repeated exposure to some chemicals, e.g., in contact dermatitis, skin proteins are altered when some chemicals are absorbed, the altered proteins act as antigens, stimulating the production of antibodies and initiating the inflammatory process.

LEUKOCYTE DISORDERS

LEUKOPENIA

This is the name of the condition in which the total blood leukocyte count is less than $4 \times 10^9/1$ (4000/mm³).

GRANULOCYTOPENIA

This is a general term used to indicate an abnormal reduction in the numbers of circulating granulocytes (polymorphonuclear leukocytes), sometimes called neutropenia because 40 to 75% of granulocytes are neutrophils. A reduction in the number of circulating granulocytes occurs when production does not keep pace with the normal removal of cells or when the life span of the cells is reduced. Extreme shortage or the absence of granulocytes is called *agranulocytosis*. A temporary reduction occurs in response to inflammation but the numbers are usually quickly restored. Inadequate granulopoiesis (production of granulocytes) may be caused by:

1. Drugs, e.g., cytotoxic drugs, phenylbutazone, phenothiazines, some sulphonamides and antibiotics
2. Irradiation damage to granulocyte precursors in the bone marrow by, e.g., X-rays, radioactive isotopes
3. Malignant disease of red bone marrow, e.g., leukaemias involving other leukocytes
4. Severe infections

LYMPHOPENIA

This relatively uncommon condition, in which there is a decrease in the number of lymphocytes in the blood, is associated with Hodgkin's disease, non-lymphocytic leukaemias and treatment of patients with, e.g., ionising radiations, cytotoxic drugs and corticosteroids. It also occurs in children in some rare congenital diseases.

LEUKOCYTOSIS

An increase in the number of circulating leukocytes occurs as a normal protective reaction in a variety of pathological conditions, especially in response to infections. When the infection subsides the leukocyte count returns to normal.

Pathological leukocytosis exists when a blood leukocyte count of more than $11 \times 10^9/1$ (11 000/mm³) is sustained and is not consistent with the normal protective function. One or more of the different types of cell is involved.

LEUKAEMIA

Leukaemia is a *malignant myeloproliferative disease* of the bone marrow that results in the uncontrolled increase in the production of leukocytes and/or their precursors. The total leukocyte count is usually raised but in some cases it may be normal or even low. The proliferation of leukaemic cells crowds out other blood cells formed in bone marrow, causing anaemia, thrombocytopenia and leukopenia.

Causes of leukaemia

Some causes of leukaemia are known but many cases cannot be accounted for. Some people have a genetic predisposition that is triggered by environmental factors.

Ionising radiations produced by X-rays and radioactive isotopes are known to cause malignant changes in the precursors of white blood cells. Leukaemia may develop at any time after irradiation, even 20 or more years later. The cells are affected in different ways:

1. Genetic material of the cells may be changed. Some cells die while other reproduce at an abnormally rapid rate
2. Dormant viruses already present in the cells may be activated, but conclusive evidence is not yet available

Chemical substances. Some chemicals encountered in the general or work environment are known to change the genetic make-up of the white cell precursors in the bone marrow. These include benzene and its derivatives, asbestos, cytotoxic drugs, chloramphenicol.

Types of leukaemia

Leukaemias are usually classified according to the type of cell involved and whether they are acute or chronic (see Table 3:2).

Table 3:2 Types of leukaemia and the cells involved

Type of leukaemia	Type of cell involved
Myeloid (myelocytic, myelogenous, myeloblastic)	Granulocytes
Lymphocytic (lymphoblastic)	Lymphocytes
Monocytic	Monocytes

In chronic leukaemia there tends to be a higher leukocyte count than in the acute form and the cells in circulation are more mature. During its course there may be sudden exacerbations similar to the corresponding acute type.

Acute myeloid and lymphocytic leukaemias

These types usually have a sudden onset, reaching a climax within a few weeks or months. Small haemorrhages may occur due to reduced platelet production.

Acute myeloid leukaemia. This occurs at any age, but most commonly, between 25 and 60 years.

Acute lymphocytic leukaemia. This form of the disease occurs most commonly in children under 10 years although it may occur up to about 40 years of age.

Chronic myeloid leukaemia

There is a gradual increase in the number of immature granulocytes in the blood. In the later stages anaemia, secondary haemorrhages, infections and fever become increasingly severe. It is slightly more common in men than women and usually occurs between the ages of 20 and 40 years. Although treatment may appear to be successful, death usually occurs within about 5 years.

Chronic lymphocytic leukaemia

There is enlargment of the lymph nodes and hyperplasia of the aggregates of lymphoid tissue throughout the body. The lymphocyte count is considerably higher than normal. Lymphocytes accumulate in the bone marrow and there is progressive anaemia and thrombocytopenia. It is three times more common in males than females and it occurs mainly between the ages of 50 and 70 years. There is great variation in survival times and death is usually due to repeated infections of increasing severity.

ERYTHROCYTES OR RED BLOOD CELLS
(Fig. 3:3)

These are circular bi-concave non-nucleated discs, about 7 micrometres* (μm) in diameter.

Each cell contains *haemoglobin* which combines with oxygen in the lungs and transports it to all body cells.

Erythrocyte count. This is the number of red cells per litre or cubic millimetre of whole blood (1 mm^3 = 10^{-3}/l).

Normal: males 5×10^{12}/l to 5.5×10^{12}/l
(5 to 5.5 million/mm^3)
females 4.5×10^{12}/l to 5×10^{12}/l
(4.5 to 5 million/mm^3)

Haemocrit or packed cell volume (PCV). This is the volume of red blood cells in 1 litre or 100 mm^3 of whole blood.

Normal: 0.44 to 0.5 litres (44 to 50 mm^3)

* micrometre = 10^{-6} metre

Mean cell volume.
Normal: 85 to 95 femtolitres
(1 fl = 10^{-15} litres)

Haemoglobin.
Normal: males   13 to 18 grams per decilitre
(1 dl = 10^{-1} litres)
females 11.5 to 16.5 g/dl)

Mean cell haemoglobin (MCH). This is the average weight of haemoglobin per cell.
Normal: 25 to 30 picograms per cell
(1 pg = 10^{-12} gram)

Mean cell haemoglobin concentration (MCHC). This is the amount of haemoglobin in 1 decilitre or 100 mm^3 of red cells.
Normal: 30 to 35 grams per decilitre of red cells

DEVELOPMENT AND LIFE-SPAN OF ERYTHROCYTES

Erythrocytes are formed in red bone marrow and pass through several stages of development before entering the blood. Their life-span in the circulation is about 120 days (Fig. 3:4).

The process of development of red blood cells is *erythropoiesis*. They originate from *haemocytoblasts* and follow two main lines of development to the stage of maturity when they enter the blood stream. One is *maturation of the cell* and the other, the *formation of haemoglobin* inside the cell. Normal erythrocytes develop satisfactorily along both lines.

Maturation of the cell. This depends on a number of factors, especially the presence of *vitamin B$_{12}$* and *folic acid*. These are present in sufficient quantity in a normal diet containing dairy products, meat and green vegetables. If the diet contains more than is needed it is stored in the liver. Absorption of vitamin B$_{12}$ depends on a glycoprotein called *intrinsic factor* secreted by *parietal cells* in the *gastric glands*. Together they form the *intrinsic factor-vitamin B$_{12}$ complex* (IF–B$_{12}$) absorbed by cells in the wall of the distal part of the ileum. Only the vitamin B$_{12}$ part enters the blood. *Folic acid* is absorbed by cells in the wall of the jejunum where it undergoes change before entering the blood.

Haemoglobin. This is a complex protein, consisting of globin and an iron-containing substance called *haem*, synthesised inside developing erythrocytes in red bone marrow. A normal diet containing meat, eggs, wholemeal bread and green vegetables contains more iron than the daily requirement. Women lose blood during menstruation and need more iron throughout the reproductive years and during pregnancy.

Haemoglobin in erythrocytes combines with oxygen to form *oxyhaemoglobin*. In this way the bulk of oxygen absorbed from the lungs is conveyed to the cells of the body. Haemoglobin is also involved in the transport of the waste product, carbon dioxide, from the body cells to the lungs for excretion.

Erythropoiesis

The number of red cells remains fairly constant, which means that the bone marrow produces erythrocytes at the rate they are destroyed. The primary stimulus to increased erythropoiesis is *hypoxia*, i.e., deficient oxygen supply to body cells. This occurs when, e.g.:

1. The oxygen-carrying power of blood is reduced by, e.g., haemorrhage or excessive haemolysis due to disease
2. The oxygen tension in the air is reduced, as at high altitudes

Hypoxia stimulates the production of the hormone *erythropoietin*, mainly by the kidneys. Erythropoietin stimulates an increase in the production of pro-erythroblasts and the release of increased numbers of reticulocytes into the blood. It is assumed the erythropoietin regulates normal red cell replacement, i.e., in the absence of hypoxia.

Destruction of erythrocytes

The life-span of erythrocytes is about 120 days and their breakdown, or *haemolysis*, is carried out by *phagocytic reticuloendothelial cells*. These cells are found in many tissues but the main sites of haemolysis are the spleen, bone marrow and liver. As erythrocytes age, changes in their walls make them gradually more susceptible to haemolysis. Iron released by haemolysis is retained in the body and re-used in the bone marrow to form haemoglobin. *Biliverdin* is formed from the protein part of the erythrocytes. It is almost completely reduced to the yellow pigment *bilirubin*, before it is bound to plasma proteins, mainly albumin, and transported to the liver. In the liver it is changed from a fat-soluble to a water-soluble form before it is excreted as a constituent of bile.

BLOOD GROUPS

Different blood groups are associated with genetically determined differences in antigens on red blood cell membranes and antibodies in blood serum. There are two main systems used to classify blood donated for administration by transfusion. If the donor's blood does not *match* that of the recipient the *incompatibility* results in agglutination and lysis of donated red cells after transfusion.

ABO SYSTEM

In some people there are genetically determined antigens on red cell membranes and *natural antibodies* in serum.

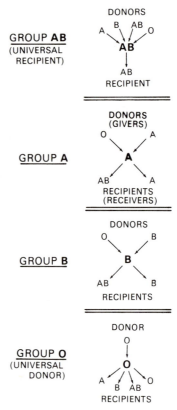

Figure 3:12   Blood groups showing their compatibility.

The antibodies are not associated with acquired immunity, described above. Table 3:3 and Figure 3:12 summarise the system.

Table 3:3   Summary of the ABO system

Blood group	Red cell antigens	Antibodies in serum	Can donate to groups	Can receive from groups
AB	A and B	None	AB	All groups
A	A	anti-B	A and AB	A and O
B	B	anti-A	B and AB	B and O
O	None	anti-A and anti-B	All groups	O

RHESUS SYSTEM

In over 80% of people the *rhesus factor* (antigen) is present on red blood cell membranes, i.e., they are *rhesus positive* (Rh+ve). Administration of Rh+ve blood to Rh−ve recipients stimulates an immune response with the production of antibodies that cause heamolysis of the transfused red cells. As in active immunisation, second and subsequent encounters with Rh+ve red cells lead to a sharp increase in antibody production.

Rhesus incompatibility of maternal and fetal blood is described on page 48.

ERYTHROCYTE DISORDERS

ANAEMIA

In anaemia there is not enough haemoglobin available to carry sufficient oxygen from the lungs to supply the needs of body tissues. A number of different types occur due to shortage of red cells, abnormal development of cells or abnormal formation of haemoglobin.

NORMOCYTIC ANAEMIA (Normochromic)

The cells are normal (MCV 90) but the numbers are reduced and the proportion of reticulocytes in circulation may be raised. It occurs when:
1. Erythropoiesis has been stimulated by loss of red cells through, e.g., haemorrhage, excess haemolysis
2. There is chronic inflammation and infection

HYPOPLASTIC AND APLASTIC ANAEMIA

This occurs when erythropoiesis is unable to keep pace with the continuous normal removal of aged red cells from the circulation. The degree of bone marrow depression varies from mild hypoplasia to aplasia. The known causes include:
1. Drugs, e.g., cytotoxic drugs, some anti-inflammatory and anti-epileptic drugs, some sulphonamides and antibiotics
2. Irradiation
3. Some chemicals, e.g., benzene and its derivatives
4. Chronic nephritis
5. Virus diseases, including hepatitis
6. Destruction of bone marrow by, e.g., malignant disease

MEGALOBLASTIC ANAEMIA (Macrocytic)

The development of erythrocytes stops at the *megaloblast* stage due to deficiency of vitamin B_{12} and/or folic acid. This reduces the rate of DNA and RNA synthesis and delays cell division. The cells continue to grow but remain immature. Circulating cells are larger than normal and some are nucleated (MCV>94). The haemoglobin content of each cell is normal or raised. The cells are fragile and their life span is reduced to between 40 and 50 days. Depressed production and early lysis cause the anaemia.

PERNICIOUS ANAEMIA (Vitamin B_{12} deficiency anaemia)

Dietary deficiency
Intake of less than the required $1\mu g$ of vitamin B_{12} per day

is rare, even in a strictly vegetarian diet. It occurs only when the diet generally is deficient.

Intrinsic factor (IF) deficiency or absence
This occurs when the production of IF by the parietal cells in gastric glands is deficient, e.g.,
1. After gastrectomy
2. In chronic gastritis, malignant disease, after exposure to irradiation
3. In autoimmune disease when antibodies destroy the intrinsic factor and parietal cells

Deficiency of intrinsic factor–vitamin B_{12} complex (IF–B_{12}). Microbes may colonise the small intestine and use or destroy IF–B_{12} before it reaches the terminal ileum where it is absorbed. This occurs when the contents are slow moving or static, especially in diverticuli or a blind loop of bowel left after surgery. It is sometimes called *blind loop syndrome.*

Malabsorption of intrinsic factor–vitamin B_{12} complex. This may follow resection of terminal ileum or inflammation of ileum, e.g., in Crohn's regional ileitis or tropical sprue.

Complications of pernicious anaemia
These may appear before the signs of anaemia. They include:
1. Subacute combined degeneration of the spinal cord in which nerve fibres in the posterior and lateral columns of white matter become demyelinated, causing ataxia, spastic paralysis, loss of sensation and pain. The link between this condition and pernicious anaemia is not clear
2. Atrophy and ulceration of the tongue and glossitis

FOLIC ACID DEFICIENCY ANAEMIA

Deficiency in the bone marrow causes a form of megaloblastic anaemia not associated with degeneration of the spinal cord. It may be due to:
1. Dietary deficiency, e.g., in infants if there is delay in establishing a mixed diet, or in pregnancy when the normal requirement is raised
2. Malabsorption from the jejunum caused by, e.g., coeliac disease, tropical sprue, anticonvulsant drugs
3. Interference with utilisation by, e.g., cytotoxic and anticonvulsant drugs

HYPOCHROMIC ANAEMIA (Iron deficiency, microcytic)

The amount of haemoglobin in each cell is regarded as below normal when the MCH is less than 27 pg/cell. It is caused by deficiency of iron in the bone marrow and may be due to dietary deficiency, excessively high requirement or malabsorption.

DIETARY DEFICIENCY

Normal iron requirements with low intake

This occurs in:
1. Babies fed only on milk because it contains very little iron and, until a mixed diet is established, the baby is dependent on iron stored in the liver before birth
2. Some ethnic groups where there are dietary restrictions for religious or cultural reasons

High iron requirements with normal or low intake

This type of anaemia is usually due to chronic blood loss and may be exacerbated by low iron intake. The causes include:

Chronic peptic ulcers	Haemorrhoids
Menorrhagia	Carcinoma
Intestinal ulceration	

MALABSORPTION OF IRON

Iron absorption is usually increased following haemorrhage but may be reduced in abnormalities of the stomach, duodenum or jejunum, including:
1. Resection of stomach or upper part of small intestine
2. Hypochlorhydria, e.g., in malignant disease, pernicious anaemia

HAEMOLYTIC ANAEMIA

This occurs when red cells are destroyed while in circulation or are removed from the circulation prematurely because the cells are abnormal or the spleen is overactive.

CONGENITAL HAEMOLYTIC ANAEMIAS

In these diseases genetic abnormality leads to the synthesis of abnormal haemoglobin and increased cell membrane friability, reducing cell oxygen-carrying capacity and life-span. The most common forms are sickle cell anaemia and thalassaemia.

Sickle cell anaemia

The abnormal haemoglobin molecules become mis-shapen when deoxygenated, making the erythrocytes sickle shaped. A high proportion of abnormal molecules makes the sickling permanent. The life-span of cells is reduced by early haemolysis. Sickle cells do not move smoothly through the small blood vessels. This tends to increase the viscosity of the blood, reducing the rate of blood flow and leading to intravascular clotting, ischaemia and infarction. The anaemia is due to early haemolysis of irreversibly sickled cells.

Negroes are affected more than other races but in this race, some affected individuals have a degree of immunity to malaria because the life-span of the sickled cells is less than the time needed for the parasites to mature.

Thalassaemia

There is reduced globin synthesis with resultant reduced haemoglobin production together with increased friability of the cell membrane and early haemolysis. Severe cases may cause death in infants or young children.

ACQUIRED HAEMOLYTIC ANAEMIA

In this context acquired means haemolytic anaemia in which no familial or racial factors have been identified. There are several causes.

Chemical agents

These substances cause early or excessive haemolysis, e.g.:
1. Some drugs, especially when taken long-term in large doses, e.g., phenacetin, primaquine, sulphonamides
2. Chemicals encountered in the general or work environment, e.g., lead and arsenic compounds
3. Toxins produced by microbes, e.g., *Streptococcus pyogenes*, *Clostridium welchii*

AUTOIMMUNE HAEMOLYTIC ANAEMIA

In this disease antibodies are formed which react with red cell antigens, causing haemolysis. It may be acute or chronic and primary or secondary to other diseases, e.g., carcinoma, rheumatoid arthritis and other connective tissue diseases, virus infections, Raynaud's disease, other autoimmune conditions.

Cold antibody type. The antibodies react with red cells at temperatures below 30° C, causing agglutination and haemolysis, usually in blood vessels in hands and feet on exposure to cold.

Warm antibody type. The antibodies are active at normal temperature (37° C). The haemolysis is usually chronic but there may be fluctuations with periodic acute exacerbations.

INCOMPATIBLE BLOOD TRANSFUSION

In some people there are antigens on the erythrocyte cell membrane that react with natural antibodies in the plasma of others. Incompatibility of blood occurs when the antigens on donor's red cells react with the antibodies in recipient's plasma, causing erythrocyte agglutination and haemolysis (see Table 3:3, p. 46 and Fig. 3.12, p. 46).

HAEMOLYTIC DISEASE OF THE NEWBORN

This is caused by the presence of the *Rhesus (Rh) factor* antigens on the erythrocyte membrane. Incompatibility occurs when the Rh factor is absent from the mother's cells and the child has inherited it from the father. A small number of fetal erythrocytes cross the placental barrier and enter the mother's blood. Her immune system treats the fetal Rh factor as antigen and develops specific antibodies against it. These antibodies then pass from mother to fetus and haemolyse fetal erythrocytes before birth and for some time afterwards. It is believed that fetal erythrocytes are able to cross the placenta as it becomes more permeable near the end of pregnancy. The severity of the condition depends on the concentration, or titre, of maternal antibodies. In second and subsequent pregnancies red cells from an Rh+ve fetus stimulate a dramatic increase in antibody production by the mother unless she is immunised with serum containing antibodies.

OTHER CAUSES OF HAEMOLYTIC ANAEMIA

These include:

1. Parasitic diseases, e.g., malaria
2. Irradiation, e.g., X-rays, radioactive isotopes
3. Destruction of blood trapped in tissues in, e.g., severe burns, crushing injuries

HAEMORRHAGIC ANAEMIA

Severe loss of blood leads to immediate anaemia. In chronic haemorrhage the anaemia develops gradually when erythropoiesis does not keep pace with the rate of cell loss.

Table 3:4   Summary of the types of anaemia

Type of anaemia	Bone marrow	Blood	Spleen
Normocytic	Erythropoiesis inadequte	Reduced numbers	
Hypoplastic	Reduced erythropoiesis	Reduced number of normal cells	
Megaloblastic	Deficiency vitamin B_{12} and/or folic acid	Large cells, some nucleated	
Hypochromic	Deficiency of iron	Reduced haemoglobin	
Haemolytic	Genetic abnormality	Abnormal cells	Early lysis
	Immune reaction		Early lysis

POLYCYTHAEMIA

There are an abnormally large number of erythrocytes in the blood. This increases blood viscosity, slows the rate of flow and increases the risk of intravascular clotting, ischaemia and infarction.

RELATIVE INCREASE IN ERYTHROCYTE COUNT

This occurs when the erythrocyte count is normal but the blood volume is reduced by fluid loss, e.g., excessive serum exudate from extensive superficial burns.

TRUE INCREASE IN ERYTHROCYTE COUNT

Physiological
Prolonged hypoxia stimulates erythropoiesis and the number of cells released into the normal volume of blood is increased. This occurs in people living at high altitudes where the oxygen tension in the air is low and the partial pressure of oxygen in the alveoli of the lungs is correspondingly low. Each cell carries less oxygen so more cells are needed to meet the body's oxygen needs.

Pathological
The reason for this increase in circulating red cells, sometimes to twice the normal, is not known. It may be due to hypoxia of the red bone marrow caused by, e.g., pulmonary disease or bone marrow cancer.

POLYCYTHAEMIA VERA

In this condition there is abnormal excessive production of the erythrocyte precursors, i.e., *myeloproliferation*. This raises the haemoglobin level and the haemocrit (relative proportions of cells to plasma). The blood viscosity is increased and may lead to cerebral, coronary and mesenteric thrombosis. Aplastic anaemia and leukaemia may also be present.

THROMBOCYTES OR PLATELETS

These are very small non-nucleated discs derived from the cytoplasm of megakaryocytes in red bone marrow (see Fig. 3:4). They contain a variety of substances in inert form including *platelet factors III and IV*, serotonin (5-hydroxytryptamine), *platelet fibrinogen* and *adenosine triphosphate* (ATP). When platelets encounter damaged epithelium they undergo a 'release reaction' during which their inert contents are activated and initiate the clotting mechanism.

The normal blood platelet count is between $200 \times 10^9/1$ and $350 \times 10^9/1$ (200 000 to 350 000/mm³). The control of platelet production is not yet entirely clear but it is believed that one stimulus is a fall in platelet count and that a substance called *thrombopoietin* is involved. The life-span of platelets is between 8 and 11 days and those not used in the process of *haemostasis* (cessation of bleeding) are destroyed by macrophages, mainly in the spleen.

Haemostasis
When a blood vessel is damaged, loss of blood is stopped by a series of closely linked processes. Platelets play an essential part, e.g.,

1. Their surface becomes sticky, they adhere to the blood vessel wall around the damaged area and clump together, forming a *haemostatic plug*

2. They stimulate constriction of the blood vessel, reducing the size of the opening
3. They undergo 'release reaction' which initiates the process of blood clot formation

Blood clot formation. This is a complex process and only a few stages are included here. The factors involved are listed in Table 3:5. Their numbers represent the order in which they were discovered and not the order of participation in the clotting process.

Damage to endothelium and platelets triggers a series of events in which *prothrombin* is activated to *thrombin*. Thrombin acts on *fibrinogen*, converting it to an insoluble thread-like mesh of *fibrin* which traps blood cells. Together fibrin and entrapped blood cells form the blood clot. After a time it shrinks, squeezing out *serum*, i.e., a clear sticky fluid that consists of plasma from which fibrinogen has been removed.

Fibrinolysis. After the clot has formed the process of removing it and healing the damaged blood vessel begins. The breakdown of the clot, or fibrinolysis, is the first stage. An inactive substance called *plasminogen* is converted to *plasmin* by activators released from the damaged endothelial cells. Plasmin initiates the breakdown of fibrin to soluble products and these are treated as waste material and removed by phagocytosis. As the clot is removed the healing process restores the integrity of the blood vessel wall.

Plasminogen + activators → Plasmin
Plasmin + fibrin   → Waste material

Table 3:5   Blood clotting factors

I	Fibrinogen
II	Prothrombin
III	Tissue factor
IV	Calcium
V	Labile factor, proaccelerin, Ac-globulin
VII	Stable factor, proconvertin
VIII	Antihaemophilic globulin (AHG), antihaemophilic factor A
IX	Christmas factor, plasma thromboplastin component (PTA), antihaemophilic factor B
X	Stuart–Power factor
XI	Plasma thromboplastin antecedent (PTA), antihaemophilic factor C
XII	Hageman factor
XIII	Fibrin stabilising factor

(There is no Factor VI)

Vitamin K is essential for synthesis of Factors II, VII, IX and X

HAEMORRHAGIC DISEASES

THROMBOCYTOPENIA

This is defined as a blood platelet count below $150 \times 10^9/l$ (150 000/mm^3) but spontaneous capillary bleeding does not usually occur unless the count falls below $30 \times 10^9/l$ (30 000/mm^3). It may be due to reduced rate of platelet production or increased rate of destruction.

Reduced platelet production
This is usually due to:
1. Platelets being crowded out of the bone marrow in bone marrow diseases, e.g., leukaemias, pernicious anaemia, malignant tumours
2. X-radiation, radioactive isotopes, cytotoxic and other drugs, e.g., chloramphenicol, chlorpromazine, phenylbutazone, sulphonamides

Increased platelet destruction
A reduced platelet count occurs when production of new cells does not keep pace with destruction of damaged and worn out cells.

IDIOPATHIC THROMBOCYTOPENIC PURPURA

This is an autoimmune disease, possibly triggered by a virus infection, in which anti-platelet antibodies are formed that lead to the destruction of platelets by phagocytosis. The severity of the disease varies but when the platelet count is very low there may be severe bruising, haematuria, gastrointestinal or cranial haemorrhages.

SECONDARY THROMBOCYTOPENIC PURPURA

This may occur in association with red bone marrow diseases, excessive irradiation and some drugs, e.g., digoxin, chlorothiazides, quinine, sulphonamides. When the spleen is enlarged and congested, platelets are trapped in it and are destroyed by macrophages.

VITAMIN K DEFICIENCY
HAEMORRHAGIC DISEASE OF THE NEWBORN

Spontaneous haemorrhage from the umbilical cord and intestinal mucosa occurs in babies when the stored vitamin K obtained from the mother before birth has been used up and the intestinal bacteria needed for its synthesis in the infant's bowel are not yet established.

DEFICIENT ABSORPTION IN ADULTS

Vitamin K is fat soluble and bile salts are required in the colon for absorption. Deficiency may occur when there is prolonged obstruction to the biliary tract or in any other disease where fat absorption is impaired, e.g., coeliac disease.

DIETARY DEFICIENCY

This is rare because a sufficient supply is usually synthesised in the intestine by bacterial action. However, deficiency may occur during treatment with drugs that sterilise the bowel.

FIBRINOGEN DEFICIENCY (Hypofibrinogenaemia)

Fibrinogen is produced by the liver and synthesis may be deficient in any condition which causes progressive fibrosis of liver or liver failure.

DISSEMINATED INTRAVASCULAR COAGULATION (DIC)

In this disease the complex mechanisms that control intravascular clotting break down and clots form in many small blood vessels in a wide range of internal organs. It occurs in association with a number of conditions in which substances that promote clotting are produced, e.g.:

1. Severe shock, especially when due to microbial infection
2. Septicaemia when endotoxins are released by Gram-negative bacteria
3. Premature separation of placenta when amniotic fluid enters maternal blood
4. Acute pancreatitis when digestive enzymes are released into the blood
5. Malignant tumours with widely dispersed metastases

GENETIC ABNORMALITIES

In each body cell there are 46 chromosomes arranged in 23 pairs. One pair are sex chromosomes. In the female, the two sex chromosomes are identical and are called X chromosomes. In the male, each cell has one X chromosome and one Y chromosome.

Female — XX   Male — XY

Each sex cell, the ovum or spermatazoon, has only 23 chromosomes, one from each of the 23 pairs. This means that each ovum has an X chromosome and each spermatazoon has either an X or a Y chromosome.

The gene that promotes the antihaemophilic activity of Factors VIII and IX in the clotting mechanism are carried on X chromosomes only. When this gene is abnormal or absent, *haemophilia* occurs in males. It is transmitted to the next generation by symptom-free females and by affected males.

HAEMOPHILIA A

In this disease, Factor VIII is abnormal and less biologically active. For descriptive purposes abnormal X chromosomes are designated X'. Thus the distribution of chromosomes is:

Female: XX' or XX   Ovum: X or X'
Male: X'Y or XY   Spermatazoon: X' or X or Y

The system of inheritance of haemophilia is as follows (see Fig. 3:13):

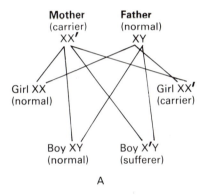

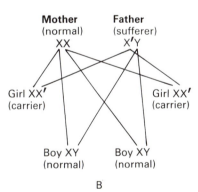

Figure 3:13 Diagram showing the pattern of inheritance of abnormal chromosomes in haemophilia. A. Mother is a carrier. B. Father is a sufferer.

1. When an X' ovum combines with a Y sperm the male child (X'Y) is a haemophiliac
2. When an X' ovum combines with an X sperm the female child (X'X) is a carrier. The disease potential is present in her cells but the normal X chromosome provides the healthy gene that promotes clotting.
3. When an X' sperm combines with an X ovum the female child (X'X) is a carrier

Sufferers from haemophilia may experience severe and prolonged bleeding at any site with little evidence of trauma. Recurrent bleeding into joints is common, causing severe pain and, in the long term, cartilage is damaged (see p. 336).

HAEMOPHILIA B (Christmas disease)

This is the less common sex-linked genetic haemorrhagic disease. Factor IX is not activated, resulting in deficiency of thromboplastin. The abnormal gene is on the X chromosome and the distribution and characteristics of the disease are the same as those in haemophilia A.

VON WILLEBRAND'S DISEASE

In this disease Factor VIII is abnormal and, as the inheritance is not sex-linked, haemorrhages due to defective clotting occur in either males or females.

PURPURA

This is generalised bleeding due to increased capillary wall fragility and/or thrombocytopenia. It is seen as small haemorrhagic spots (petechiae) on the skin and mucous membranes or as large blotchy areas. Purpura occurs as a complication of some diseases, e.g., leukaemia, thrombocytopenia, severe anaemia, severe acute infections, scurvy. It may also result from some drug treatment, e.g., antibiotics, corticosteroids, sulphonamides.

Electrolytes and Body Fluids

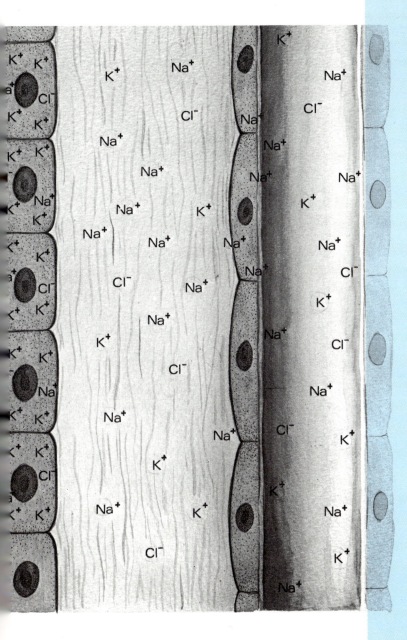

4. Electrolytes and Body Fluids

Before going on to consider the importance and nature of body fluids it is necessary to understand the meaning of such terms as electrolytes, isotopes, pH and buffer substances. To do so it is essential to survey some of the principles of the physics and chemistry of atoms, elements and compounds. Compounds which contain carbon are classified as organic and all others as inorganic. The body fluids contain both.

Before discussing the compounds found in solution in the water of the body, the atom and its main constituent parts have to be considered.

ATOMS, MOLECULES AND IONS

The atom is described as the smallest particle of an element which can exist as a stable entity. All atoms of an element are identical, but the atoms of each element are different from those of all other elements. This will be more clearly understood when the structure of the atom has been described.

Structure
Atoms are made up of a number of types of particles, but only three are considered here.

Protons are particles present in the nucleus or central part of the atom. Each proton has *one unit of positive electrical charge* and *one unit of mass*.

Neutrons are also found in the nucleus of the atom. They have *no electrical charge* and *one unit of mass*.

Electrons are particles which revolve in orbit around the nucleus of the atom at a distance from it, as the planets revolve round the sun. Each electron carries *one unit of negative electrical charge* and its mass is so small that it can be disregarded when compared with the mass of the other particles.

Table 4:1 summarises the characteristics of these subatomic particles.

Table 4:1 Characteristics of subatomic particles

Particle	Mass	Electric charge
Proton	1 unit	1 positive
Neutron	1 unit	neutral
Electron	negligible	1 negative

In all atoms the number of positively charged protons in the nucleus is *equal to* the number of negatively charged electrons in orbit around the nucleus.

The difference between elements is in the *numbers* of these essential particles which make up their atoms. The planetary electrons revolve in rings or shells around the nucleus and there is an optimum number of electrons in each shell (Fig. 4:1).

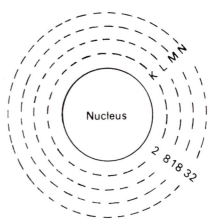

Figure 4:1 Diagram of the atom showing the nucleus and four electron shells.

Atomic number
The atomic number of an element is the number of protons in the nuclei, or electrons in orbit around the nuclei of the atoms of that element. For example, the atomic number of oxygen is 8 because it has 8 protons and 8 planetary electrons. Hydrogen is 1 and sodium 12 (Fig. 4:2).

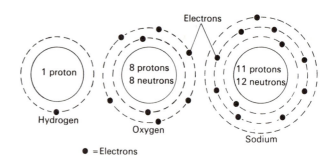

Figure 4:2 Diagram of the structure of atoms of different elements showing protons, neutrons and orbiting electrons.

Formation of compounds

It was mentioned earlier that the atoms of each element have a specific number of electrons in orbit around the nucleus. When the number of electrons in the outer shell of an element is the optimum number (Fig. 4:1), the element is described as inert or chemically unreactive, i.e. it will not easily combine with other elements to form compounds. These elements include the inert gases — neon, argon, krypton, xenon and radon.

Elements which have incomplete outer shells of electrons are reactive and will combine with other elements which also have incomplete outer electron shells. In the formation of *electrovalent* or *ionic compounds*, electrons are transferred from one element to another. For example, when sodium (Na) combines with chlorine (Cl) to form sodium chloride (NaCl) there is a transfer of the only electron in the outer shell of the sodium atom to the outer shell of the chlorine atom. This makes the outer shell (the M shell) of the chlorine up to its full capacity of 8 electrons and leaves the sodium part of the compound with a complete L shell and no electrons in the M shell (Fig. 4:3).

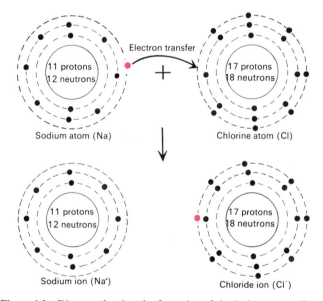

Figure 4:3 Diagram showing the formation of the ionic compound sodium chloride.

The number of electrons is the only change which occurs in the atoms in this type of reaction. There is no change in the number of protons or neutrons in the nuclei of the atoms. The chloride part now has *18 electrons*, each with one negative electrical charge, and *17 protons*, each with one positive charge. The sodium part has lost one electron, leaving *10 electrons* orbiting round the nucleus with *11 protons*. When sodium chloride is dissolved in water the two atoms separate, i.e., they *ionise*, and the imbalance of protons and electrons leads to the formation of two *charged particles* called *ions*. Sodium, with the positive charge, is a *cation*, written Na$^+$, and chloride is an *anion*, written Cl$^-$. By convention the number of elec-

trical charges carried by an ion is indicated by the superscript plus or minus signs.

The nature of electrical charge, which is used daily in lighting, heating and so on, is closely related to the structure of the atom. Power initiated by water falling from a height or by steam under pressure is used to drive a dynamo to generate electricity, which is the movement of electrons. The electric current is then led away along a metal wire called a *conductor*. A substance which prevents the flow of electricity, or electrons, is a *non-conductor* or *insulator*.

An ionic compound, e.g. sodium chloride, in solution in water is called an *electrolyte* because it can conduct electricity. Non-ionising compounds are not electrolytes, e.g. carbohydrates.

In this discussion, sodium chloride has been used as an example of the formation of an ionic compound and to illustrate electrolyte activity. There are, however, many other electrolytes within the human body which, though in relatively small quantities, are equally important. Although these substances may enter the body in the form of compounds, such as sodium bicarbonate, they are usually discussed in the ionic form, that is as sodium ions (Na$^+$) and bicarbonate ions (HCO$_3^-$).

The bicarbonate part of sodium bicarbonate is derived from carbonic acid (H$_2$CO$_3$). All inorganic acids contain hydrogen combined with another element, or with a group of elements called a *radical* which acts like a single element. Hydrogen combines with chlorine to form hydrochloric acid (HCl) and with the *phosphate radical* to form phosphoric acid (H$_3$PO$_4$). When these two acids ionise they do so thus:

$$HCl \rightarrow H^+ \; Cl^-$$
$$H_3PO_4 \rightarrow 3H^+ \; PO_4^{---}$$

In the second example, three atoms of hydrogen have each lost one electron, all of which have been taken up by one unit, the phosphate radical, to make a phosphate ion with three negative charges.

A large number of compounds present in the body are not ionic and therefore have no electrical properties when dissolved in water, e.g., carbohydrates.

Isotopes

In an oversimplification it was stated earlier that all atoms of an element are identical. This is true as far as the numbers or protons and electrons are concerned, but some atoms of the same element have a *different number of neutrons in the nucleus*. This does not affect the chemical properties of these atoms but it does affect their weight, e.g., there are three forms of the hydrogen atom, all with one electron. The most common form has one proton in the nucleus, a second form has one proton and *one neutron* in the nucleus and a third form has one proton and *two neutrons* in the nucleus. These three forms of hydrogen are called *isotopes* (Fig. 4:4).

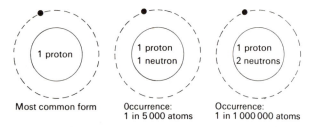

Most common form   Occurrence: Occurrence:
                    1 in 5 000 atoms   1 in 1 000 000 atoms

Figure 4:4 Diagram of the isotopes of hydrogen.

Atomic weight

The atomic weight of an element is the sum of the protons and neutrons in the nucleus of the atoms of the element. Taking into account the isotopes of hydrogen and the proportions in which they occur, the atomic weight of hydrogen is 1.008, although for many practical purposes it can be taken as 1.

Oxygen has an atomic weight of 16. This means that a specific number of atoms of oxygen weigh 16 times more than the same number of atoms of hydrogen.

Chlorine has an atomic weight of 35.5, because it exists in two isotopic forms, one form has 18 neutrons in the nucleus, and the other 20. Because the proportion of these two forms are not equal, the *average atomic weight* which emerges is 35.5.

Atomic weight is expressed simply as a number until a standard scale of measurement of weight is applied, e.g. grams, milligrams, micrograms.

Molecules and molecular weight

Molecules are the units of substances consisting of two or more atoms, e.g., a molecule of sodium chloride (NaCl) consists of one atom of chlorine and one atom of sodium; and a molecule of atmospheric oxygen (O_2) consists of two atoms of oxygen.

The molecular weight of a compound is the sum of the atomic weights of the elements which form its molecules. For example:

Water (H.OH)

2 hydrogen atoms (atomic weight 1)	2	
1 oxygen atom (atomic weight 16)	16	
Molecular weight	=18	

Sodium bicarbonate ($NaHCO_3$)

1 sodium atom (atomic weight 23)	23	
1 hydrogen atom (atomic weight 1)	1	
1 carbon atom (atomic weight 12)	12	
3 oxygen atoms (atomic weight 16)	48	
Molecular weight	= 84	

Molecular weight, like atomic weight, is expressed simply as a figure until a scale of measurement of weight is applied.

Equivalent weight (Chemical equivalent)

Equivalent weight is the weight of an element that combines with, or displaces, 16 parts by weight of oxygen, i.e., 1 atomic weight. Other elements may be used to establish equivalence, e.g., chlorine (atomic weight 35.5) or hydrogen (atomic weight 1).

If someone in a laboratory wished to synthesise water he would find that he required twice as much hydrogen as oxygen and would need to provide the appropriate conditions for their chemical combination to form water. Thus two measures of hydrogen are *equivalent to* one of oxygen.

To make 36.5 grams of hydrochloric acid (HCl) it would be necessary to take 1 gram of hydrogen and 35.5 grams of chlorine. One gram of hydrogen *contains the same number of atoms* as 35.5 grams of chlorine. Provided these two gases were given the appropriate conditions for chemical combination, hydrochloric acid would be formed with no hydrogen or chlorine left. It can be said, therefore, that 1 atomic weight of hydrogen is *equivalent* to 35.5 atomic weights of chlorine.

Using hydrochloric acid as an example, it can be seen that the equivalent weight of both elements is their atomic weight, but if hydrochloric acid is combined with calcium it is found that one atomic weight of calcium combines with two atomic weights of chlorine releasing two atomic weights of hydrogen. The equation for this reaction is:

$$Ca + 2HCl \rightarrow CaCl_2 + H_2$$

The weight of calcium, therefore, which is equivalent to one atomic weight of chlorine, is half an atomic weight of calcium (calcium: atomic weight = 40.8, therefore its equivalent weight = 20.4).

Any scale of measurement of the weight of these elements can be used provided the same scale is used for all the elements involved in the reaction.

The electrolytes found within the human body may be expressed in terms of *milliequivalents per litre of body fluids (mEq/l)*. This is *the equivalent weight in milligrams in each litre* of body fluid.

Molar concentration

This is the term used for expressing the concentration of substances present in the body fluids recommended in the *Système Internationale* which has now been adopted in the United Kingdom.

The mole (mol) is the molecular weight in grams of a substance (formerly called 1 gram molecule). One mole of all substances contains 6.023×10^{23} molecules, or atoms.

A molar solution is a solution in which 1 mole of a substance is dissolved in 1 litre of solvent. In the human body the solvent is water or fat.

Molar concentration may be used to measure quantities of electrolytes, non-electroytes, ions and atoms, e.g., molar solutions of the following substances mean:

1 mole of sodium chloride molecules (NaCl) = 58.5 g per litre
1 mole of sodium ions (Na^+) = 23 g per litre
1 mole of carbon atoms (C) = 12 g per litre
1 mole of atmospheric oxygen (O_2) = 32 g per litre

In physiology this system has the advantage of being a measure of the number of particles (molecules, atoms, ions) of substances present because molar solutions of different substances contain the same number of particles. It has the advantage over the measure milliequivalents per litre because it can be used for non-electrolytes, in fact for any substance of known molecular weight.

Many of the chemical substances present in the body are in very low concentrations so it is more convenient to use smaller metric measures, e.g., *millimoles per litre* (mmol/l) or *micromoles per litre* (μmol/l) as a biological measure (see Table 4:2)

Table 4:2 Examples of normal plasma levels

Substance	Amount in SI units	Amount in other units
Chloride	95–105 mmol/l	95–105 mEq/l
Sodium	138–148 mmol/l	1 38–148 mEq/l
Glucose	3.5–5.5 mmol/l	60–100 mg/100 ml
Iron	14.28 μmol/l	80–160 μg/100 ml

HYDROGEN ION CONCENTRATION (pH)

The number of hydrogen ions present in a solution is a measure of the acidity of the solution. The maintenance of the normal hydrogen ion concentration within the body is an important factor in the environment of the cells.

A standard scale for the measurement of the hydrogen ion concentration in solution has been developed. All acids do not ionise completely when dissolved in water. The hydrogen ion concentration is a measure, therefore, of the amount of *dissociated acid* (ionised acid) rather than of the total amount of acid present. Strong acids dissociate more freely than weak acids, e.g., hydrochloric acid dissociates freely into H^+ and Cl^-, while carbonic acid dissociates much less freely into H^+ and HCO^-. The number of *free hydrogen ions* in a solution is a *measure of its acidity* rather than an indication of the type of molecule from which the hydrogen ions originated.

The alkalinity of a solution depends on the number of hydroxyl ions (OH^-), or other negatively charged ions present. Water is a neutral solution because every molecule contains one hydrogen ion and one hydroxyl radical. For every molecule of water (H.OH) which dissociates, one hydrogen ion (H^+) and one hydroxyl ion (OH^-) are formed, neutralising each other.

The scale for measurement of pH was developed taking water as the standard. It was found that 1 molecule in 550 000 000 molecules of water ionises into a hydrogen ion

and a hydroxyl ion. This is the same proportion as 1 gram hydrogen ion in 10 000 000 litres of water. Therefore 1 litre of water contains $\dfrac{1}{10\,000\,000}$ of a gram of hydrogen ion. In order to make these figures more manageable this can be written:

$$1 \text{ litre of water contains } \frac{1}{10^7} \text{ g hydrogen ion}$$

$$(10 = 10^1; \ 100 = 10^2; \ 10\,000 = 10^4)$$

The fraction $\dfrac{1}{10^7}$ may be written 10^{-7}, the negative power indicating that this is a fraction. $\dfrac{1}{10\,000\,000}$ may be written $\dfrac{1}{10^7}$ or 10^{-7}.

For every-day use, only the '*power*' figure without the negative sign is used and the symbol pH placed before it.

Thus in a neutral solution such as water, where the number of hydrogen ions is balanced by the same number of hydroxyl ions, the pH = 7. The range of this scale is from 0 to 14. If the pH is 0 it means that 1 litre contains $\dfrac{1}{1} = 1$ gram of hydrogen ion; or at the other end of the scale if there are no hydrogen ions present it may be written $\dfrac{1}{10^{14}}$ or pH 14. It will be noted that a change of pH of *one* at any level in the scale means an increase or decrease by a factor of 10 in hydrogen ion concentration.

A pH reading *below* 7 indicates an *acid solution*, while readings *above* 7 indicates an *alkaline solution* (Fig. 4:5).

Ordinary litmus paper indicates whether a solution is

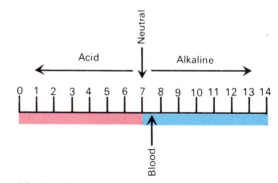

Figure 4:5 The pH scale.

acid or alkaline by colouring blue for alkaline and red for acid. Other specially treated absorbent papers give an approximate measure of pH by a colour change. When accurate measurements of pH are required, sensitive pH meters are used.

pH values of the body fluids

Body fluids have pH values that must be maintained within relatively narrow limits for normal cell activity. The pH values are not the same in all parts of the body, e.g., the normal range of pH values of the following secretions are:

Blood	7.4 to 7.5
Saliva	6.4 to 7.4
Gastric juice	1.5 to 3.5
Pancreatic juice	7.5 to 8.2
Urine	4.5 to 8.0

The pH value in an organ is produced by its secretion of acids or alkalis which will establish the optimum level. The highly acid pH of the gastric juice is maintained by hydrochloric acid secreted by the parietal cells in the walls of the gastric glands. The low pH value in the stomach provides the environment best suited to the functioning of the enzyme pepsin that begins the digestion of dietary protein. Saliva has a pH of between 6.4 and 7.4 — the optimum value for the action of ptyalin, the enzyme present in saliva which initiates the digestion of carbohydrates. Ptyalin action is inhibited when food containing it reaches the stomach and is mixed with acid gastric juice.

The blood has a pH value of between 7.4 and 7.5. This, therefore, is the general pH level in the body. The range of pH of the blood compatible with life is 7.0 to 7.8. The metabolic activity of the body cells produces certains acids and alkalis which tend to alter the pH of the tissue fluid and blood. To maintain the pH within the normal range, there are substances present in blood that act as *buffers*.

Buffers

Buffers are chemical substances such as phosphates, bicarbonates and some proteins which are able to 'bind' free hydrogen ions or hydroxyl ions and so prevent a change in pH. For example, if there is sodium hydroxide (NaOH) and carbonic acid (H_2CO_3) present, both will ionise to some extent, but they will also react together to form sodium bicarbonate ($NaHCO_3$) and water (H.OH). One of the hydrogen ions from the acid has been 'bound' in the formation of the bicarbonate radical and the other by combining with the hydroxyl radical to form water.

$$NaOH \quad + \quad H_2CO_3 \quad \rightarrow \quad NaHCO_3 \quad + \quad H.OH$$

sodium hydroxide	carbonic acid	sodium bicarbonate	water

The ability of the complex buffer systems of the blood to 'bind' hydrogen ions, or neutralise acids, is called the *alkali reserve* of the blood. If the pH of the blood falls below 7.4, the reserve of alkali is reduced by an increase in production of hydrogen ions, and the condition of *acidosis* exists. When the reverse situation pertains and the pH is raised above 7.5, the increased alkali produced uses up the *acid reserve*, and the state of *alkalosis* exists.

The buffer system serves to prevent dramatic changes in the pH values in the blood, but it can only function effectively if there is some means by which excess acid or alkali can be excreted from the body. The organs most active in this way are the lungs and the kidneys. When respiration is decreased there is an accumulation of carbon dioxide in the body which uses up the alkali reserve of the blood resulting in the development of acidosis. On the other hand, if there is 'over-breathing', which results in excessive excretion of carbon dioxide, the condition of alkalosis may develop.

The kidneys have the ability to form ammonia which combines with the acid products of protein metabolism which are then excreted in the urine.

The buffer and excretory systems of the body together maintain the *acid–base balance* so that the pH range of the blood remains within normal, but narrow, limits.

BODY FLUIDS

The total body water in adults of average build is about 60% of body weight, i.e., about 42 litres. This proportion is higher in young people and in adults below average weight. It is lower in the elderly and in obesity in all age groups. About 22% of body fluid is extracellular and the remainder, intracellular.

The extracellular fluid consists of fluid in the blood and lymph vessels, cerebrospinal fluid and fluid in the interstitial spaces of the body. Interstitial fluid, or tissue fluid, bathes all the cells of the body. The constituents of blood which do not pass into the tissue fluid are those with molecules too large to pass through the semipermeable walls of the capillaries. Substances which remain in the blood vessels are plasma proteins, erythrocytes, thrombocytes and leukocytes, except those which are amoeboid. Electrolytes, enzymes, hormones, antibodies, nutritional materials, oxygen, carbon dioxide and water have molecules small enough to pass through capillary walls. This means that their concentration in tissue fluid is similar to the concentrations in plasma. Differences are due to the uptake of substances in tissue fluid by cells and the addition of cellular waste materials.

Cells obtain the materials they require from interstitial fluid, but their membranes are more complex than capillary walls. Some substances pass through by *diffusion*, water crosses cell membranes by diffusion and *osmosis* and some other substances cross by *active transfer*, involving the expenditure of energy. Measurement of the blood concentration of substances gives an indication of their intracellular concentration.

Diffusion

Diffusion is the physical process in which *dissolved substances* cross a semipermeable membrane in order to establish equality of concentration on the two sides of the membrane. It is a fairly slow process as it takes time for equilibrium to be established. This type of movement of dissolved substances is not the deliberate movement of a specific number of particles (as known schematically in

A. Diffusion   B. Osmosis

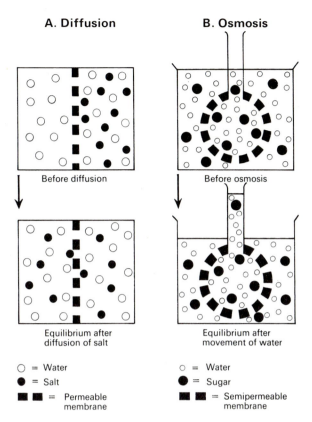

Figure 4:6   A. Before and after diffusion. B. Before and after osmosis.

Fig. 4:6) but results from the constant random movement characteristic of all substances in solution. The particles bombard the dividing membrane and some pass through. The net result is equality of concentration of substances on both sides of the membrane. In diffusion the net movement of dissolved substances is from the higher concentration to the lower, i.e., down the concentration gradient. Diffusible substances and those with molecules smaller than the membrane pores.

Osmosis

Osmosis is the process of the transfer of the *water* of the solution across a semipermeable membrane when equilibrium cannot be achieved by diffusion of solute molecules, i.e., when they are too large to pass through the pores. The force with which this occurs is called the *osmotic pressure*.

It would be easier to understand this phenomenon if osmotic pressure was considered as *osmotic pull*, because water crosses the semipermeable membrane from the low concentration side to the high concentration side. By taking water away from the side of lower concentration its concentration increases, and by adding it to the high concentration solution its concentration is reduced. The process will continue until the concentrations on each side of the membrane are equal. When equilibrium is established the solutions are called *isotonic*. The strength of osmotic pressure depends on the difference between the

ratio of water molecules to non-diffusible particles on the two sides of the membrane.

Osmosis is demonstrated in the following example: a dilute solution of gelatine and water is prepared and placed in a beaker and a more concentrated solution is placed inside a semipermeable membrane such as a cellophane bag, as shown in Figure 4:6. Water passes from the *solution of lower concentration to the solution of higher concentration* and the number of molecules of gelatine on each side remains unchanged.

The processes of diffusion and osmosis have been described separately, but within the body they occur at the same time.

Osmosis and diffusion cannot explain the transfer of all substances which are known to cross living cell membranes. Some substances:

1. *Dissolve in the fat* of the cell membrane and diffuse through it until they enter the cytoplasm of the cell.
2. *Are actively transferred* through the walls of the cell. For example, for each chemical transferred (substances X), there is a carrier (substance Y) in the cell membrane. Y combines with X to form XY. The combined substances, XY, then cross the cell wall and at its inner surface, X is released into the cell cytoplasma and Y remains in the cell membrane. From this it can be seen that the amount of X which can be transferred will depend on the amount of Y present and the rate at which the transfer occurs. This process requires the use of energy.

SHOCK

Shock occurs when the metabolic needs of cells are not being met because of inadequate blood flow. In effect, there is a reduction in blood circulating, in blood pressure and in cardiac output. This causes hypoxia, an inadequate supply of nutrients and the accumulation of waste products. A number of different types of shock are described:

Hypovolaemic   Neurogenic
Cardiogenic   Peritoneal
Septic

HYPOVOLAEMIC SHOCK

The volume of blood in circulation is reduced sufficiently to causes shock following:

1. Severe haemorrhage — whole blood is lost
2. Extensive superficial burns — serum is lost and blood cells at the site of the burn are destroyed
3. Severe vomiting and diarrhoea — water and electrolytes are lost
4. Severe injury — water is lost via the kidneys when excess waste material from cell breakdown is excreted

CARDIOGENIC SHOCK

This occurs in acute heart disease when the cardiac output is suddenly reduced, e.g., in myocardial infarction.

SEPTIC SHOCK (Bacteraemic, Endotoxic)

This resembles hypovolaemic shock but is more prolonged and does not respond so readily to treatment. It is caused by severe infections in which endotoxins are released from dead Gram-negative bacteria, e.g., *Enterobacteria, Pseudomonas.*

The mode of action of the toxins is not clearly understood. It may be that they cause an apparent reduction in the blood volume because of vasodilatation and pooling of blood in the large veins. This reduces the venous return to the heart and the cardiac output.

NEUROGENIC SHOCK (Vaso-vagal attack, Fainting)

The causes include sudden acute pain and severe emotional experience. Nerve impulses reduce the heart rate, reducing the cardiac output. The venous return may also be reduced by the pooling of blood in dilated veins. These changes effectively reduce the blood supply to the brain, causing fainting. The period of unconsciousness is usually of short duration.

PERITONEAL SHOCK

This occurs when there is a break in the wall of an abdominal organ and its contents enter the peritoneal cavity. The peritoneum becomes acutely inflamed, the blood vessels dilate and excess serous fluid is secreted. It may occur following:

1. Perforation of gastric or duodenal ulcer
2. Haemorrhagic pancreatitis (non-microbial)
3. Perforation of uterine tube
4. Rupture of spleen, liver, uterus

CHANGES TAKING PLACE DURING SHOCK

In the short term these are associated with the physiological attempts to restore an adequate blood circulation. If the state of shock persists, the longer term changes may be irreversible.

IMMEDIATE CHANGES

As the blood pressure falls, a number of reflexes are stimulated and hormone secretions increased. These raise the blood pressure by increasing the blood volume and the cardiac output. The changes include:
1. Vasoconstriction, following:
   a. stimulation of the baroreceptors in the aorta and carotid sinuses
   b. sympathetic stimulation of the adrenal glands which causes increased secretion of adrenaline and noradrenaline
   c. stimulation of the renin/angiotensin system by diminished blood flow to the kidneys
2. Increased heart rate, following sympathetic stimulation
3. Water retention by the kidney, following increased release of antidiuretic hormone by the posterior lobe of the pituitary gland

In shock of moderate severity the circulation to the heart and brain is maintained, in the short term. If shock is very severe there may not be time for the above changes to be effective. The severe hypoxia that occurs disrupts cell metabolism. Large amounts of lactic acid are formed and hydrogen ions accumulate, reaching dangerous levels in a few minutes. These are the changes that lead to the severe metabolic acidosis which occurs immediately following cardiac arrest.

LONG-TERM CHANGES ASSOCIATED WITH SHOCK

If the state of shock is not reversed, hypoxia and low blood pressure cause irreversible brain damage and a vicious circle of events is established.

Hypoxia. When this persists there is cell damage and a release of chemical substances that increase the permeability of the capillaries. More fluid enters the interstitial spaces, leading to further hypovolaemia, further reduction in blood pressure and increased hypoxia.

Low blood pressure. As the blood pressure continues to fall, cerebral and myocardial hypoxia becomes progressively more marked and the reduced blood flow encourages the formation of thrombi and infarcts. There is a marked reduction in the secreton of urine, leading to the retention of damaging metabolic waste products. If effective treatment is not possible these irreversible changes become progressively more severe and eventually may cause death.

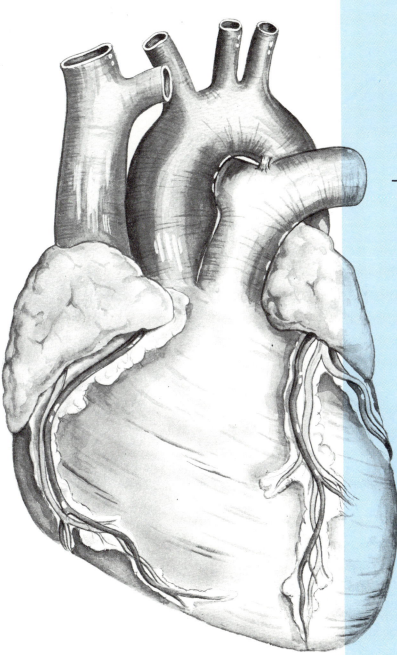

The Circulatory System

5. The Circulatory System

The circulatory or vascular system is divided for descriptive purposes into two main parts.

1. *The blood circulatory system*, consisting of *the heart*, which acts as a pump, and the *blood vessels* through which the *blood* circulates.

2. *The lymphatic system*, consisting of *lymph nodes* and *lymph vessels* through which colourless *lymph* flows.

The two systems communicate with one another and are intimately associated.

BLOOD VESSELS

There are several types:

Arteries	Veins
Aterioles	Venules
Capillaries	

ARTERIES AND ARTERIOLES

These are the blood vessels that transport blood away from the heart. They vary considerably in size, but have much the same structure (Fig. 5:1). They consist of three layers of tissue:

Tunica adventitia or outer layer consists of fibrous tissue
Tunica media or middle layer consists of smooth muscle and elastic tissue
Tunica intima or inner lining consists of squamous epithelium called *endothelium*

The amount of muscular and elastic tissue varies in the arteries depending upon their size. In the large arteries the tunica media consists of more elastic tissue and less muscle. These proportions gradually change as the arteries become smaller until in the *arterioles* (the smallest arteries) the tunica media consists almost entirely of smooth muscle.

Anastomoses and end arteries

Anastomoses are arteries that form a link between main arteries supplying an area, e.g., the arterial supply to the palms of the hand and soles of the feet, the brain, the joints and, to a limited extent, the heart muscle. If one

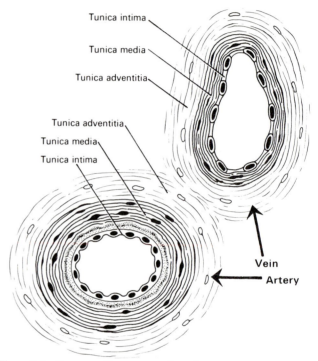

Figure 5:1   Structure of an artery and a vein.

artery supplying the area is occluded anastomotic arteries provide a *collateral circulation*. This is most likely to provide an adequate blood supply when the occlusion occurs gradually, giving the anastomotic arteries time to dilate.

End arteries are the arteries with no anastomoses or those beyond the most distal anastomosis, e.g., the branches from the circle of Willis in the brain, the central artery to the retina of the eye. When an end artery is occluded the tissues it supplies die because there is no alternative blood supply.

VEINS AND VENULES

The veins are the blood vessels that transport blood to the heart. The walls of the veins are thinner than those of arteries but have the same three layers of tissue (Fig. 5:1). They are thinner because there is less muscle and elastic tissue in the tunica media. When cut, the veins collapse while the thicker-walled arteries remain open.

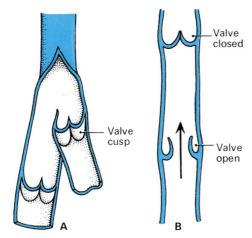

Figure 5:2 Interior of a vein. A. Showing the valves and cusps. B. Arrow showing the direction of blood flow through a valve.

Some veins possess *valves*, which ensure that blood flows towards the heart (Fig. 5:2). Valves are abundant in the veins of the limbs, especially the lower limbs, and are absent in very small and very large veins in the thorax and abdomen. They are formed by a fold of tunica intima strengthened by connective tissue. They are *semilunar* in shape with the concavity towards the heart.

The smallest veins are called *venules*.

CAPILLARIES AND SINUSOIDS

The smallest arterioles break up into a number of minute vessels called *capillaries*. Capillary walls consist of a single layer of endothelial cells through which water and other small-molecule substances can pass. Blood cells and large-molecule substances such as plasma proteins do not normally pass through capillary walls. The capillaries form a vast network of tiny vessels which link the smallest arterioles to the smallest venules. Their diameter is approximately that of an erythrocyte (7 μm).

Sinusoids are wider than capillaries and have extremely thin walls separating the blood from the neighbouring cells. In some there are distinct spaces between the endothelial cells. In the liver some of the sinusoid lining cells are phagocytic macrophages (Kupffer cells). Sinusoids are also found in bone marrow, endocrine glands and the spleen. Because of their larger calibre the blood pressure in sinusoids is lower than in capillaries and there is a slower rate of blood flow.

NERVOUS CONTROL OF THE BLOOD VESSELS

All blood vessels except capillaries have muscle fibres in the tunica media which are supplied by nerves of the *autonomic nervous system*. These nerves arise from the *vasomotor centre* in the *medulla oblongata* and they change the calibre

of the vessels, controlling the amount of blood circulating to any part of the body. Medium-sized and small arteries have more muscle than elastic tissue in their walls. In large arteries, such as the aorta, the middle layer is almost entirely elastic tissue. This means that small arteries and arterioles respond to nerve stimulation whereas the calibre of large arteries varies according to the amount of blood they contain.

The nerves which reduce the lumen of the blood vessels are *vasoconstrictors*, and those which increase the lumen are *vasodilators* (see Ch. 12).

BLOOD SUPPLY

The outer layers of tissue of thick-walled blood vessels receive their blood supply via a network of blood vessels called the *vaso vasorum*. Vessels with thin walls and the endothelium of the others are supplied by the blood passing through them.

CELL RESPIRATION (Fig. 5:3)

Internal or *cell respiration* is the interchange of gases between the capillary blood and neighbouring body cells.

Oxygen is carried from the lungs to the tissues in chemical combination with haemoglobin as *oxyhaemoglobin*. The exchange in the tissues takes place between blood at the arterial end of the capillaries and the tissue fluid and then between the tissue fluid and the cells. The process involved is that of *diffusion from a higher concentration of oxygen in the blood to a lower concentration in the cells*, i.e., down the concentration gradient.

Oxyhaemoglobin is an unstable compound and breaks up (dissociates) easily to liberate oxygen. Factors that increase dissociation include raised carbon dioxide content of tissue fluid, raised temperature and a substance present in red blood cells. In active tissues there is an increased production of carbon dioxide and heat which leads to an increased availability of oxygen. In this way oxygen is available to the tissues in greatest need.

Carbon dioxide is one of the waste products of cell metabolism and, towards the venous end of the capillary,

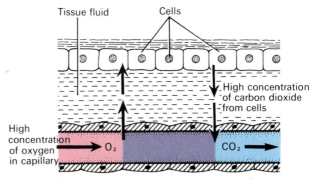

Figure 5:3 Diagram of the exchange of gases in cell respiration.

it diffuses into the blood down the concentration gradient. Blood transports carbon dioxide to the lungs of excretion by three different mechanisms:

1. Some dissolved in the water of the blood plasma
2. Some in chemical combination with sodium in the form of sodium bicarbonate
3. The remainder in combination with haemoglobin (about 25%)

CELL NUTRITION

The nutritive materials required by the cells of the body are transported round the body in the blood plasma. In the process of passing from the blood to the cells the nutritive materials pass through the semipermeable capillary walls into the tissue fluid which bathes the cells, then through the cell wall into the cell. The mechanism of the transfer of water and other substances from the blood capillaries depends mainly upon diffusion, osmosis and active transport.

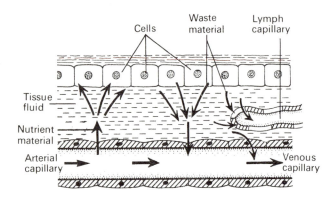

Figure 5:4 Diagram of the exchange of nutrient materials and waste products between the capillaries and the cells.

Diffusion
The capillary walls consist of a single layer of epithelial cells that constitutes a *semipermeable membrane* which allows substances with small molecules to pass through into tissue fluid, and retains large molecules in the blood. Diffusable substances, i.e., those with small molecules, pass down the concentration gradient from a high concentration in blood to a lower concentration in tissue fluid. Glucose, amino acids, fatty acids, glycerol, vitamins, minerals salts and water, needed by body cells, either diffuse through capillary walls or are actively transported.

Osmotic pressure
This is the pressure across a semipermeable membrane that draws water from a dilute to a more concentrated solution in an attempt to establish a state of equilibrium. The force of the osmotic pressure depends on *the number of non-diffusible* particles in the solutions separated by the

membrane. The main substances responsible for the osmotic pressure between blood and tissue fluid are the plasma proteins, especially albumin.

At the arterial end of the capillaries the blood pressure is about *40 mm of mercury (mmHg)*. This is the hydrostatic pressure that forces substances into the tissue spaces. The osmotic pressure in the capillaries, exerted mainly by the plasma proteins, is about *25 mmHg*. This tends to retain water within the blood vessels. The net outward pressure is the difference between the two pressures, i.e., *15 mmHg* (Fig. 5:4).

At the venous end of the capillaries the blood pressure drops to about *10 mmHg*; thus there is a net pressure drawing fluid into the blood vessel of about *15 mmHg*.

This transfer of substances, including water, to the tissue spaces is a dynamic process. Blood flows slowly through the large network of capillaries from the arterial to the venous end and there is constant change. All the water and cell waste products do not return to the blood capillaries. The excess is drained away from the tissue spaces in the minute *lymph capillaries* which originate as blind-end tubes with walls similar to, but more permeable than, those of the blood capillaries (Fig 5:4). Extra tissue fluid and some cell waste materials enter the lymph capillaries and are eventually returned to the blood stream (see Ch. 6).

HEART

The heart is a roughly cone-shaped hollow muscular organ. It is about 10 cm long and is about the size of the owner's fist. It weighs about 225 g in women and is heavier in men.

POSITION (Fig. 5:5)

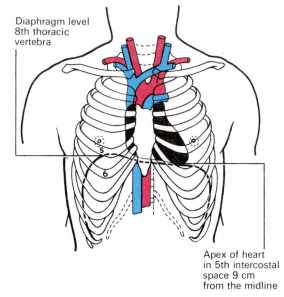

Figure 5:5 Position of the heart in the thorax.

The heart is in the thoracic cavity in the middle mediastinum between the lungs. It lies obliquely, a little more to the left than the right, and presents a *base* above and an *apex* below. The apex is about 9 cm to the left of the midline at the level of the 5th intercostal space, i.e., a little below the nipple and slightly nearer the midline. The base extends to the level of the 2nd rib.

Organs associated with the heart (Fig. 5:6)

Inferiorly — the apex rests on the central tendon of the diaphragm

Superiorly — the great blood vessels, i.e., the aorta, superior vena cava, pulmonary artery and pulmonary veins

Posteriorly — the oesophagus, trachea, left and right bronchus, descending aorta, inferior vena cava and thoracic vertebrae

Laterally — the lungs — the left lung overlaps the left side of the heart

Anteriorly — the sternum, ribs and intercostal muscles

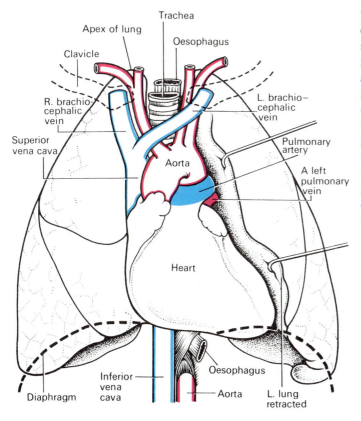

Figure 5:6 Organs associated with the heart.

STRUCTURE

The heart is composed of three layers of tissue: pericardium, myocardium and endocardium.

Pericardium

The pericardium is made up of two sacs. The outer sac consists of fibrous tissue and the inner of a double layer of serous membrane.

The outer fibrous sac is continuous with the tunica adventitia of the great blood vessels above and is adherent to the diaphragm below. Its fibrous nature prevents overdistension of the heart.

The outer layer of the serous membrane, the *parietal pericardium*, lines the fibrous sac, and the inner layer, the *visceral pericardium*, is reflected on to the heart and is adherent to the heart muscle.

The serous membrane is made up of *flattened epithelial cells*. It secretes serous fluid into the space between the visceral and parietal layers which allows smooth movement between them when the heart beats. The space between the parietal and visceral pericardium is only a *potential space*. In life the two layers are in close association, with only a thin film of serous fluid between them.

Myocardium (Fig 5:7)

The myocardium is composed of specialised *cardiac muscle* tissue found only in the heart. It is not under the control of the will but, like voluntary muscle, cross stripes can be seen on microscopic examination. Each fibre (cell) has a nucleus and one or more branches. The ends of the cells and their branches are in very close contact with the ends and branches of adjacent cells. Microscopically these 'joints', or *intercalated discs*, can be seen as thicker, darker lines than the ordinary cross stripes. This arrangement gives cardiac muscle the appearance of being a sheet of muscle rather than a very large number of individual cells. Because of the end-to-end contiguity of the fibres, each one does not need to have a separate nerve supply. When an impulse of contraction is initiated it spreads from cell to cell over the whole 'sheet' of muscle.

The myocardium is thickest at the apex and thins out

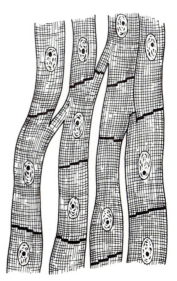

Figure 5:7 Cardiac muscle with fibres separated.

towards the base. The atria and the ventricles are separated by *a ring of fibrous tissue*. Consequently, when a wave of contraction passes over the atrial muscle, it can only spread to the ventricles through the conducting system (see p. 67).

Endocardium

This forms a lining to the myocardium and is a thin, smooth, glistening membrane consisting of flattened epithelial cells, continuous with the lining of the blood vessels.

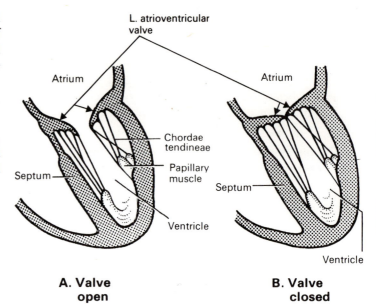

A. Valve open **B. Valve closed**

Figure 5:9 Diagram of the left atrioventricular valve.

movement is limited by tendinous cords, called *chordae tendineae*, which extend from the inferior surface of the cusps to little projections of myocardium covered with endothelium, called *papillary muscles* (Fig. 5:9).

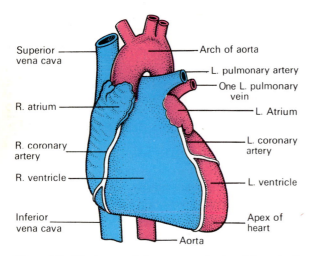

Figure 5:8 The heart and the great vessels viewed from the front.

INTERIOR OF THE HEART (Figs 5:9 and 5:10)

The heart is divided into a right and left side by *the septum*, a partition consisting of myocardium covered by endocardium. After birth blood cannot cross the septum from one side to the other. Each side is divided by an *atrioventricular valve* into an upper chamber, *the atrium*, and a lower chamber, *the ventricle*. The atrioventricular valves are formed by double folds of endocardium strengthened by a little fibrous tissue. The *right atrioventriculr valve* (tricuspid valve) has three flaps or *cusps* and the *left atrioventricular valve* (mitral valve) has two cusps.

The valves between the atria and ventricles open and close when pressure in the chambers changes. During ventricular systole (contraction) the pressure in the ventricles rises to a level higher than that in the atria, closing the valves and preventing backward flow of blood. The valves do not open upwards into the atria because their

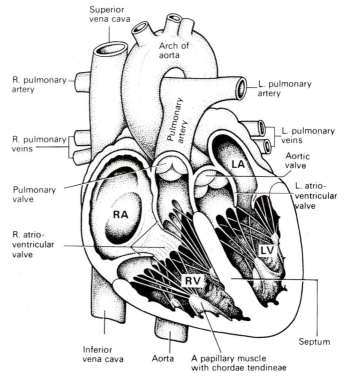

Figure 5:10 Interior of the heart.

FLOW OF BLOOD THROUGH THE HEART
(Fig. 5:11)

The two largest veins of the body, the *superior* and *inferior venae cavae*, empty their contents into the right atrium. This blood passes via the right atrioventricular valve into the right ventricle, and from there it is pumped into the *pulmonary artery* or *trunk* (the only artery in the body which carries deoxygenated blood). The opening of the pulmonary artery is guarded by the *pulmonary valve*, formed by three *semilunar cusps*. This valve prevents the back flow of blood into the right ventricle when the ventricular muscle relaxes. After leaving the heart the pulmonary artery divides into *left* and *right pulmonary arteries* which carry the venous blood to the lungs where the interchange of gases takes place: carbon dioxide is excreted and oxygen is absorbed.

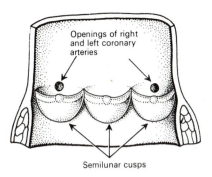

Figure 5:12 The aorta cut open to show the cusps of the semilunar valve.

with the amount of work it does. The atria, usually assisted by gravity, only propel the blood through the atrioventricular valve into the ventricles, whereas the ventricles pump the blood to the lungs and round the whole body. The muscle layer is thickest in the wall of the left ventricle.

The pulmonary trunk leaves the heart from the upper part of the right ventricle, and the aorta leaves from the upper part of the left ventricle.

Summary
The *right side* of the heart deals with *deoxygenated blood*.
The *left side* of the heart deals with *oxygenated blood*.
The vessels *carrying blood to* the heart are *veins*.
The vessels *carrying blood away* from the heart are *arteries*.

BLOOD SUPPLY TO THE HEART

The heart is supplied with arterial blood by the *right* and *left coronary arteries*. These are the first branches from the aorta immediately distal to the aortic valve.

The venous return is by the *coronary sinus* which empties into the right atrium and by some small channels that open directly into the chambers of the heart.

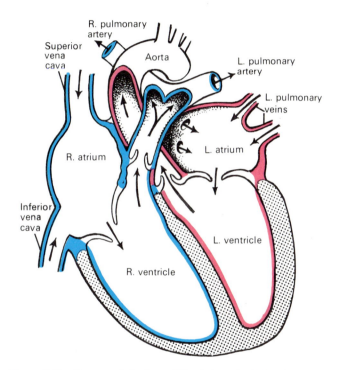

Figure 5:11 Diagram of the flow of blood through the heart.

The arterial or oxygenated blood is carried from each lung in two pulmonary veins that empty their contents into the *left atrium*. It then passes through the left atrioventricular valve into the left ventricle, and from there it is pumped into the aorta, the first artery of the general circulation. The opening of the aorta is guarded by the *aortic valve*, formed by three *semilunar cusps* (Fig. 5:12).

From this sequence of events it can be seen that the blood passes from the right to the left side of the heart via the lungs. However, it should be noted that both atria contract at the same time and this is followed by the simultaneous contraction of both ventricles.

The muscle layer of the walls of the atria is very thin in comparison with that of the ventricles. This is consistent

CONDUCTING SYSTEM OF THE HEART
(Fig. 5:13)

The heart has an intrinsic system whereby the muscle is stimulated to contract without the need for a nerve supply from the brain. However, the intrinsic system can be stimulated or depressed by nerve impulses initiated in the brain and by some hormones.

There are small groups of specialised neuromuscular cells in the myocardium which initiate and conduct impulses of contraction over the heart muscle.

Sinuatrial node (SA node)
This small mass of specialised cells is in the wall of the right atrium near the opening of the superior vena cava.

The SA node is often described as the 'pace-maker' of the heart because it initiates impulses of contraction more rapidly than other groups of neuromuscular cells.

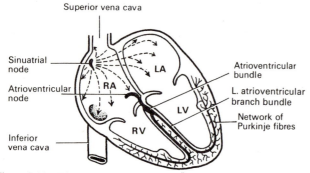

Figure 5:13   Diagram of the conducting system in the heart.

Atrioventricular node (AV node)

This mass of neuromuscular tissue is situated in the wall of the atrial septum near the atrioventricular valves. Normally the AV node is stimulated by the contraction which sweeps over the atrial myocardium. However, it too is capable of initiating impulses of contraction but at a slower rate than the SA node.

Atrioventricular bundle (AV bundle or bundle of His)

This consists of a mass of specialised fibres that originate from the AV node. The AV bundle crosses the fibrous ring that separates atria and ventricles then, at the upper end of the ventricular septum, it divides into *right and left branch bundles*. Within the ventricular myocardium the branches break up into fine fibres, called *the Purkinje fibres*. The AV bundle, branch bundles and Purkinje fibres convey impulses of contraction from the AV node to the apex of the myocardium where the wave of ventricular contraction begins then sweeps upwards, pumping blood into the pulmonary artery and the aorta.

NERVE SUPPLY TO THE HEART

In addition to the intrinsic stimulation of the myocardium described above the heart is influenced by nerves originating in the *cardiac centre* in the *medulla oblongata* which reach it through the autonomic nervous system. These are the *parasympathetic* and *sympathetic nerves* and they are antagonistic to one another.

The *vagus nerves* (parasympathetic) supply mainly the SA and AV nodes and atrial muscle. Parasympathetic stimulation tends to reduce the rate at which impulses are produces, *decreasing* the rate and force of the heart beat.

The *sympathetic nerves* supply the SA and AV nodes and the myocardium of atria and ventricles. Sympathetic stimulation tends to *increase* the rate and force of the heart beat. Adrenaline, a hormone secreted by the medulla of the adrenal gland, has the same effect as sympathetic stimulation. Noradrenaline depresses the heart.

The rate at which the heart beats is the result of a fine balance of sympathetic and parasympathetic effects. Cardiac activity usually decreases during rest and increases during excitement, exercise and when the blood volume is decreased.

FUNCTION

The function of the heart is to maintain a constant circulation of blood throughout the body. The heart acts as a pump and its action consists of a series of events known as the *cardiac cycle*.

Cardiac cycle (Fig. 5:14)

The normal number of cardiac cycles per minute ranges from 60 to 80. Taking 74 as an example each cycle lasts about *0.8 of a second* and consists of:

Atrial systole — contraction of the atria
Ventricular systole — contraction of the ventricles
Complete cardiac diastole — relaxation of the atria and ventricles

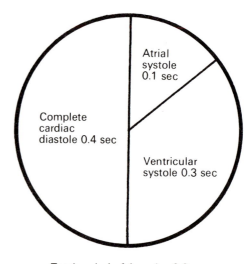

Total period of 1 cycle=0.8 sec

Figure 5:14   Diagram of the stages of one cardiac cycle.

It does not matter at which stage of the cardiac cycle a description starts. For convenience the period when the atria are filling has been chosen.

The superior vena cava and the inferior vena cava pour deoxygenated blood into the right atrium *at the same time* as the four pulmonary veins pour oxygenated blood into the left atrium. The atrioventricular valves are open and blood flows through to the ventricules. The SA node emits an impulse of contraction. This stimulates the wave of contraction that spreads over the myocardium of both atria, emptying the atria and completing ventricular filling (atrial systole 0.1 sec). When the wave of contraction reaches the AV node it is stimulated to emit an impulse of contraction which spreads to the ventricular muscle via the AV bundle, branch bundles and Purkinje fibres. This results in a wave of contraction which sweeps upwards from the apex of the heart and pushes the blood into the pulmonary artery and the aorta (ventricular systole 0.3 sec).

After contraction of the ventricles there is *complete cardiac diastole*, a period of *0.04 seconds*, when atria and ventricles are relaxed.

The valves of the heart and of the great vessels open and close according to the pressure within the chambers of the heart. The AV valves are open while the ventricular muscle is relaxed during atrial filling and systole. When the ventricles contract there is a gradual increase in the pressure in these chambers, and when it rises above atrial pressure the atrioventricular valves close. When the ventricular pressure rises above that in the pulmonary artery and in the aorta, the pulmonary and aortic valves open and blood flows into these vessels. When the ventricles relax and the pressure within them falls, the reverse process occurs. First the pulmonary and aortic valves close, then the atrioventricular valves open and the cycle begins again. This sequence of opening and closing valves ensures that the blood flows in only one direction.

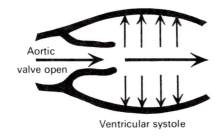

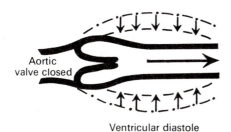

Figure 5:15   Diagram showing the elasticity of the walls of the aorta.

HEART SOUNDS

The individual is not usually conscious of his heart beat, but if the ear or the receiver of a stethoscope is placed on the chest wall a little below the left nipple and slightly nearer the midline the heartbeat can be heard.

Two sounds, separated by a short pause, can be clearly distinguished. They are described in words as 'lub dup'.

The first sound, 'lub', is fairly loud and is due to the contraction of the ventricular muscle and the closure of the atrioventricular valves.

The second sound, 'dup', is softer and is due to the closure of the aortic and pulmonary valves.

ELECTRICAL CHANGES IN THE HEART

When muscles contract there is a change in the electrical potential across the membrane of muscle fibres. As the body fluids and tissues are good conductors of electricity, the electrical changes which occur in the contracting myocardium can be detected by attaching electrodes to the surface of the body. The pattern of electrical activity may be displayed on an oscilloscope screen or traced on paper. The apparatus used is an *electrocardiograph* and the tracing is an *electrocardiogram* (ECG).

The normal ECG tracing shows five waves which, by convention, have been named P, Q, R, S and T (Fig. 5:16).

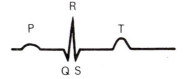

Figure 5:16   Electrocardiogram of one cardiac cycle.

The P wave is caused by the impulse of contraction as it sweeps over the atria.

The QRS wave shows the spread of the impulse of contraction from the AV node through the AV bundle and the Purkinje fibres and the contraction of the ventricular muscle.

The T wave represents the relaxation of the ventricular muscle.

By examining the pattern of waves and the time interval between cycles and parts of cycles the physician obtains valuable information about the state of the myocardium and the conducting system within the heart.

BLOOD PRESSURE

Blood pressure may be defined as the force or pressure which the blood exerts on the walls of the blood vessels. As there is some delay in the movement of blood through the arteriolar and capillary systems, the blood pressure in the arteries is higher than that in the veins.

The arterial blood pressure is the result of the discharge of blood from the left ventricle into the *already full aorta*.

When the left ventricle contracts and pushes blood into the aorta the pressure produced is called the *systolic blood pressure*. In adults it is about 120 mmHg (millimetres of mercury) or 16 kPa (kilopascals).

When *complete cardiac diastole* occurs and the heart is resting following the ejection of blood, the pressure within the arteries is called *diastolic blood pressure*. In an adult this is about 80 mmHg or 11 kPa.

These figures vary according to the time of day, the posture, sex and age of the individual. During bedrest at night the blood pressure tends to be lower. It increases with age and is usually higher in women than in men.

Arterial blood pressure is measured by the use of a sphygmomanometer and is usually expressed in the following manner:

$$BP = \frac{120}{80} \text{ mmHg or } BP = \frac{16}{11} \text{ kPa}$$

MAINTENANCE OF NORMAL BLOOD PRESSURE

The blood pressure is maintained within normal limits by fine adjustments, involving a number of factors, including:

Cardiac output
Blood volume
Peripheral resistance
Elasticity of the artery walls
Venous return

Cardiac output
The cardiac output may be considered as the amount of blood ejected from the heart by each contraction of the ventricles (*stroke volume*), or the amount ejected each minute (*minute volume*). The minute volume takes into consideration the rate *and* force of cardiac contraction. An increase in minute volume raises both the systolic and diastolic pressure. An increase in the stroke volume increases the systolic pressure more than it does the diastolic pressure.

Blood volume
A sufficient amount of blood must be circulating in the vessels to maintain the normal blood pressure, e.g., if haemorrhage has occurred and a large amount of blood is lost, there is an accompanying fall in the blood pressure.

Peripheral or arteriolar resistance
Arterioles are the smallest arteries and they have a tunica media composed almost entirely of smooth muscle which responds to nerve stimulation. Nerve impulses reach the muscle layer in blood vessel walls via vasoconstrictor nerves (sympathetic nerves), originating in the vasomotor centre in the medulla oblongata (see Ch. 12). Nerve endings sensitive to stretch, called *baroreceptors*, are located in the arch of the aorta and in the coratid sinuses. A rise in blood pressure in these arteries stimulates the baroreceptors and initiates vasodilatation and a slowing of the heart rate. A fall in blood pressure stimulates vasoconstriction and an increase in heart rate. Normally the vasomotor centre maintains artery walls in a state of slight constriction, or *tone*.

Dilatation and constriction of arterioles occurs selectively around the body, resulting in changes in the blood flow through organs according to their needs. The highest priorities are the blood supply to the brain and the heart muscle, and in an emergency, supplies to other parts of the body are reduced in order to ensure an adequate supply to these organs. Generally, changes in the amount of blood flowing to any organ depend on how active it is. A very active organ needs more oxygen and nutritional materials than a resting organ and it produces more waste materials for excretion.

The elasticity of the artery walls
There is a considerable amount of elastic tissue in the arterial walls, especially in large arteries. Therefore, when the left ventricle ejects blood into the already full aorta, it distends, then the elastic recoil pushes the blood onwards. This distension and recoil occurs all through the arterial system. During cardiac diastole the elastic recoil of the arteries maintains the diastolic pressure (Fig. 5:15).

Venous return
The amount of blood returned to the heart through the superior and inferior venae cavae plays an important part in cardiac output. The force of contraction of the left ventricle ejecting blood into the aorta and subsequently into the arteries, arterioles and capillaries, is not sufficient to return the blood through the veins back to the heart. Other factors are involved in assisting the venous return.

The position of the body. Gravity assists the venous return from the head and neck when the individual is standing or sitting and offers less resistance to venous return from the lower parts of the body when the individual is lying flat.

Muscular contraction. The contraction of muscles, particularly skeletal muscles, puts pressure on the veins.

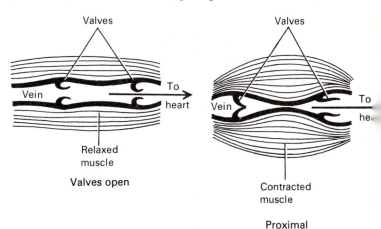

Figure 5:17   Diagram of the flow of blood through a vein aided by the contraction of skeletal muscle.

This squeezing or milking action has the effect of pushing the blood towards the heart. When the pressure in deep veins in lowered during muscle relaxation blood flows into them from superficial veins through *communicating veins*. Backward flow of blood is prevented by valves (Fig. 5:17).

Effects of respiratory movements. During inspiration the expansion of the chest creates a negative pressure within the thorax, assisting the flow of blood towards the heart. In addition, when the diaphragm descends during inspiration, the increased intra-abdominal pressure squeezes blood towards the heart.

PULSE

The pulse is described as a wave of distension and elongation felt in an artery wall due to the contraction of the left ventricle forcing about 40 to 80 millilitres of blood into the already full aorta. When the aorta is distended, a wave passes along the walls of the arteries and can be felt at any point where an artery can be pressed gently against a bone. The number of pulse beats per minute varies considerably in different people and in the same person at different times. An average of 60 to 80 is common.

Some information may be obtained by taking the pulse.
1. *The rate* at which the heart is beating.
2. *The rhythm*, or regularity with which the heart beats occur, i.e., the length of time between beats should be the same.
3. *The volume or strength* of the beat. It should be poss-ible to compress the artery with moderate pressure, stopping the flow of blood. The compressibility of the blood vessel gives some indication of the blood pressure and the state of the blood vessel wall.
4. *The tension.* The artery wall should feel soft and pliant under the fingers.

Factors affecting the pulse rate

Position. When the individual is standing up the pulse rate is usually more rapid than when he is lying down.

Age. The pulse rate in children is more rapid than in adults.

Sex. The pulse rate tends to be more rapid in females than in males.

Exercise. Any exercise — walking, running or playing games — will increase the rate of the pulse. The resting rate should be restored soon after exercise has stopped.

Emotion. In strong emotional states the pulse rate is increased, e.g., excitement, fear, anger, grief.

CIRCULATION OF THE BLOOD

Although circulation of blood round the body is continuous it is convenient to describe it in three parts:
Pulmonary circulation
Systemic or general circulation
Portal circulation

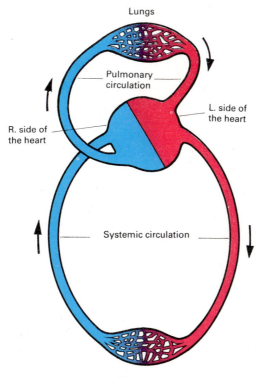

Figure 5:18 Diagram showing the relationship between the systemic and pulmonary circulation.

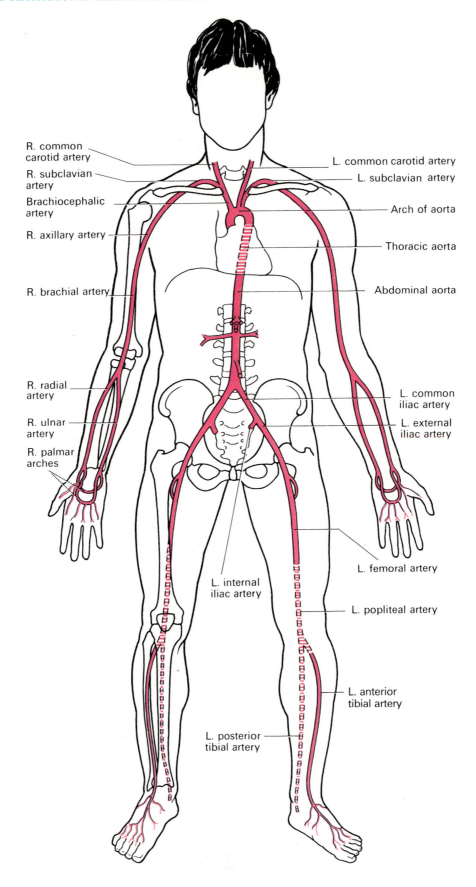

Figure 5:19 Aorta and the main arteries of the limbs.

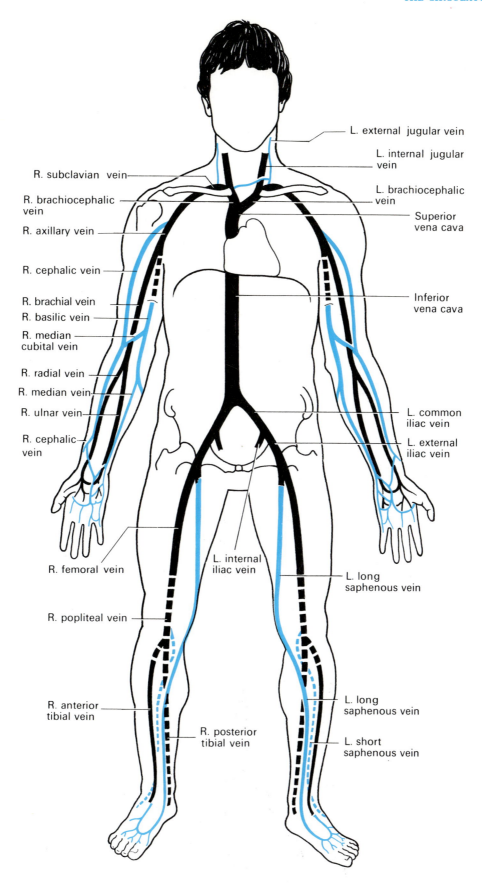

Figure 5:20   Venae cavae and the main veins of the limbs. Deep veins in black, superficial veins in blue.

PULMONARY CIRCULATION (Fig. 5:18)

This consists of the circulation of blood from the right ventricle of the heart to the lungs and back to the left atrium. In the lungs, carbon dioxide is excreted and oxygen is absorbed.

The pulmonary artery or trunk, carrying *deoxygenated blood*, leaves the upper part of the right ventricle of the heart. It passes upwards and divides into left and right pulmonary arteries at the level of the 5th thoracic vertebra.

The left pulmonary artery runs to the root of the left lung where it divides into two branches, one passing into each lobe.

The right pulmonary artery passes to the root of the right lung and divides into two branches. The larger branch carries blood to the middle and lower lobes, and the smaller branch to the upper lobe.

Within the lung these arteries divide and subdivide into smaller arteries, arterioles and capillaries. The interchange of gases takes place between capillary blood and air in the alveoli of the lungs (see p. 114). In each lung the capillaries containing oxygenated blood join up and eventually form two veins.

Two pulmonary veins leave each lung, returning oxygenated blood to the left atrium of the heart. During atrial systole this blood passes into the left ventricle, and during ventricular systole it is forced into the aorta, i.e., the first artery of the general circulation.

SYSTEMIC OR GENERAL CIRCULATION

The blood pumped out from the left ventricle is carried by the *branches of the aorta* around the body and is returned to the heart by the *superior* and *inferior venae cavae*. Figure 5:19 (p. 72) gives a general impression of the positions of the aorta and the main arteries to the limbs. Figure 5:20 (p. 73) provides an overview of the venae cavae and the veins of the limbs.

The circulation of blood to the different parts of the body will be described in the order in which their arteries branch off the aorta.

AORTA (Fig. 5:21)

The aorta begins at the upper part of the left ventricle and, after passing upwards for a short distance, it arches backwards and to the left. It then descends behind the heart through the thoracic cavity a little to the left of the thoracic vertebrae. At the level of the 12th thoracic vertebra it passes behind the diaphragm then downwards in the abdominal cavity to the level of the 4th lumbar vertebra, where it divides into the *right* and *left common iliac arteries*.

Throughout its length the aorta gives off numerous branches. Some of the branches are *paired*, i.e., there is a right and left branch of the same name and some are single or *unpaired*.

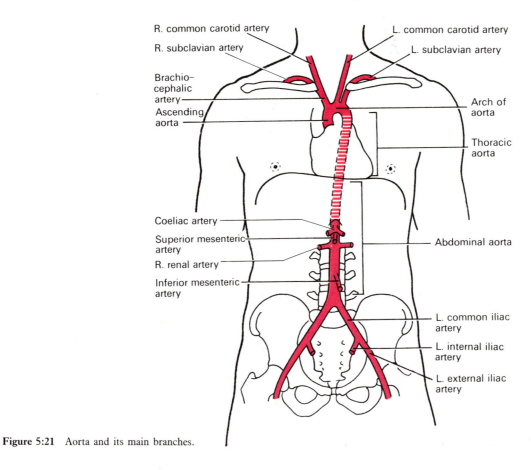

Figure 5:21   Aorta and its main branches.

THORACIC AORTA

This part of the aorta is above the diaphragm and is described in three parts:
Ascending aorta
Arch of the aorta
Descending aorta

Ascending aorta

This is about 5 cm long and lies behind the sternum.

The right and left coronary arteries (Fig. 5:22) arise from the aorta just above the level of the aortic valve. They supply the tissues of the heart with oxygenated blood. As they traverse the heart they break up, eventually becoming capillaries. Most of the venous blood is collected into several small veins which join up to form the *coronary sinus* which opens into the right atrium. The remainder passes directly into the chambers of the heart through little venous channels.

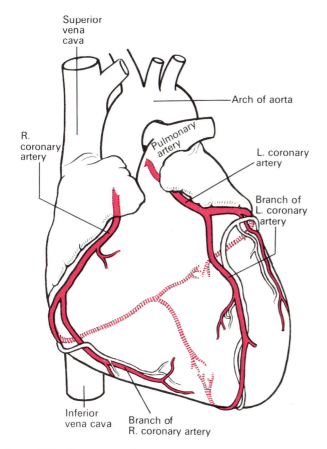

Figure 5:22   Coronary arteries.

Arch of the aorta

The arch of the aorta is a continuation of the ascending aorta. It begins behind the manubrium of the sternum and runs upwards, backwards and to the left in front of the trachea. It then passes downwards to the left of the trachea and is continuous with the descending aorta.

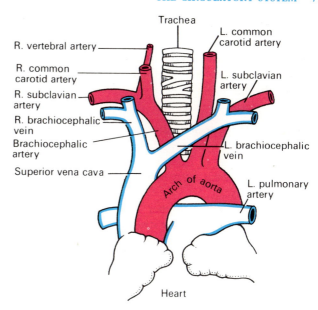

Figure 5:23   The arch of the aorta and its branches.

Three branches are given off from its upper aspect (Fig. 5:23):
Brachiocephalic artery or trunk
Left common carotid artery
Left subclavian artery

The brachiocephalic artery is about 4 to 5 cm long and passes obliquely upwards, backwards and to the right. At the level of the sternoclavicular joint it divides into the *right common carotid artery* and the *right subclavian artery*.

CIRCULATION OF BLOOD TO THE HEAD AND NECK

The *right common carotid artery* is a branch of the brachiocephalic artery. The *left common carotid artery* arises directly from the arch of the aorta. They pass upwards on either side of the neck and have the same distribution on each side. The common carotid arteries are embedded in fascia, called the *carotid sheath*. At the level of the upper border of the thyroid cartilage they divide into:
External carotid artery
Internal carotid artery

The carotid sinuses are slight dilatations at the point of division of the common carotid arteries into their internal and external branches, i.e., their bifurcation. The walls of the sinuses are thin and contain numerous nerve endings of the glossopharyngeal nerves. These nerve endings, or *baroreceptors*, are stimulated by changes in *blood pressure* in the carotid sinuses. The resultant nerve impulses initiate reflex adjustments of blood pressure through the vasomotor centre in the medulla oblongata.

The carotid bodies are two small groups of specialised cells, one lying in close association with each common carotid artery at its bifurcation. They are supplied by the glossopharyngeal nerves and their cells are stimulated by changes in the carbon dioxide and oxygen content of blood. The resultant nerve impulses initiate reflex adjustments of respiration through the respiratory centre in the medulla oblongata.

Branches of the external carotid artery (Fig. 5:24)

This artery supplies the superficial tissues of the head and neck, via a number of branches:

The superior thyroid artery supplies the thyroid gland and adjacent muscles.

The lingual artery supplies the tongue, the lining membrane of the mouth, the structures in the floor of the mouth, the tonsil and the epiglottis.

The facial artery passes outwards over the mandible just in front of the angle of the jaw and supplies the muscles of facial expression and structures in the mouth. The pulse may be felt where the artery crosses the jaw bone.

The occipital artery supplies the posterior part of the scalp.

The temporal artery passes upwards over the zygomatic process in front of the ear and supplies the frontal, temporal and parietal parts of the scalp. The pulse may be felt in front of the upper part of the ear.

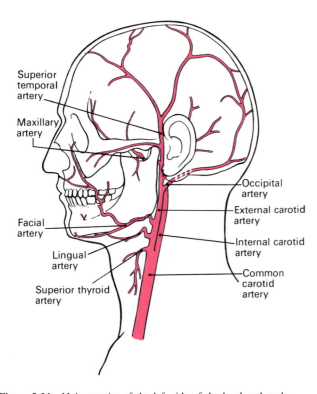

Figure 5:24 Main arteries of the left side of the head and neck.

The maxillary artery supplies the muscles of mastication and a branch of this artery, the *middle meningeal artery*, runs deeply to supply structures in the interior of the skull.

Branches of the internal carotid artery

The internal carotid artery is a major contributor to the *circle of Willis* which supplies the greater part of the brain. It also has branches that supply the eyes, forehead and nose. It ascends to the base of the skull and passes through the carotid foramen in the temporal bone.

Circulus arteriosus (circle of Willis) (Fig. 5:25)

The greater part of the brain is supplied with arterial blood by an arrangement of arteries called the *circulus arteriosus* or the *circle of Willis*. Four large arteries contribute to its formation: two *internal carotid arteries* and two *vertebral arteries*. The arrangement is such that the brain as a whole receives an adequate blood supply even when a contributing artery is damaged.

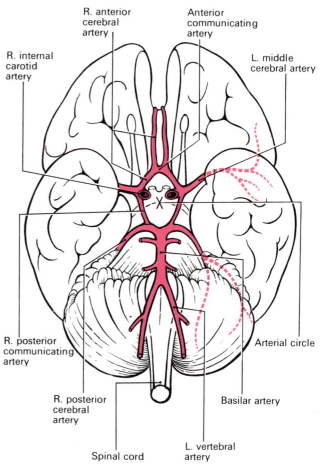

Figure 5:25 Arteries forming the circulus arteriosus (circle of Willis) and its main branches to the brain.

Anteriorly, *two anterior cerebral arteries* arise from the internal carotid arteries and are joined together by an artery known as the *anterior communicating artery*.

Posteriorly, *two vertebral arteries* join together to form the *basilar artery*. After travelling for a short distance the basilar artery divides to form the *two posterior cerebral arteries*, each of which is joined to the corresponding internal carotid artery by a *posterior communicating artery*. The *circulus arteriosus* is therefore formed by:

2 anterior cerebral arteries
2 internal carotid arteries
1 anterior communicating artery
2 posterior communicating arteries
2 posterior cerebral arteries
1 basilar artery

From this circle, the *anterior cerebral arteries* pass forward to supply the anterior part of the brain, the *middle cerebral arteries* pass laterally to supply the sides of the brain, and the *posterior cerebral arteries* supply the posterior part of the brain.

Branches of the basilar artery supply parts of the brain stem.

The vertebral arteries arise from the subclavian arteries, pass upwards through the foramina in the transverse processes of the cervical vertebrae, enter the skull through the foramen magnum, then join to form the *basilar artery* (Fig. 5:26).

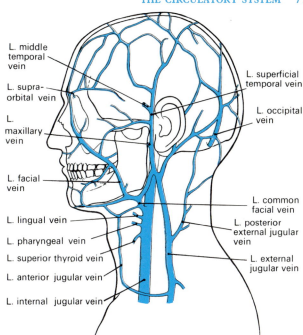

Figure 5:27   Veins of the left side of the head and neck.

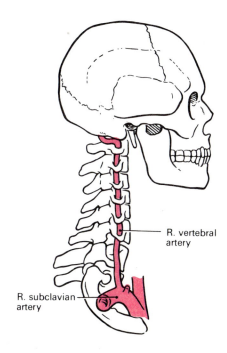

Figure 5:26   Right vertebral artery.

VENOUS RETURN FROM THE HEAD AND NECK
(Figs 5:27, 5:28 and 5:29)

The venous blood from the head and neck is returned by *deep and superficial veins*.

Superficial veins with the same names as the branches of the external carotid artery return venous blood from the superficial structures of the face and scalp and unite to form the external jugular vein.

The external jugular vein begins in the neck at the level of the angle of the jaw. It passes downwards in front of the sternocleidomastoid muscle, then behind the clavicle before entering the *subclavian vein*.

The venous blood from the deep areas of the brain is collected into channels called *sinuses*.

Venous sinuses of the brain (Figs 5:28 and 5:29)
The walls of the venous sinuses are formed by layers of *dura mater* lined with endothelium. The dura mater is the outer protective covering of the brain (see p. 214). The main venous sinuses are:

1 superior sagittal sinus
1 inferior sagittal sinus
1 straight sinus
2 transverse or lateral sinuses

The *superior sagittal sinus* carries the venous blood from the superior part of the brain. It begins in the frontal region and passes directly backwards in the midline of the skull to the occipital region where it turns to the right side and continues as the *right transverse sinus*.

The *inferior sagittal sinus* lies deep within the brain and passes backwards to form the *straight sinus*.

The *straight sinus* runs backwards and downwards to become the *left transverse sinus*.

The *transverse sinuses* begin in the occipital region. The one on the right side is a continuation of the superior sagittal sinus and the one on the left, of the straight sinus. They run forward and medially in a curved groove of the skull, to become continuous with the *internal jugular veins* in the middle cranial fossa.

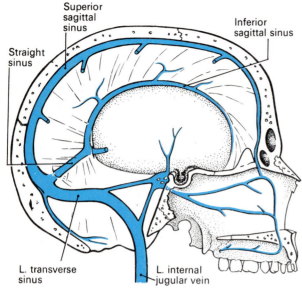

Figure 5:28   Venous sinuses of the brain viewed from the right side.

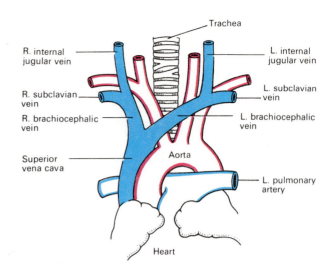

Figure 5:30   Superior vena cava and the veins which form it.

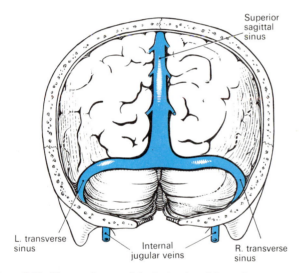

Figure 5:29   Venous sinuses of the brain viewed from above.

The *internal jugular veins* begin at the jugular foramina in the middle cranial fossa and each is the continuation of a transverse sinus. They run downwards in the neck behind the sternocleidomastoid muscles. Behind the clavicle they unite with the *subclavian veins* to form the *brachiocephalic veins*.

The *brachiocephalic veins* are situated one on each side in the root of the neck. Each is formed by the union of the internal jugular and the subclavian veins. The left brachiocephalic vein is longer than the right and passes obliquely behind the manubrium of the sternum, where it joins the right brachiocephalic vein to form the *superior vena cava* (Fig. 5:30).

The superior vena cava, which drains all the venous blood from the head, neck and upper limbs, is about 7 cm long. It passes downwards along the right border of the sternum and ends in the right atrium of the heart.

CIRCULATION OF BLOOD TO THE UPPER LIMB

Arterial supply (Figs 5:19 and 5:31)

The subclavian arteries. The right subclavian artery arises from the brachiocephalic artery, the left branches from the arch of the aorta. They are slightly arched and pass behind the clavicles and over the first ribs before entering the axillae, where they continue as the *axillary arteries*.

Before entering the axilla each subclavian artery gives off two branches: the *vertebral artery*, which passes upwards to supply the brain and the *internal mammary artery*, which supplies the breast and a number of structures in the thoracic cavity.

The *axillary artery* is a continuation of the subclavian artery and lies in the axilla. The first part lies deeply, then it runs more superficially to become the *brachial artery*.

The *brachial artery* is a continuation of the axillary artery. It runs down the medial aspect of the upper arm, passes to the front of the elbow and extends to about 1 cm below the joint, where it divides into *radial* and *ulnar arteries*.

The *radial artery* passes down the radial or lateral side of the forearm to the wrist. Just above the wrist it lies superficially and can be felt in front of the radius, i.e., where the radial pulse is palpable. The artery then passes between the first and second metacarpal bones and enters the palm of the hand.

The *ulnar artery* runs downwards on the ulnar or medial aspect of the forearm to cross the wrist and pass into the hand.

There are anastomoses between the radial and ulnar arteries, called, the *deep* and *superficial palmar arches*, from which *palmar metacarpal* and *palmar digital arteries* arise to supply the structures in the hand and fingers.

Branches from the axillary, brachial, radial and ulnar arteries supply all the structures in the upper limb.

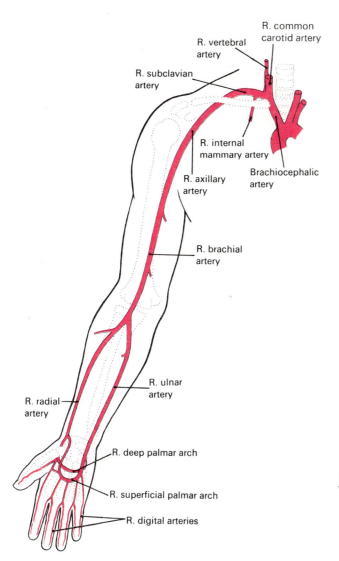

Figure 5:31   Main arteries of the right arm.

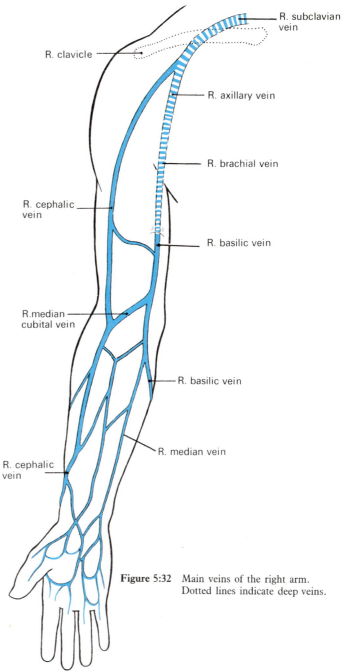

Figure 5:32   Main veins of the right arm. Dotted lines indicate deep veins.

Venous return from the upper limb (Figs 5:20 and 5:32)

The veins of the upper limb are divided into two groups: deep and superficial veins.

The deep veins follow the course of the arteries and have the same names:

Palmar metacarpal veins
Deep palmar venous arch
Ulnar and radial veins
Brachial vein
Axillary vein
Subclavian vein

The superficial veins begin in the hand and consist of the following:

Cephalic vein
Basilic vein
Median vein
Median cubital vein

The *cephalic vein* begins at the back of the hand where it collects blood from a complex of superficial veins, many of which can be easily seen. It then winds round the radial side to the anterior aspect of the forearm. In front of the elbow it gives off a large branch, the *median cubital vein*, which slants upwards and medially to join the *basilic vein*. After crossing the elbow joint the cephalic vein passes up the lateral aspect of the arm and in front of the shoulder joint to end in the axillary vein. Throughout its length it receives blood from the superficial tissues on the lateral aspects of the hand, forearm and arm.

The *basilic vein* begins at the back of the hand on the ulnar aspect. It ascends on the medial side of the forearm and upper arm then joins the axillary vein. It receives blood from the medial aspect of the hand, forearm and arm. There are many small veins which link the cephalic and basilic veins.

The *median vein* is a small vein that is not always present. It begins at the palmar surface of the hand, ascends on the front of the forearm and ends in the basilic vein or the median cubital vein.

The *brachiocephalic vein* is formed when the subclavian and internal jugular veins unite. There is one on each side.

The *superior vena cava* is formed when the two brachiocephalic veins unite. It drains all the venous blood from the head, neck and upper limbs and terminates in the right atrium. It is about 7 cm long and passes downwards along the right border of the sternum.

Descending aorta (Figs 5:21 and 5:33)

The descending aorta is continuous with the arch of the aorta and begins at the level of the 4th thoracic vertebra. It extends downwards on the anterior surface of the bodies of the thoracic vertebrae to the level of the 12th thoracic vertebra, where it passes *behind* the diaphragm to become the abdominal aorta.

The descending aorta in the thorax gives off many *paired branches* which supply the walls of the thoracic cavity and the organs within the cavity, including:

1. *Bronchial arteries* that supply the bronchi and their branches, connective tissue in the lungs and the lymph nodes at the root of the lungs

2. *Oesophageal arteries* that supply the oesophagus.

3. *Intercostal arteries* that run along the inferior border of the ribs and supply the intercostal muscles, some muscles of the thorax, the ribs, the skin and its underlying connective tissues.

Venous return from the thoracic cavity (Fig. 5:34)

Most of the venous blood from the organs in the thoracic cavity is drained into the *azygos vein* and the *hemiazygos vein*. Some of the main veins which join them are the *bronchial*, *oesophageal* and *intercostal veins*. The azygos vein joins the superior vena cava and the hemiazygos vein joins the left brachiocephalic vein. At the distal end of the oesophagus some oesophageal veins join the azygos vein and others, the left gastric vein. A venous plexus is formed by anastomoses between the veins joining the azygos and the left gastric veins, linking the general and portal circulations.

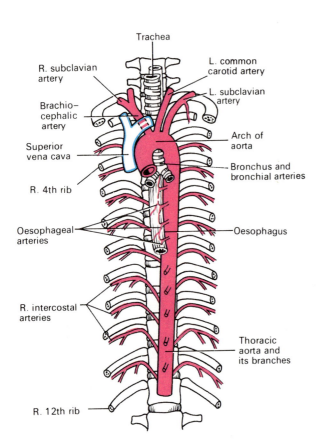

Figure 5:33 Aorta and its main branches in the thorax.

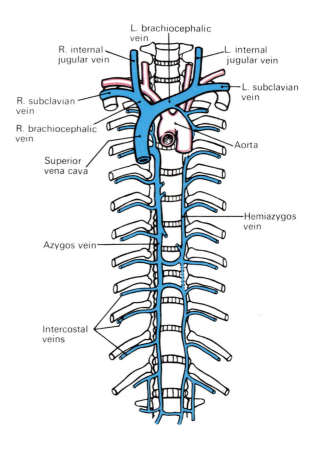

Figure 5:34 Superior vena cava and the main veins of the thorax.

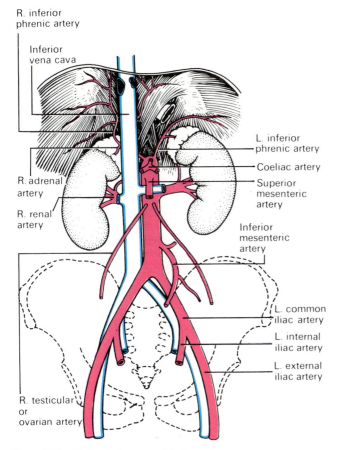

R. inferior
phrenic artery

Inferior
vena cava

R. adrenal
artery

R. renal
artery

L. inferior
phrenic artery

Coeliac artery

Superior
mesenteric
artery

Inferior
mesenteric
artery

L. common
iliac artery

L. internal
iliac artery

L. external
iliac artery

R. testicular
or
ovarian artery

Figure 5:35 Abdominal aorta and its branches.

ABDOMINAL AORTA (Fig. 5:35)

The abdominal aorta is a continuation of the thoracic aorta. The name changes when the aorta enters the abdominal cavity by passing behind the diaphragm at the level of the 12th thoracic vertebra. It descends in front of the bodies of the vertebrae to the level of the 4th lumbar vertebra, where it divides into the *right* and *left common iliac arteries*.

When a branch of the abdominal aorta supplies an organ it is only named here and is described in more detail in association with the organ. However, illustrations showing the distribution of blood from the coeliac, superior and inferior mesenteric arteries are presented here (Figs 5:36 and 5:37).

Many branches arise from the abdominal aorta some of which are *paired* and some *unpaired*.

Paired branches

Inferior phrenic arteries supply the diaphragm.

Renal arteries supply the kidneys and give off branches to supply the adrenal glands.

Testicular arteries supply the testes in the male.

Ovarian arteries supply the ovaries in the female.

The testicular and ovarian arteries are much longer than the other paired branches. This is because the testes and the ovaries begin their development in the region of the kidneys. As they grow they descend into the scrotum and the pelvis respectively and are accompanied by their blood vessels.

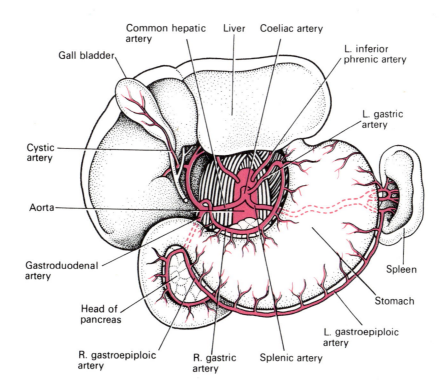

Common hepatic
artery

Gall bladder

Cystic
artery

Aorta

Gastroduodenal
artery

Head of
pancreas

R. gastroepiploic
artery

Liver   Coeliac artery

L. inferior
phrenic artery

L. gastric
artery

Spleen

Stomach

L. gastroepiploic
artery

R. gastric
artery

Splenic artery

Figure 5:36 Coeliac artery and its branches and the inferior phrenic arteries.

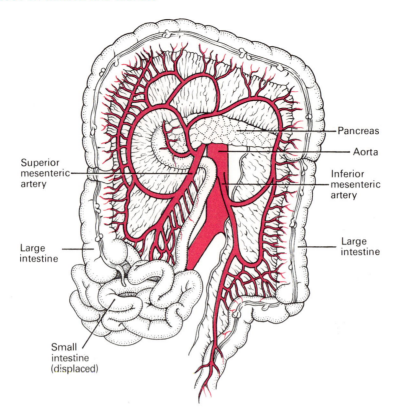

Figure 5:37 Superior and inferior mesenteric arteries and their branches.

Unpaired branches (Figs 5:36 and 5:37)
The *coeliac artery* is a short thick artery about 1.25 cm long. It arises immediately below the diaphragm and divides into three branches:

Left gastric artery: supplies the stomach
Splenic artery: supplies the pancreas and the spleen
Common hepatic artery: supplies the liver, gall bladder and parts of the stomach, duodenum and pancreas

The *superior mesenteric artery* branches from the aorta between the coeliac artery and the renal arteries. It supplies the whole of the small intestine and the proximal half of the large intestine.

The *inferior mesenteric artery* arises from the aorta about 4 cm above its division into the common iliac arteries. It supplies the distal half of the large intestine and part of the rectum.

Venous return from the abdominal organs (Fig. 5:38)
The *inferior vena cava* is formed when *right* and *left common iliac veins* join at the level of the body of the 5th lumbar vertebra. This is the largest vein in the body and it conveys blood from all parts of the body below the diaphragm to the right atrium of the heart. It passes through the central tendon of the diaphragm at the level of the 8th thoracic vertebra.

Paired testicular, ovarian, renal and adrenal veins join the inferior vena cava.

Blood from the remaining organs in the abdominal cavity passes through the liver before entering the inferior vena cava via the *portal circulation*.

PORTAL CIRCULATION

In all the parts of the circulation which have been described previously, venous blood passes from the tissues to the heart by the most direct route. In the portal circulation, venous blood passes from the capillary bed of *the abdominal part of the digestive system, the spleen and pancreas* to the *liver*. It passes through a second capillary bed in the liver before entering the general circulation via the inferior vena cava. In this way blood with a high concentration of nutrient materials goes to the liver first, where certain modifications take place, including the regulation of nutrient supply to other parts of the body.

Portal vein (Figs 5:38 and 5:39)
This is formed by the union of the following veins, each of which drains blood from the area supplied by the corresponding artery:

Splenic vein
Inferior mesenteric vein
Superior mesenteric vein
Gastric veins
Cystic vein

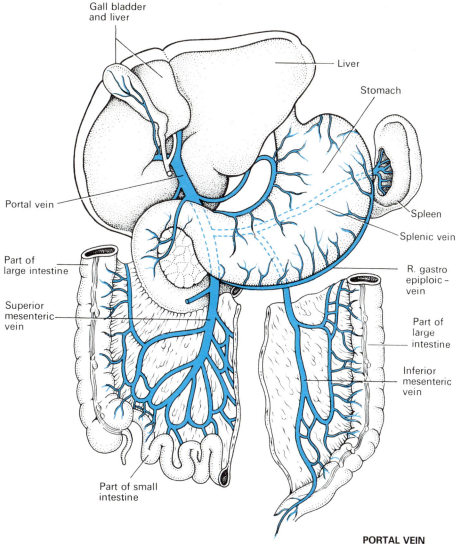

Gall bladder and liver

Liver

Stomach

Portal vein

Spleen

Splenic vein

Part of large intestine

R. gastro epiploic - vein

Superior mesenteric vein

Part of large intestine

Inferior mesenteric vein

Part of small intestine

Figure 5:38 Venous drainage from the abdominal organs and the formation of the portal vein.

The *splenic vein* drains blood from the spleen, the pancreas and part of the stomach.

The *inferior mesenteric vein* returns the venous blood from the rectum, pelvic and descending colon of the large intestine. It joins the splenic vein.

The *superior mesenteric vein* returns venous blood from the small intestine and the proximal parts of the large intestine, i.e., the caecum, ascending and transverse colon.

The *superior mesenteric vein* unites with the *splenic vein* to form the *portal vein*.

The *gastric veins* drain blood from the stomach and the distal end of the oesophagus then join the portal vein.

The *cystic vein* which drains venous blood from the gall bladder joins the portal vein.

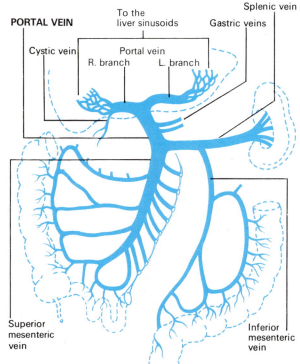

PORTAL VEIN

To the liver sinusoids

Splenic vein

Gastric veins

Cystic vein

Portal vein

R. branch

L. branch

Superior mesenteric vein

Inferior mesenteric vein

Figure 5:39 Portal vein — formation and termination.

CIRCULATION OF BLOOD TO THE PELVIS AND LOWER LIMB (Figs 5:19, 5:20, 5:40, 5:41 and 5:42)

Arterial supply

Common iliac arteries. The right and left common iliac arteries are formed when the abdominal aorta divides at the level of the 4th lumbar vertebra. In front of the sacro-iliac joint each divides into:

Internal iliac artery

External iliac artery

The *internal iliac artery* runs medially to supply the organs within the pelvic cavity. In the female, one of the main branches is the *uterine artery* which provides the main arterial blood supply to the reproductive organs.

The *external iliac artery* runs obliquely downwards and passes behind the inguinal ligament into the thigh where it becomes the *femoral artery*.

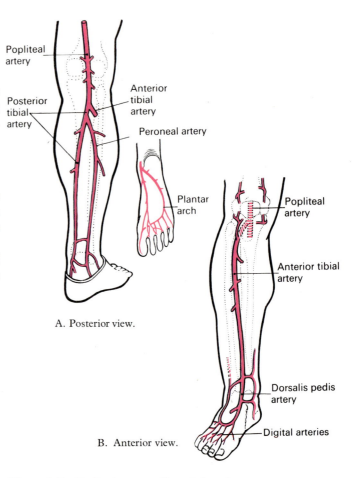

A. Posterior view.

B. Anterior view.

Figure 5:41   Popliteal artery and its main branches.

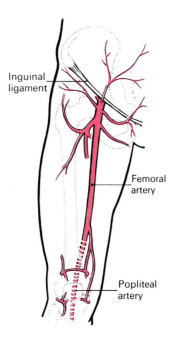

Figure 5:40   Femoral artery and its main branches.

The *femoral artery* (Fig. 5:40) begins at the mid-point of the inguinal ligament and extends downwards in front of the thigh then it turns medially, and eventually passes round the medial aspect of the femur to enter the popliteal space where it becomes the *popliteal artery*. It supplies blood to the structures of the thigh and some superficial pelvic and inguinal structures.

The *popliteal artery* (Fig. 5:41) passes through the popliteal fossa behind the knee. It supplies the structures in this area, including the knee joint. At the lower border of the popliteal fossa it divides into the anterior and posterior tibial arteries.

The *anterior tibial artery* (Fig. 5:41) passes forwards between the tibia and fibula and supplies the structures in the front of the leg. It lies on the tibia, runs in front of

the ankle joint and continues over the dorsum (top) of the foot as the *dorsalis pedis artery*.

The *dorsalis pedis artery* is a continuation of the anterior tibial artery and passes over the dorsum of the foot, supplying arterial blood to the structures in this area. It ends by passing between the first and second metatarsal bones into the sole of the foot where it contributes to the formation of the plantar arch.

The *posterior tibial artery* (Fig. 5:41) runs downwards and medially on the back of the leg. Near its origin it gives off a large branch called the *peroneal artery* which supplies the lateral aspect of the leg. In the lower part it becomes superficial and passes medial to the ankle joint to reach the sole of the foot where it continues as the *plantar artery*.

The *plantar artery* supplies the structures in the sole of the foot. This artery and its branches form an arch from which the *digital branches* arise to supply the toes.

Venous return from the lower limb and pelvis

There are some deep and some superficial veins in the leg (Fig. 5:20). Blood entering the superficial veins passes to the deep veins through *communicating veins*. Movement of blood towards the heart is partly dependent on contraction of skeletal muscles. Backward flow is prevented by a large number of valves. Superficial veins receive less support by surrounding tissues than deep veins.

The deep veins

The deep veins accompany the arteries and their branches and have the same names. They are:

Digital veins
Plantar venous arch
Posterior tibial vein
Anterior tibial vein
Popliteal vein
Femoral vein
External iliac vein
Internal iliac vein
Common iliac vein

The *femoral vein* ascends in the thigh to the level of the inguinal ligament where it becomes the external iliac vein.

The *external iliac vein* is the continuation of the femoral vein where it enters the pelvis lying close to the femoral artery. It passes along the brim of the pelvis and at the level of the sacroiliac joint it is joined by the *internal iliac vein* to form the *common iliac vein*.

The *internal iliac vein* receives tributaries from several veins which drain the organs of the pelvic cavity.

The *two common iliac veins* begin at the level of the sacroiliac joints. They ascend obliquely and end a little to the right of the body of the 5th lumbar vertebra by uniting to form the *inferior vena cava*.

Superficial veins (Fig. 5:42)

The two main superficial veins draining blood from the superficial structures of the lower limbs are:

Short saphenous vein
Long saphenous vein

The *short saphenous vein* begins behind the ankle joint where many small veins which drain the dorsum of the foot join together. It ascends superficially along the back of the leg and in the popliteal space it joins the *popliteal vein* — a deep vein.

The *long saphenous vein* is the longest vein in the body. It begins at the medial half of the dorsum of the foot and runs upwards, crossing the medial aspect of the tibia and up the inner side of the thigh. Just below the inguinal ligament it joins the *femoral vein*.

Many communicating veins join the superficial and deep veins of the lower limb.

SUMMARY OF THE SYSTEMIC CIRCULATION

THORACIC AORTA

Ascending aorta
Coronary arteries — paired

Arch of aorta

Brachiocephalic { right common carotid artery / right subclavian artery
Left common carotid artery
Left subclavian artery

Descending aorta
Bronchial arteries
Oesophageal arteries } paired arteries
Intercostal arteries

ABDOMINAL AORTA

(branches in the order in which they leave the aorta)
Phrenic arteries — paired

Coeliac artery { gastric arteries / splenic artery / hepatic artery } unpaired

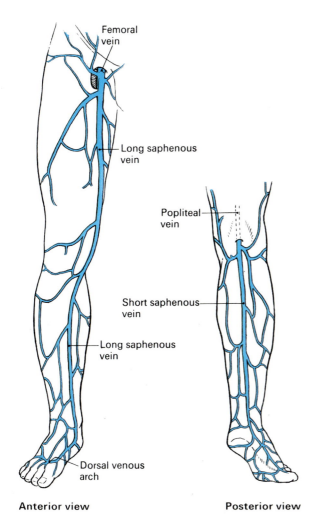

Anterior view   **Posterior view**

Figure 5:42   Superficial veins of the leg.

Labels in figure: Femoral vein, Long saphenous vein, Popliteal vein, Short saphenous vein, Long saphenous vein, Dorsal venous arch

Superior mesenteric artery — unpaired
Renal arteries — paired
Ovarian or testicular arteries — paired
Inferior mesenteric artery — unpaired
Common iliac arteries { external iliac artery / internal iliac artery } paired

ARTERIES SUPPLYING THE HEAD AND NECK

Common carotid arteries — paired
  External carotid artery:
      thyroid artery
      lingual artery
      facial artery
      occipital artery
      temporal artery
      maxillary artery
  Internal carotid artery:
      opthalmic artery
      anterior cerebral artery
      middle cerebral artery
      posterior communicating artery
  Circulus arteriosus (circle of Willis):
      anterior cerebral arteries
      middle cerebral arteries
      posterior communicating arteries
      posterior cerebral arteries, which are branches
        from the basilar artery
  Vertebral arteries — paired
  Basilar artery — unpaired

VENOUS RETURN FROM THE HEAD AND NECK

Superficial veins
  Thyroid vein
  Facial vein   } join the external jugular vein
  Occipital vein

Deep sinuses
  Superior sagittal sinus
  Inferior sagittal sinus } unpaired
  Straight sinus
  Transverse sinuses — paired

ARTERIES SUPPLYING THE UPPER EXTREMITY

Subclavian artery
Axillary artery
Brachial artery
Radial artery
Ulnar artery
Deep and superficial palmar arches
Palmar metacarpal arteries
Palmar digital arteries

VENOUS RETURN FROM THE UPPER EXTREMITY

Deep veins
  Palmar digital veins
  Palmar metacarpal veins
  Palmar venous arch
  Radial vein
  Ulnar vein
  Brachial vein
  Axillary vein
  Subclavian vein
  Brachiocephalic vein
  Superior vena cava

Superficial veins
  Digital veins
  Palmar venous arch
  Cephalic vein
  Basilic vein
  Median cubital vein
  Median vein

ARTERIES SUPPLYING THE PELVIS AND LOWER EXTREMITY

Common iliac artery
Internal iliac artery
External iliac artery
Femoral artery
Popliteal artery
Anterior tibial artery
Posterior tibial artery
Plantar arch
Digital arteries

VENOUS RETURN FROM THE PELVIS AND LOWER EXTREMITY

Deep veins
  Digital veins
  Plantar veins
  Anterior tibial veins
  Posterior tibial veins
  Popliteal vein
  Femoral vein
  External iliac vein
  Internal iliac vein
  Common iliac vein
  Inferior vena cava

Superficial veins
  Long saphenous vein
  Short saphenous vein

DISEASES OF BLOOD VESSELS

ARTERIOSCLEROSIS AND ARTERIOLOSCLEROSIS

These are diseases of arteries, most common in men between 35 and 65 years of age and in women after the menopause.

ATHEROMA

Patchy changes (*atheromatous plaques*) develop in the tunica media of large and medium-sized arteries. These consist of accumulations of cholesterol compounds, excess smooth muscle and fibroelastic cells. As the plaques grow they spread along the artery walls forming swellings that protrude into the lumen. Eventually the whole thickness of the wall and long sections of the vessel may be affected (Fig. 5:43).

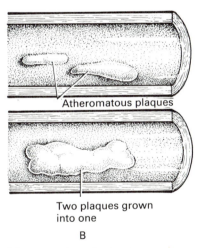

Figure 5:43 Atheroma and arteriosclerosis.

The origin of atheromatous plaques is uncertain. *Fatty streaks* present in artery walls of infants are usually absorbed. Their incomplete absorption may be the source of atheromatous plaques in later life. Arteries most commonly affected are those in the heart, brain, kidneys, small intestine and lower limbs.

CAUSES OF ATHEROMA

Why atheromatous plaques develop is not yet clearly understood but predisposing factors include:
1. Familial factors, genetic and environmental
2. Sex, they are more common in males
3. Other diseases, e.g., hypertension, hypothyroidism and diabetes mellitus, especially when it develops at an early age
4. Smoking, especially cigarettes
5. Excessive stress in work and/or home environment
6. Diet, high kilopascal (calorie) diet, rich in animal fats and sugars, soft drinking water, i.e., slightly acid and low in mineral salt content
7. Obesity
8. Sedentary life-style

COMPLICATIONS OF ATHEROMA

Thrombosis
If the epithelium lying over a plaque breaks down, platelets are stimulated by the damaged cells and a blood clot (thrombus) forms, blocking the artery and causing ischaemia and infarction. Pieces of the clot (emboli) may break off, travel in the blood stream and lodge in small arteries distal to the clot, causing small infarcts (areas of dead tissue).

Calcification
If calcium salts are deposited in the plaques, the artery walls become brittle, rigid and unresponsive to rises in blood pressure and may rupture, causing haemorrhage.

Aneurysm formation
If the artery wall is weakened by spread of the plaque between the layers of tissue, a local dilatation (aneurysm) may develop. This may lead to thrombosis and embolism, or the aneurysm may rupture causing severe haemorrhage. The most common sites affected are the abdominal and pelvic arteries.

EFFECTS OF ATHEROMA

Arteries may be partially or completely blocked by atheromatous plaques alone, or by plaques combined with a thrombus. This may reduce or completely block the blood supply. The effects depend on the site and size of the artery involved.

Narrowing of artery
The tissues distal to the narrow point become ischaemic. The cells may receive enough blood to meet their minimum needs, but not enough to cope with an increase in metabolic rate, e.g., when muscle activity is increased. This causes acute cramp-like ischaemic pain. Heart muscle and skeletal muscles of the lower limb are most commonly affected. Ischaemic pain in the heart is called *angina pectoris* and in the lower limbs *intermittent claudication*.

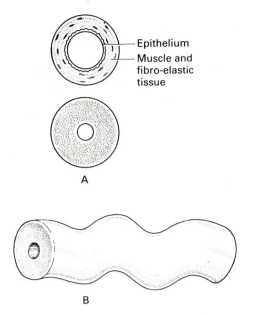

Figure 5:44 Arteriosclerotic arteries. A. cross-section B. Tortuous artery.

Occlusion of artery

When an artery is completely blocked, the tissues it supplies rapidly undergo degeneration and die, i.e., ischaemia leads to infarction. The effect on the body depends on:

1. The size of the artery occluded
2. The type of tissue involved
3. The extent of collateral circulation, e.g., in the brain the circle of Willis provides extensive collateral blood vessels while in the heart there are very few

If a coronary artery is occluded and the individual survives the initial heart attack, a blood clot may form inside the heart over the infarct. An embolus from this mural clot in the right side of the heart becomes lodged in the lung, while one from the left side may lodge in any other organ, most commonly in the brain.

THROMBANGITIS OBLITERANS

In this condition there is acute inflammation with thrombosis of the arteries and veins mainly in the lower limbs. Later, it becomes chronic and the vessel walls become fibrosed, lose their elasticity and do not dilate during exercise. The individual suffers from acute ischaemic pain and, as the disease progresses, the distance walked with comfort is gradually reduced. In the long term the skin may ulcerate and in extreme cases, gangrene may develop.

It is believed to be caused by an immune response to an antigen, possibly a tobacco protein, as the disease commonly occurs in men who are heavy smokers. No microbes have yet been isolated.

POLYARTERITIS NODOSUM

This is an inflammatory condition of the small arteries and arterioles in any part of the body. The most common sites are the heart, kidneys, alimentary tract, liver, pancreas and nervous system. It is usually acute at first but may become chronic. Necrosis and rupture of blood vessels may occur in the acute phase followed by thrombosis, ischaemia, infarction and death. It is believed to be caused by an immune reaction but the antigen is not known.

ANEURYSMS

An aneurysm is a local abnormal dilatation of an artery. It may be caused by a congenital abnormality of the tunica media or be secondary to another disease. The most common sites are the circle of Willis in the brain (berry aneurysm) and the aorta. There are three types of aneurysm (Fig. 5:45).

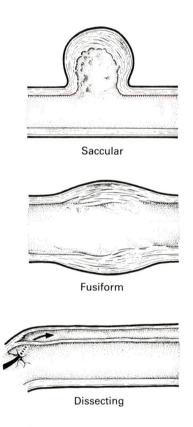

Figure 5:45 Types of aneurysm.

Saccular which bulge on only one side of the artery, e.g., berry aneurysms.

Fusiform which are spindle-shaped and occur mostly in the aorta.

Dissecting, formed by infiltration of blood between the endothelium and tunica media, are most common in the aorta.

UNDERLYING CAUSES OF ANEURYSMS

Congenital

There is a familial tendency but no genetic defect has been identified. In berry aneurysms there is impaired development of the elastic tissue at the branching points of the arteries.

Diseases of the arteries

These include:
1. Arteriosclerosis especially of the abdominal aorta
2. Degeneration of elastic tissue (Endheim's)
3. Arteritis caused by microbes in the blood or from an adjacent site

COMPLICATIONS OF ANEURYSMS

Haemorrhage

A ruptured aneurysm may cause sudden death or disability of varying severity, depending on the size and site of the artery. Berry aneurysms commonly cause subarachnoid or cerebral haemorrhage.

Pressure

This may affect adjacent tissues, e.g., an aortic aneurysm can cause displacement of the heart or it may compress:
1. The recurrent laryngeal nerve, causing laryngeal paralysis
2. Blood vessels, causing necrosis of bones, e.g., sternum, ribs, vertebrae
3. The oesophagus, causing dysphagia (difficulty in swallowing)
4. The trachea, causing dyspnoea (difficulty in breathing)

Thrombosis and embolism

A blood clot (thrombus) may form in an artery where the endothelium has been damaged by an aneurysm. A piece of clot (embolus) may lodge in a small artery distal to the aneurysm and obstruct the blood flow, causing ischaemia and infarction.

THROMBOPHLEBITIS

In this condition a thrombus forms in a vein where endothelium has been damaged by inflammation.

PREDISPOSING FACTORS

Reduced rate of blood flow which may be caused by:
1. Immobility associated with prolonged bed rest
2. Pressure on veins in the popliteal region by, e.g., a pillow under the knees in bed or sitting in a chair for long periods
3. Pressure on a vein by an adjacent tumour
4. Prolonged low blood pressure, as in cardiac failure

Changes in the blood may trigger intravascular clotting, e.g.,
1. Increased blood viscosity in, e.g., dehydration, polycythaemia
2. Increased adhesiveness of platelets, e.g., associated with the use of some contraceptive drugs, and in some malignant diseases

Traumatic injury to a vein caused by, e.g., accident, surgery or long-lasting intravenous infusion.

EMBOLISM

This is a common complication of blood vessel diseases and injuries.

An *embolus* is an abnormal mass of material in the blood stream, large enough to block a blood vessel, e.g., a fragment of blood clot, tumour or atheromatous plaque, fat from bone marrow following a fracture or a bubble of air. *Embolism* is obstruction to blood flow by an embolus.

Emboli in veins move towards the heart and lodge in the lungs or liver. Those in arteries lodge in small arteries or arterioles. The effect depends on the size and site of the blocked blood vessel.

VARICOSE VEINS

A varicosed vein is one which is so dilated that the valves do not close to prevent backward flow of blood. Such veins lose their elasticity, become elongated and tortuous and fibrous tissue replaces the tunica media.

PREDISPOSING FACTORS

Heredity. There appears to be a familial tendency but no abnormal genetic factor has been identified.

Age. There is progressive loss of elasticity in the vein walls with increasing age so that elastic recoil is less efficient.

Obesity. Superficial veins in the limbs are supported by subcutaneous areolar tissue. Excess adipose tissue may not provide sufficient support.

Gravity. Standing for long periods with little muscle contraction tends to cause pooling of blood in the lower limbs and pelvis.

Pressure. Because of their thin walls veins are easily compressed by surrounding structures, leading to increased venous pressure distal to the site of compression.

SITES OF VARICOSE VEINS (Fig. 5:46A and B)

Varicose veins of the legs

When valves in the anastomosing veins between the deep and superficial veins in the legs become incompetent the venous pressure in the superficial veins rises and varicosities develop. The long and short saphenous veins and the anterior tibial veins are most commonly affected. These dilated, inelastic veins rupture easily if injured, and haemorrhage occurs. The skin over a varicose vein may become poorly nourished due to the stasis of blood, leading to the formation of varicose ulcers usually on the medial aspect of the leg just above the ankle.

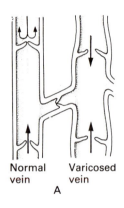

Normal
vein

Varicosed
vein

A

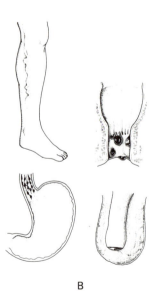

B

Figure 5:46   A. Varicosed veins. B. Common sites of varicosities: the leg, anal haemorrhoids, oesophagus, scrotal hydrocele.

Haemorrhoids

Sustained pressure on the veins at the junction of the rectum and anus leads to increased venous pressure, valvular incompetence and the development of haemorrhoids. The most common causes are chronic constipation,

and the pressure in the pelvis towards the end of pregnancy. Slight bleeding may occur each time stool is passed and may cause anaemia. Severe haemorrhage is rare.

Oesophageal varices

The veins involved are at the lower end of the oesophagus. When the venous pressure in the liver rises, there is a rise in pressure in the anastomosing veins between the left gastric vein and the azygos vein. Sustained pressure causes varicosities to develop. The commonest causes of increased portal vein pressure are cirrhosis of the liver and right-sided cardiac failure. If the pressure continues to rise, the inelastic varicosed veins may rupture causing severe haemorrhage, and possibly death.

Varicocele

Each spermatic cord is surrounded by a plexus of veins that may become varicosed, especially in men whose work necessitates standing for long periods. If the varicocele is bilateral the increased temperature due to venous congestion may cause depressed spermatogenesis.

TUMOURS OF BLOOD AND LYMPH VESSELS

ANGIOMAS

Angiomas are benign tumours of either blood vessels (haemangiomas) or lymph vessels (lymphangiomas). The latter rarely occur so angioma is usually taken to mean haemangioma.

HAEMANGIOMAS

These are not true tumours but are sufficiently similar to be classified as such. They consist of an excessive growth of blood vessels arranged in an uncharacteristic manner and interspersed with collagen fibres.

Capillary haemangiomas

Excess capillary growth interspersed with collagen in a localised area makes a dense, plexus-like network of tissue. Each haemangioma is supplied by only one blood vessel and if it thromboses the haemangioma atrophies and disappears.

Capillary haemangiomas are usually present at birth and are seen as a purple or red mole or birthmark. They may be quite small at birth but grow at an alarming rate in the first few months, keeping pace with the growth of the child. After 1 to 3 years, atrophy may begin and by the end of 5 years in about 80% of cases the tumours have disappeared.

Cavernous haemangiomas

Blood vessels larger than capillaries grow in excess of normal needs in a localised area and are interspersed with collagen fibres. They are dark red in colour and may be present in the skin, though more commonly in the liver. They grow slowly, do not regress and may become large and unsightly.

HYPERTENSION

The term hypertension is used to describe blood pressure that is sustained at a higher than the 'normal' maximum level for a particular age group, e.g., at 20 years 140/90 mmHg, at 75 years 170/105 mmHg.

CHANGES IN THE BLOOD VESSELS IN HYPERTENSION

At first the amount of muscle and elastic tissue in the walls of the arteries increases. Later it is replaced by fibrous tissue. The walls of small and medium arteries become thick, hard and inflexible (*arteriosclerotic*) and the lumen is reduced but large arteries lose their elasticity and dilate. When elasticity is lost, smooth blood flow changes to pulsating flow, i.e., there are fluctuations in blood pressure in the arteries that reflect the pressure difference between cardiac systole and diastole. At bends and bifurcations in small arteries there is a tendency for platelets and fibrin to be deposited, increasing the risk of thrombosis. Aneurysms tend to form in cerebral arteries and when the pressure continues to rise they may rupture, especially in elderly people (cerebrovascular accident). Patchy changes in blood vessel walls are more common in the spleen, liver, kidneys and pancreas.

Senile arteriosclerosis is a condition affecting elderly people in which the progressive loss of elasticity and reduced arterial lumen leads to cerebral ischaemia and loss of function. There may or may not be evidence of hypertension.

Hypertension is described as *essential* (primary, idiopathic) or *secondary to other diseases*.

ESSENTIAL HYPERTENSION

This means hypertension of unknown cause. It accounts for 85 to 90% of all cases and is subdivided according to the rate at which the disease progresses.

BENIGN OR CHRONIC HYPERTENSION

The rise in pressure is usually slight to moderate and continues to rise slowly. Sometimes complications are the first indication of hypertension, e.g., heart failure, cerebrovascular accident, myocardial infarction. Occasionally the rate of progress increases and the hypertension becomes malignant.

MALIGNANT (ACCELERATED) HYPERTENSION

The blood pressure is very high and continues to rise rapidly. Diastolic pressure in excess of 120 mmHg is common. The effects quickly become apparent, e.g., haemorrhages into the retina, papilloedema (oedema around the optic disc), encephalopathy (cerebral irritation), renal failure.

SECONDARY HYPERTENSION

Hypertension resulting from other diseases accounts for 10 to 15% of all cases.

Kidney diseases

Raised blood pressure is a complication of many kidney diseases. The vasoconstrictor effect of excess *renin* released by damaged kidneys is one causative factor but there may be others, as yet unknown.

Adrenal cortex diseases

Secretion of excess *aldosterone* and *cortisol* stimulate the retention of excess sodium and water by the kidneys, raising the blood volume and pressure. Oversecretion of aldosterone (Conn's syndrome) is due to a hormone-secreting tumour. Oversecretion of cortisol may be due to overstimulation of the gland by *adrenocorticotrophic hormone* secreted by the pituitary gland, or to a hormone-secreting tumour.

Stricture of the aorta

Hypertension develops in branching arteries proximal to the site of a stricture. In *congenital coarctation* the stricture is between the ductus arteriosus and the left subclavian artery causing hypertension in the head, neck and right arm. Compression of the aorta by an adjacent tumour may cause hypertension proximal to the stricture (see Fig. 5:47, p. 95).

EFFECTS OF HYPERTENSION

The effects of longstanding and progressively rising blood pressure are best considered in relation to essential hypertension, as it is not association with other diseases.

Heart

The rate and force of cardiac contraction is increased to maintain the cardiac output against a sustained rise in arterial pressure. The left ventricle hypertrophies and begins to fail when compensation has reached its limit. This is followed by stasis of blood in the lungs, hypertrophy of the right ventricle and eventually to right ventricular failure.

Brain

Cerebral haemorrhage is common, the effects depending on the position and size of the ruptured vessel. When a series of small blood vessels rupture at different times there is progressive disability. Rupture of a large vessel causes extensive loss of function or possibly death.

Kidneys

Essential hypertension causes kidney damage. If sustained for only a short time recovery may be complete. Otherwise the kidney damage causes further hypertension, progressive loss of kidney function and kidney failure.

PULMONARY HYPERTENSION

Raised blood pressure in the pulmonary circulation is secondary to:
1. Changes in blood vessels, described above
2. Diseases of the respiratory system
3. Diseases of the heart, e.g., congenital defects of the septum, stenosis and incompetence of the mitral or aortic valve
4. Diseases of other organs that cause raised pressure in the left side of the heart, e.g., cirrhosis of the liver, thrombosis of the portal vein

HYPOTENSION

This usually occurs as a complication of other conditions, e.g.:

Shock	Myocardial infarction
Addison's disease	Haemorrhage

Low blood pressure leads to inadequate blood supply to the brain. Depending on the cause, unconsciousness may be brief (fainting) or more prolonged, possibly causing death.

HEART DISEASES

CARDIAC FAILURE

The heart is described as failing when the myocardium of the ventricles is unable to maintain the circulation of sufficient blood to meet the needs of the body.

ACUTE CARDIAC FAILURE

A sudden diminution in output of blood from both ventricles causes acute reduction in the oxygen supply to all the tissues. Recovery from the acute phase may be followed by chronic failure, or death may occur due to anoxia of vital centres in the brain. The commonest causes are:
1. Severe damage to an area of cardiac muscle due to ischaemia caused by sudden occlusion of one of the larger coronary arteries by atheroma or atheroma with thrombosis
2. Pulmonary embolism
3. Acute toxic myocarditis
4. Severe cardiac arrhythmia
5. Rupture of a heart chamber or valve cusp
6. Severe malignant hypertension

CHRONIC CARDIAC FAILURE

This develops gradually and in the early stages there may be no symptoms because the heart compensates by increasing the rate and force of contraction and the ventricles dilate. Myocardial cell hypertrophy increases the strength of the muscle. When further compensation is not possible there is a gradual decline in myocardial efficiency. During the development of chronic failure hypoxia and venous congestion cause changes in other systems, making still greater demands on the heart, e.g., renal, endocrine, respiratory. Underlying causes include:
1. Chronic hypertension, myocardial fibrosis, valvular disease, lung diseases, anaemia
2. Previous acute failure
3. Degenerative changes of old age

RIGHT-SIDED OR CONGESTIVE CARDIAC FAILURE

The right ventricle fails when pressure developed within it by the contracting myocardium is less than the force needed to push blood through the lungs. This discrepancy may be caused by increased resistance in the lungs, weakness of the myocardium and/or stenosis and incompetence of valves in the heart or great vessels.

When compensation has reached its limit, and the ventricle is not emptying completely, the right atrium and venae cavae become congested with blood and this is followed by congestion throughout the systemic circulation. The organs affected first are the liver, spleen and kidneys. Oedema of the lower limbs and ascites usually follow.

Resistance to the flow of blood through the lungs
When this is increased the right ventricle has more work to do. It may be caused by:
1. The formation of fibrous tissue following inflammation

2. Back pressure of blood from the left side of the heart, e.g., in left ventricular failure, when the mitral valve is stenosed and/or incompetent.

Weakness of the myocardium
This may be caused by ischaemia following numerous small myocardial infarcts.

STENOSIS AND INCOMPETENCE OF VALVES

Stenosis is narrowing of a valve. Inflammation and crustation roughen the edges of the cusps so that they stick together, narrowing the valve opening. When healing occurs fibrous tissue is formed which shrinks as it ages, increasing the stenosis. Stenosis is accompanied by valvular incompetence.

Incompetence is a functional defect caused by failure of a valve to close completely, allowing blood to flow back into the ventricle when it relaxes.

The most common causes of valve defects are rheumatic fever, fibrosis following inflammation and congenital abnormalities.

LEFT-SIDED OR LEFT VENTRICULAR FAILURE

This occurs when the pressure developed in the left ventricle by the contracting myocardium is less than the pressure in the aorta. Causes include:
1. Excessively high systemic blood pressure
2. Incompetence of the mitral and/or the aortic valve
3. Aortic valve stenosis
4. Myocardial weakness

Failure of the left ventricle leads to dilatation of the atrium and an increase in pulmonary blood pressure. This is followed by a rise in the blood pressure in the right side of the heart and eventually general venous congestion.

The congestion in the lungs leads to pulmonary oedema and dyspnoea, often most severe at night. This *paroxysmal nocturnal dyspnoea* may be due to raised blood volume as fluid from peripheral oedema is reabsorbed when the patient slips down in bed during sleep.

ISCHAEMIC HEART DISEASE

Ischaemic heart disease is due to narrowing or occlusion of one or more branches of the coronary arteries. The narrowing is caused by atheromatous plaques. Occlusion may be by plaques alone, or plaques complicated by thrombosis. The overall effect depends on the size of the coronary artery involved and whether it is narrowed or occluded. Narrowing of the artery leads to *angina pectoris*, occlusion to *myocardial infarction*, i.e., an area of dead tissue.

COLLATERAL ARTERIAL BLOOD SUPPLY

When atheroma is gradually developing, a collateral arterial blood supply may have time to develop and effectively supplement or replace the original. This consists of the dilatation of normally occurring anastomotic arteries joining adjacent branch arteries. When sudden severe narrowing or occlusion of an artery occurs the anastomotic arteries dilate but may not be able to supply enough blood to meet the needs of the myocardium.

ANGINA PECTORIS

This is sometimes called *angina of effort* because physical effort causes severe ischaemic pain. A narrowed artery may supply sufficient blood to an area of heart wall to meet its needs during rest or moderate exercise but not when effort is increased, e.g., walking may be tolerated but not running. The thick, inflexible atheromatous artery wall is unable to dilate to allow for the increased blood flow needed by the more active myocardium. In the early stages of development of the disease the pain stops soon after the extra effort stops.

MYOCARDIAL INFARCT

An *infarct* in an area of tissue that has died because of lack of blood. It may occur in any part of the body and the effect depends on the size and site of the infarct.

In *chronic ischaemic heart disease* many small infarcts form and collectively they lead to myocardial weakness and possibly heart failure.

In *acute ischaemic heart disease* one or more large arteries are occluded and the atheroma is usually complicated by thrombosis. The infarct is correspondingly large and disrupts the heart activity. Death may occur suddenly for a number of reasons:
1. Acute cardiac failure due to shock
2. Severe arrhythmias, especially ventricular fibrillation due to disruption of the conducting system
3. Rupture of a ventricle wall, occurring usually within about 2 weeks of the infarction
4. Pulmonary or cerebral embolism, originating from, e.g., a mural clot or a venous thrombosis

RHEUMATIC HEART DISEASE
RHEUMATIC FEVER

This autoimmune disease occurs 2 to 4 weeks after a throat infection, caused by *Streptococcus pyogenes* (beta-haemolytic Group A). The antibodies develop to combat the infection damage to the heart. The microbes are not present in the heart lesion and the same infection in other parts of the body is very rarely followed by rheumatic fever. How the antibodies damage the heart is not yet understood. Children and young adults are most commonly affected.

Effects on the endocardium

The endocardium becomes inflamed and oedematous and tiny pale areas called *Aschoff's bodies* appear which, when they heal, leave thick fibrous tissue. Thrombotic fibrous nodules consisting of platelets and fibrin form on the free borders of the cusps of the heart valves. When healing occurs the fibrous tissue formed shrinks as it ages, distorting the shape of the cusps and causing stenosis and incompetence of the valve. The mitral and aortic valves are commonly affected, the tricuspid valve sometimes and the pulmonary valve rarely.

Effects on the myocardium

Aschoff's bodies form on the connective tissue between the cardiac muscle fibres. As in the endocardium, healing is accompanied by fibrosis which may interfere with myocardial contraction.

Effects on the pericardium

Inflammation leads to the accumulation of exudate in the pericardial cavity. Healing is accompanied by fibrous thickening of the pericardium and adhesions form between the two layers. In severe cases the layers may fuse, obliterating the cavity. Within this inelastic pericardium the heart may not be able to expand fully during diastole, leading to reduced cardiac output, generalised venous congestion and oedema.

SUBCLINICAL RHEUMATIC HEART DISEASE

Valvular incompetence developing in older people who have a history of rheumatic fever many years previously is believed to be due to repeated subclinical attacks. These attacks are not associated with repeated episodes of sore throat so it is assumed that the original disease has remained active in a subclinical form. In some cases there is no history of rheumatic fever.

INFECTIVE ENDOCARDITIS

Infecting organisms may colonise any part of the endocardium but most commonly occur on or near the heart valves and round the margins of congenital septal defects. These areas are susceptible to infection because they are exposed to fast-flowing blood that may cause mild trauma.

The main predisposing factors are bacteraemia, depressed immune response and heart abnormalities.

Bacteraemia

Microbes may or may not multiply while in the blood stream and, if not destroyed by phagocytes or antibodies, they tend to adhere to platelets and form tiny infected emboli. Inside the heart the emboli are most likely to settle on already damaged endocardium. Vegetations consisting of platelets and fibrin surround the microbes and seem to protect them from normal body defences and antibiotics. Because of this, infection may be caused by a wide range of microbes, including some of low pathogenicity, e.g.,

1. Non-haemolytic streptococci, e.g., following tooth extraction, tonsillectomy
2. *Escherichia coli* and other normal bowel inhabitants, e.g., following intestinal surgery
3. *Staphylococcus aureus.* e.g., from boils and carbuncles
4. Microbes from infections of, e.g., the biliary, urinary, respiratory tracts
5. Microbes introduced during medical and nursing procedures, e.g., cystoscopy, bladder catheterisation, arterial and venous catheterisation, surgery, wound dressing
6. Low virulence microbes that cause infection in people with reduced immune response

Depressed immune response

This enables low virulence bacteria, viruses, yeasts and fungi to become established and cause infection. These are organisms always present in the body and the environment. Depression of the immune systems may be caused by:

1. Cytotoxic drugs
2. X-rays used in, e.g., cancer treatment or to prevent the rejection of transplanted tissues by depressing the immune system
3. Anti-inflammation drugs, e.g., corticosteroids
4. Some diseases, e.g., leukaemia, tumours of lymphoid tissue, cancer
5. Use of unsterile syringes by, e.g., drug addicts

Heart abnormalities

The sites most commonly infected are those where there is already an abnormality, e.g., valve cusps damaged by earlier attacks of rheumatic fever, endothelium damaged by the fast flow of blood through a narrow opening, such as a stenosed valve or congenital septal defect.

ACUTE INFECTIVE ENDOCARDITIS

This is usually caused by high-virulence microbes. Vegetations grow rapidly and pieces may break off, becoming infected emboli. These settle in other organs where the microbes grow, destroying tissue and forming pus. The effects depends on the organ involved, e.g., brain or kidney infection may cause death in a few·days.

SUBACUTE INFECTIVE ENDOCARDITIS

This is usually caused by low-virulence microbes, e.g., non-haemolytic streptococci, some staphylococci. Infected emboli may settle in any organ. They do not cause suppuration and rarely cause death. Microbes in the vegetations seem to be protected by surrounding platelets and fibrin from normal body defences and antibiotics. Healing by fibrosis further distorts the shape of the valve cusps, increasing the original stenosis and incompetence. Heart failure may develop later.

CARDIAC ARRHYTHMIAS

FIBRILLATION

This is the contraction of the cardiac muscle fibres in a disorderly sequence. The chambers do not contract as a whole and the pumping action is disrupted.

Atrial fibrillation
When the atria fibrillate, contraction is unco-ordinated and rapid, pumping is ineffective and stimulation of the AV node is disorderly. Ventricular contraction becomes rapid and rhythm and force irregular, although an adequate circulation may be maintained. The causes of increased excitability and disorganised activity are not always clear but predisposing conditions include:
1. Ischaemic heart disease
2. Degenerative changes in the heart due to old age
3. Thyrotoxicosis
4. Rheumatic heart disease

Ventricular fibrillation
The pumping action of the ventricles is defective when the sequence of contraction of the muscle fibres is disrupted and blood is not pumped into either the pulmonary or the systemic circulation. If normal heart action cannot be restored quickly, death follows due to cerebral anoxia. There is usually a considerable increase in the heart rate. *Tachycardia* is a rate of more than 160 beats per minute.

HEART BLOCK

Heart block occurs when the delay between atrial and ventricular contraction is increased. The severity depends on the extent of loss of stimulation of the AV node. Heart block is complete when ventricular contraction is entirely dependent on impulses initiated by the AV node, i.e., about 40 per minute. *Bradychardia* is a rate of less than 35 per minute. In this state the heart is unable to respond quickly to a sudden increase in demand by, e.g., muscular exercise. The most common causes are:
1. Acute ischaemic heart disease

2. Myocardial fibrosis following repeated infarctions or myocarditis
3. Drugs used to treat heart disease, e.g., digitalis, propanolol

When heart block develops gradually there is some degree of adjustment in the body to reduced cardiac output but, if progressive, it eventually leads to death from cardiac failure and cerebral anoxia.

CONGENITAL ABNORMALITIES

Abnormalities in the heart and great vessels at birth may be due to intrauterine developmental errors or to the failure of the heart and blood vessels to adapt to extra-uterine life. In many cases there are no symptoms in early life and the abnormality is recognised only when complications appear.

PERSISTENT DUCTUS ARTERIOSUS

Before birth the ductus arteriosus, joining the arch of the aorta and the pulmonary artery, allows blood to pass from the pulmonary artery to the aorta (Fig. 5:47). At birth, when the pulmonary circulation is established, the ductus arteriosus should close completely. If it remains patent, blood regurgitates from the aorta to the pulmonary artery, reducing the volume entering the systemic circulation. This leads to pulmonary congestion and eventually cardiac failure.

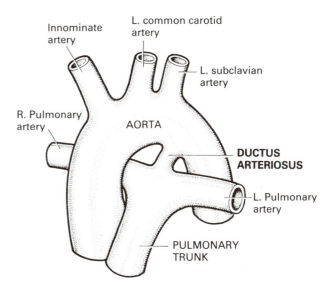

Figure 5:47 Diagram showing the position of the ductus arteriosus in the fetus.

ATRIAL SEPTAL DEFECT

Before birth most blood enters the left atrium from the right atrium through the *foramen ovale* in the septum. There is a valve-like structure across the opening consisting of two partly overlapping membranes. The 'valve' is open when the pressure in the right atrium is higher than in the left. After birth, when the pulmonary circulation is established and the pressure in the left atrium is the higher, the two membranes come in contact, closing the 'valve'. Later the closure becomes permanent due to fibrosis.

When the membranes do not overlap an opening between the atria remains patent after birth. In many cases it is too small to cause symptoms in early life but they may appear later. In severe cases blood flows back to the right atrium from the left. This increases the right ventricular and pulmonary pressure, causing hypertrophy of the myocardium and eventually cardiac failure. As pressure in the right atrium rises, blood flow through the defect may be reversed, but this is not an improvement because deoxygenated blood gains access to the general circulation.

COARCTATION OF THE AORTA

The most common site of coarctation (narrowing) of the aorta is between the left subclavian artery and ductus arteriosus. This leads to hypertension in the upper body because increased force of contraction of the heart is needed to push the blood through the coarctation. There is hypotension in the rest of the body.

FALLOT'S TETRALOGY

Four concurrent abnormalities that interfere with the flow of blood from the left ventricle cause severe general hypoxia and heart failure. They are:

1. Stenosis of the pulmonary artery at its point of origin
2. Ventricular septal defect just below the atrioventricular valves
3. Aortic misplacement, i.e. the origin of the aorta is displaced to the right so that it is immediately above the septal defect
4. Right ventricular hypertrophy to counteract the pulmonary stenosis

The Lymphatic System

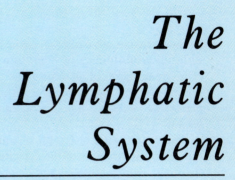

6. The Lymphatic System

All body tissues are bathed in tissue fluid, consisting of the diffusible constituents of blood and waste materials from cells. Some tissue fluid returns to the capillaries at their venous end and the remainder diffuses through the more permeable walls of the *lymph capillaries* and becomes *lymph*.

The composition of lymph is similar to that of plasma, but the concentrations of the constituents are different and there are some additional substances that are too large to pass through blood capillary walls, e.g., particles from areas of infection and cells damaged by disease. Lymph passes through vessels of increased size and a varying number of *lymph nodes* before returning to the blood. The lymphatic system consists of:

Lymph vessels
Lymph nodes and other lymphatic tissue
Spleen
Thymus gland

LYMPH VESSELS

LYMPH CAPILLARIES

These originate as blind-end tubes in the interstitial spaces. They have the same structure as blood capillaries, i.e., a single layer of endothelial cells, but their walls are more permeable (Fig. 6:1). The tiny capillaries join up to form larger lymph vessels.

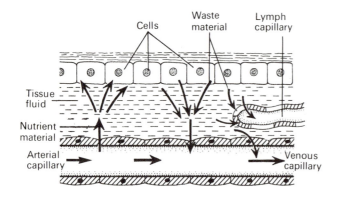

Figure 6:1   Diagram of the beginning of a lymph capillary.

LARGER LYMPH VESSELS

The walls of lymph vessels are about the same thickness as those of small veins and have the same layers of tissue, i.e., a fibrous covering, a middle layer of smooth muscle and elastic tissue and an inner lining of endothelium.

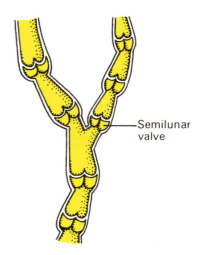

Figure 6:2   A lymph vessel cut open to show semilunar valves.

Lymph vessels have numerous cup-shaped valves that prevent backward flow of lymph (Fig. 6:2). There is no 'pump', like the heart, involved in the onward movement of lymph but factors believed to assist include:

1. Tissue fluid pressure
2. Contraction of surrounding muscles
3. Sympathetic nerve stimulation of the smooth muscle in lymph vessel walls
4. Pressure caused by the pulsation of adjacent arteries
5. Negative pressure in the thorax during inspiration

Lymph vessels become larger as they join together, eventually forming two large ducts, the *thoracic duct* and *right lymphatic duct*, that empty lymph into the *subclavian veins*.

THORACIC DUCT

This duct begins at the *cisterna chyli*, which is a lymph vessel dilatation situated in front of the bodies of the first two lumbar vertebrae. The duct is about 40 cm long and

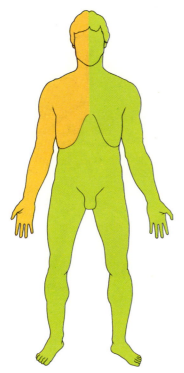

Figure 6:3 Diagram of lymph drainage. Green area drained by the thoracic duct. Orange area drained by the right lymphatic duct.

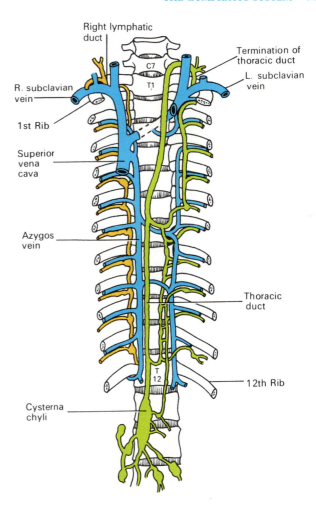

Figure 6:4 Origin and position of the thoracic duct and right lymphatic duct.

opens into the *left subclavian vein* in the root of the neck. It drains lymph from both legs, the pelvic and abdominal cavities, the left half of the thorax, head and neck and the left arm (see Figs 6:3 and 6:4).

RIGHT LYMPHATIC DUCT

This is a dilated lymph vessel about 1 cm long. It lies in the root of the neck and opens into the *right subclavian vein*. It drains lymph from the right half of the thorax, head and neck and the right arm.

DISORDERS ASSOCIATED WITH LYMPH VESSELS

The main involvements of lymph vessels are in relation to:
1. The spread of disease in the body
2. The effects of lymphatic obstruction

SPREAD OF DISEASE

The materials most commonly spread via the lymph vessels from their original site to the circulating blood are fragments of tumours and infected material.

FRAGMENTS OF TUMOURS

Tumour cells may enter a lymph capillary draining a tumour, or larger vessel when a tumour has eroded its wall. Cells from a malignant tumour, if not phagocytosed, settle and multiply in the first lymph node they encounter. Later, there may be further spread to other lymph nodes, to the blood and to other parts of the body via the blood. In this sequence of events, each new metastatic tumour becomes a source of malignant cells that may spread by the same routes.

INFECTED MATERIAL

Infected material may enter lymph vessels either at their origin in the interstitial spaces, or through the walls of larger vessels invaded by microbes when an infection spreads locally. If phagocytosis is not effective the infection may spread from node to node, and eventually reach the blood stream.

Lymphangitis (infection of lymph vessel walls)
This occurs in some acute pyogenic infections in which the microbes in the lymph draining from the area infect and spread along the walls of lymph vessels, e.g., in acute *Streptococcus pyogenes* infection of the hand, a red line may be seen extending from the hand to the axilla. This is caused by an inflamed superficial lymph vessel and adjacent tissues. The infection may be stopped at the first lymph node or spread through the lymph drainage network to the blood.

LYMPHATIC OBSTRUCTION

When a lymph vessel is obstructed there is an accumulation of lymph distal to the obstruction, called *lymphoedema*. The amount of resultant swelling and the size of the area affected depends on the size of the vessel involved. Lymphoedema usually leads to low-grade inflammation and fibrosis of the lymph vessel and further lymphoedema.

CAUSES OF LYMPHATIC OBSTRUCTION

Tumours
A tumour may grow into, and include, a lymph vessel or node, and obstruct the flow of lymph. A large tumour outside the lymphatic system may cause sufficient pressure to stop the flow of lymph.

Surgery
In some surgical procedures lymph nodes are removed because cancer cells may have already spread to them. This is done to prevent growth of secondary tumours in local lymph nodes and further spread of the disease via the lymphatic system, e.g., axillary nodes may be removed during mastectomy.

LYMPH NODES (Fig. 6:5)

All the small and medium-sized lymph vessels open into *lymph nodes* which are situated in strategic positions throughout the body. The lymph drains through a number of nodes, usually 8 to 10, before returning to the blood. These nodes vary considerably in size: some are as small as a pin head and the largest are about the size of an almond.

Structure
Lymph nodes have a *surrounding capsule of fibrous tissue* which dips down into the node substance forming partitions, or *trabeculae*. The main substance of the node consists of *reticular and lymphatic tissue* containing many *lymphocytes* and *macrophages*.

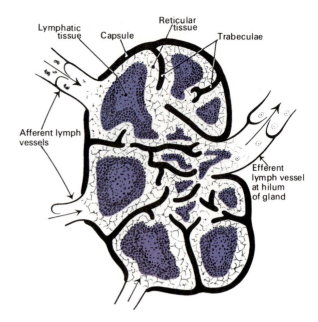

Figure 6:5   Diagram of a section of a lymph node. Arrows show the direction of flow of lymph.

As many as four or five *afferent lymph vessels* may enter a lymph node while only one *efferent vessel* carries lymph away from the node. Each node has a concave surface called the hilum where blood vessels supplying the node enter and leave and the efferent lymph vessel leaves.

The large numbers of lymph nodes situated in strategic positions throughout the body are arranged in *deep* and *superficial groups*.

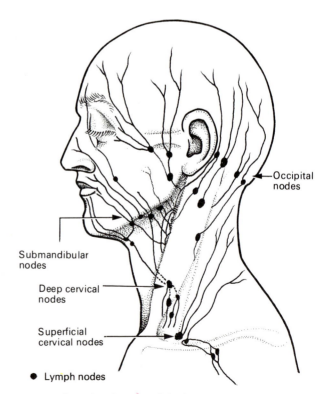

Figure 6:6   Some lymph nodes of the face and neck.

Lymph from the head and neck passes through deep and superficial *cervical nodes* (Fig. 6:6).

Lymph from the *upper limbs* passes through nodes situated in the *elbow region* then through the deep and superficial *axillary nodes* (Fig. 6:7).

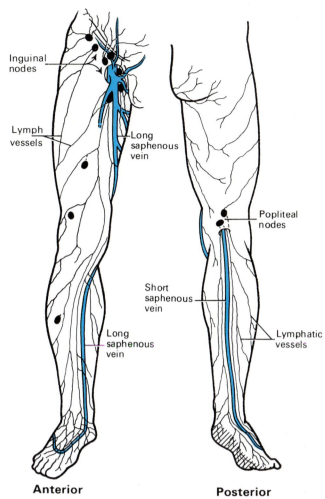

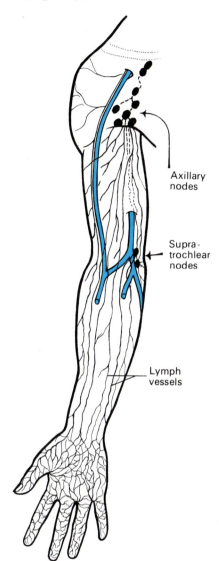

Figure 6:7 Some lymph nodes of the upper limb.

Figure 6:8 Some lymph nodes of the lower limb.

The lacteals are the lymph capillaries which drain lymph from the small intestine. The fat absorbed from the small intestine passes into the lymph capillaries and gives the lymph a milky appearance. Because of this, lymph entering the thoracic duct from the small intestine is called *chyle*.

Lymph from the organs and tissues in the *thoracic cavity* drains through many groups of nodes, including: *parasternal, intercostal, brachiocephalic, mediastinal, tracheobronchial, bronchopulmonary* and *oesophageal nodes*. Most of the lymph from the breast passes through the *axillary nodes*.

Lymph from the *pelvic and abdominal cavities* passes through many lymph nodes before entering the cisterna chyli. The abdominal and pelvic nodes are situated mainly in association with the blood vessels supplying the organs and close to the main arteries, i.e., the aorta and the external and internal iliac arteries.

The lymph from the *lower limbs* drains through deep and superficial nodes *popliteal nodes* and *inguinal nodes* (Fig. 6:8).

FUNCTIONS

1. Lymph is filtered by the reticular and lymphoid tissue as it passes through lymph nodes. Particulate matter may include microbes, dead and live phagocytes with ingested microbes, worn-out tissue cells, cells damaged by disease, inhaled particles and cells from malignant tumours. This material is destroyed by macrophages and antibodies produced by lymphocytes. Material not dealt with in one node passes on to successive nodes and lymph entering the blood has usually been cleared of foreign matter and cell debris.

2. Lymphocytes multiply in lymph nodes. Antibodies produced by B-lymphocytes are added to lymph (see Ch. 3), and to blood supplying the nodes

DISEASES OF LYMPH NODES

ENLARGED LYMPH NODES

Lymph nodes become enlarged when their workload is increased by infection. They usually return to normal when the infection subsides but if there is chronic infection or repeated acute episodes they may become fibrosed and remain enlarged. Other causes include tumours (lymphomas) and excess abnormal material in lymph, especially if present for a long time, e.g., coal dust from the lungs, necrotic material from a tumour.

ACUTE LYMPHADENITIS

This is an acute infection of lymph nodes, usually caused by microbes transported in lymph from other areas of infection. The nodes become inflamed, enlarged and congested with blood and chemotaxis attracts large numbers of phagocytes. If phagocytosis and antibody activity are not effective the infection may lead to:
1. Abscess formation in the node
2. Infection of adjacent tissues
3. Spread of infected material to other nodes and then to the blood, causing septicaemia or bacteraemia

DISEASES CAUSING ACUTE LYMPHADENITIS

Wound and skin infections
The most common infecting organisms are *Staphylococcus aureus* and *Streptococcus pyogenes*.

Infectious mononucleosis (Glandular fever)
This is a virus infection, usually of young adults, spread by direct contact. During the incubation period of 7 to 10 days viruses multiply in the epithelial cells of the pharynx. They subsequently spread to cervical nodes, then to lymphoid tissue throughout the body. The viruses enter B-lymphocytes, changing their characteristics. The abnormal B-lymphocytes are antigenic and stimulate an intense T-cell response. Many of the T-cells entering the blood are atypical but they destroy the virus-infected B-cells. Clinical or subclinical infection confers life-long immunity.

Other diseases
Minor lymphadenitis accompanies many infections and indicates the mobilisation of normal protective resources. More serious infection occurs in:

Measles	Cat-scratch fever
Anthrax	Lymphogranuloma venereum
Typhoid fever	Bubonic plague

CHRONIC LYMPHADENITIS

This occurs in many chronic infections, e.g., tuberculosis and syphilis.

LYMPHOMAS

These are malignant neoplasms of lymphoid tissue that vary in rate of growth. The causes are not known.

Hodgkin's disease
This is the most common lymphoma, usually beginning in adolescence or between 50 and 70 years of age. Enlarged lymph nodes in the neck are often the first to be noticed. The rate of progress of the disease varies considerably. There may be few signs of disease for many years or death may occur in a few months.

Malignant neoplastic metastases
Metastatic tumours develop in lymph nodes in any part of the body. Lymph from a tumour may contain cancer cells that are filtered off by the lymph nodes. If not phagocytosed they multiply, forming metastatic tumours. Nodes nearest the primary tumour are affected first but there may be further spread through the sequence of nodes, eventually reaching the blood stream.

LYMPHATIC TISSUE

Lymphatic tissue is found in a number of situations in the body in addition to the lymph nodes:

Palatine tonsil	— between the mouth and the oral part of the pharynx
Pharyngeal tonsil	— on the wall of the nasal part of the pharynx
Solitary lymphatic follicles	— in the wall of the small intestine
Aggregated lymphatic follicles (Peyer's patches)	
Vermiform appendix	— an outgrowth from the caecum (first part of the large intestine)

These structures are discussed in the sections on the respiratory and digestive systems (Chs 7 and 9).

SPLEEN (Fig. 6:9)

The spleen is formed partly by lymphatic tissue and will be described here as its functions are associated with the circulatory system.

The spleen lies in the *left hypochondriac region* of the abdominal cavity between the fundus of the stomach and

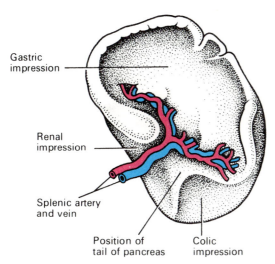

Figure 6:9   Spleen.

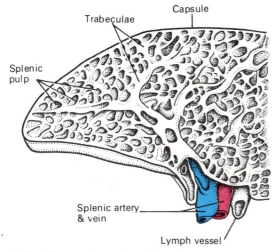

Figure 6:10   Diagram of a section of the spleen.

the diaphragm. It is purplish in colour and varies in size in different individuals, but is usually about 12 cm long, 7 cm wide and 2.5 cm thick. It weighs about 200 g.

ORGANS ASSOCIATED WITH THE SPLEEN

Superiorly and
  posteriorly — diaphragm
Inferiorly — left colic flexure of the large intestine
Anteriorly — fundus of the stomach
Medially — pancreas and the left kidney
Laterally — separated from the 9th, 10th and 11th ribs and the intercostal muscles by the diaphragm

STRUCTURE

The spleen is slightly oval in shape with the *hilum* on the lower medial border. The anterior surface is covered with peritoneum. It is enclosed in a *fibroelastic capsule* that dips

into the organ, forming *trabeculae*. The cellular material, consisting of *lymphocytes* and *macrophages*, is called *splenic pulp*, and it lies between the trabeculae.

The structures entering and leaving the spleen at the hilum are:

Splenic artery, a branch of the coeliac artery
Splenic vein, a branch of the portal vein
Lymph vessels
Nerves

Blood passing through the spleen passes through sinusoids, instead of capillaries, allowing it to come into close association with splenic pulp.

FUNCTIONS

Phagocytosis. As described previously (p. 22), erythrocytes are destroyed in the spleen and the breakdown products, bilirubin and iron, are passed to the liver via the splenic and portal veins. Other cellular material, e.g., leukocytes, platelets and microbes are phagocytosed in the spleen. Lymph does not *pass through* the spleen, as it does through lymph nodes, so it is not subject to diseases spread by lymph.

A reservoir for blood. It is known that the spleen stores blood in some animals, but the extent to which it does so in man is uncertain. It may store erythrocytes and in states of emergency, e.g., following haemorrhage, pass them into the circulation and increase the oxygen-carrying capacity of the blood.

Development of lymphocytes. The aggregates of lymphocytes respond to antigens by immune reactions (see p. 40). B- and T-lymphocytes multiply and some B-lymphocytes develop into plasma cells and produce antibodies. The spleen provides many of the lymphocytes in the blood.

DISORDERS OF THE SPLEEN

SPLENOMEGALY (Enlargement of spleen)

Enlargement of the spleen is usually secondary to other conditions, e.g., infections, circulatory disorders, blood diseases, malignant neoplasms.

INFECTIONS

The spleen may be infected by blood-borne microbes or by local spread of infection. The red pulp becomes congested with blood and there is an accumulation of phagocytes and plasma cells. *Acute infections* are rare.

Chronic infections. Some chronic non-pyogenic infections cause splenomegaly, but this is usually less severe than in the case of acute infections. The most commonly occurring primary infections include:

Tuberculosis   Brucellosis (undulant fever)
Typhoid fever   Infectious mononucleosis
Malaria

CIRCULATORY DISORDERS

Splenomegaly due to congestion of blood occurs when the flow of blood through the liver is impeded by, e.g., fibrosis in cirrhosis of liver, portal venous congestion in right-sided heart failure.

BLOOD DISEASES

Splenomegaly may *be caused by blood diseases*. The spleen enlarges to deal with the extra workload associated with removing damaged, worn out and abnormal blood cells in, e.g., haemolytic and macrocytic anaemia, polycythaemia, chronic myeloid leukaemia.

Splenomegaly may *cause blood disease*. When the spleen is enlarged for any reason, especially in portal hypertension, excessive haemolysis of red cells or phagocytosis of normal white cells and platelets leads to marked anaemia, leukopenia, thrombocytopenia.

TUMOURS

Benign and malignant tumours of the spleen are rare.

THYMUS GLAND

The thymus gland lies in the upper part of the mediastinum behind the sternum and extends upwards into the root of the neck. It weighs about 10 to 15 g at birth and grows until the individual reaches puberty, when it begins to atrophy. Its maximum weight, at puberty, is between 30 and 40 g and by middle age it has returned to approximately its weight at birth (Fig. 6:11).

ORGANS ASSOCIATED WITH THE THYMUS

Anteriorly   — sternum and upper four costal cartilages
Posteriorly — aortic arch and its branches, brachio-cephalic veins, trachea
Laterally   — lungs
Superiorly   — structures in the root of the neck
Inferiorly   — heart

STRUCTURE

The thymus consists of two lobes joined by areolar tissue. The lobes are enclosed by a fibrous capsule which dips into their substance, dividing them into lobules that consist of an irregular branching framework of epithelial cells and lymphocytes.

FUNCTIONS

Lymphocytes originate from haemocytoblasts (stem cells) in red bone marrow. Those that enter the thymus mature and develop into activated T-lymphocytes, i.e., able to respond to antigens encountered elsewhere in the body (see p. 39). They then divide into two groups:

1. Those that enter the blood, some of which remain in circulation and some lodge in lymphoid tissue
2. Those that remain in the thymus gland and are the source of future generations of T-lymphocytes

The maturation of the thymus and other lymphoid tissue is stimulated by *thymosin*, a hormone secreted by the epithelial cells that form the framework of the thymus gland. Involution of the gland begins in adolescence and, with increasing age, the effectiveness of T-lymphocyte response to antigens declines.

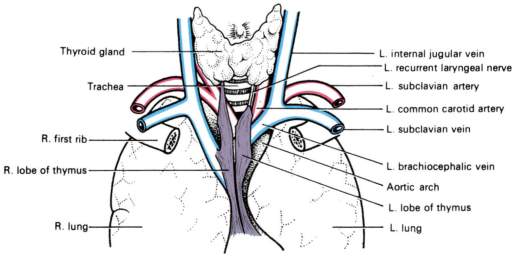

Figure 6:11 Thymus gland in the adult, and related structures.

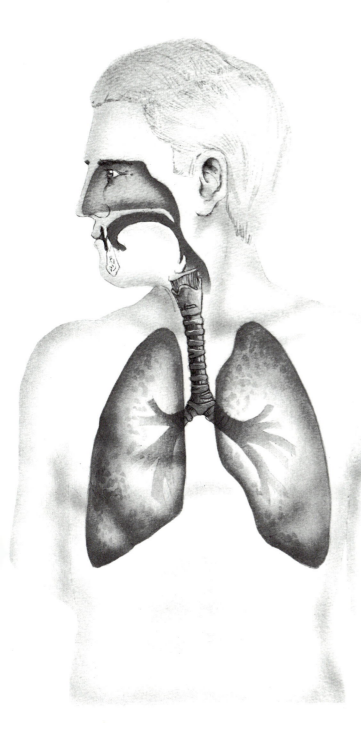

7

The Respiratory System

7. *The Respiratory System*

Most of the energy required by the cells of the body is derived from chemical reactions which can only take place in the presence of oxygen (O_2). The main waste product of these reactions is carbon dioxide (CO_2). The respiratory system provides the route by which the supply of oxygen present in the atmospheric air gains entry to the body and it provides the route of excretion of carbon dioxide.

The condition of the atmospheric air entering the body varies considerably according to the external environment, e.g., it may be dry, cold and contain dust particles or it may be moist and hot. As the air breathed in moves through the air passages to reach the lungs, it is warmed or cooled to body temperature, moistened to become saturated with water vapour and 'cleaned' as particles of dust stick to the mucus which coats the lining membrane. Blood provides the transport system for these gases between the lungs and the cells of the body. The exchange of gases between the blood and the lungs is called *external respiration* and that between the blood and the cells *internal respiration*. The organs of the respiratory systems are:

Nose
Pharynx
Larynx
Trachea
Two bronchi (one bronchus to each lung)
Bronchioles and smaller air passages
Two lungs and their coverings — the pleura
Muscles of respiration — the intercostal muscles and the diaphragm

A general view of the organs of the respiratory system is given in Figure 7:1.

NOSE AND NASAL CAVITIES

POSITION AND STRUCTURE

The nasal cavity is the first of the respiratory organs and consists of a large irregular cavity divided into two equal parts by a *septum*. The posterior bony part of the septum is formed by the perpendicular plate of the ethmoid bone and the vomer. Anteriorly it consists of hyaline cartilage (Fig. 7:2).

The roof is formed by the cribriform plate of the

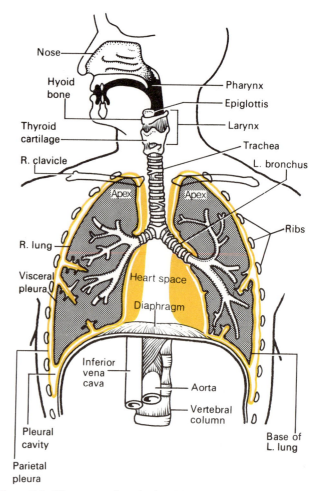

Figure 7:1   The organs of respiration. Pleura — yellow.

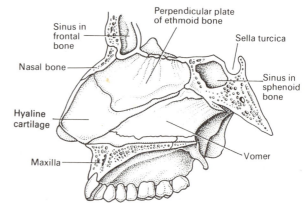

Figure 7:2   Structures forming the nasal septum.

106

ethmoid bone, and the sphenoid bone, frontal bone and nasal bones.

The floor is formed by the roof of the mouth and consists of the hard palate in front and the soft palate behind. The hard palate is composed of the maxilla and palatine bones and the soft palate consists of unstriped muscle.

The medial wall is formed by the *septum*.

The lateral walls are formed by the maxilla, the ethmoid bone and the inferior conchae (Fig. 7:3).

The posterior wall is formed by the posterior wall of the pharynx.

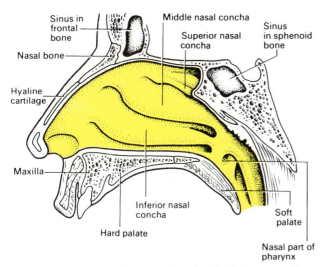

Figure 7:3   Lateral wall of right nasal cavity. Epithelium — yellow.

OPENINGS INTO THE NASAL CAVITY

The anterior nares are the openings from the exterior into the nasal cavity.

The posterior nares are the openings from the nasal cavity into the pharynx.

The sinuses are cavities in the bones of the face and the cranium which contain air. There are tiny openings between the air sinuses and the nasal cavities. They are lined with mucous membrane, continuous with that of the nasal cavities. The main sinuses are:

Maxillary sinuses in the lateral walls
Frontal and sphenoidal sinuses in the roof
Ethmoidal sinuses in the upper part of the lateral walls

The nasolacrimal ducts extend from the lateral walls of the nose to the conjunctival sacs of the eye (see p. 265). They drain tears from the eyes.

LINING OF THE NOSE

The nose is lined with very vascular *ciliated columnar epithelium* (ciliated mucous membrane) which contains mucus-secreting goblet cells (Figs 7:4 and 7:5). At the anterior nares this blends with the skin and posteriorly it extends into the nasal part of the pharynx.

RESPIRATORY FUNCTION OF THE NOSE

The nose is the first of the respiratory passages through which the incoming air passes. The function of the nose is to begin the process by which the air is *warmed, moistened and 'filtered'*.

The air is warmed as it passes over the interior surface

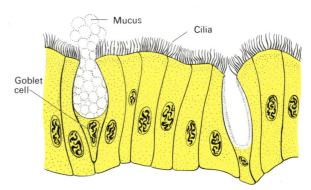

Figure 7:4   Ciliated columnar epithelium with goblet cells.

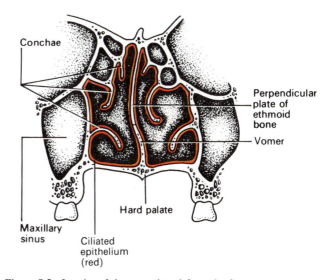

Figure 7:5   Interior of the nose viewed from the front.

of the nose; it is moistened by contact with mucus; and it is 'filtered' in the sense that particles of dust and other impurities such as microbes stick to the mucus.

The cilia of the mucous membrane waft the mucus towards the throat and it is swallowed or expectorated. The projecting conchae cause turbulence in the air passing through the nose, spreading it over the whole nasal surface (see Fig. 7:6).

OLFACTORY FUNCTION OF THE NOSE

The nose is the organ of the sense of smell. There are nerve endings and fibres that detect smell, located in the

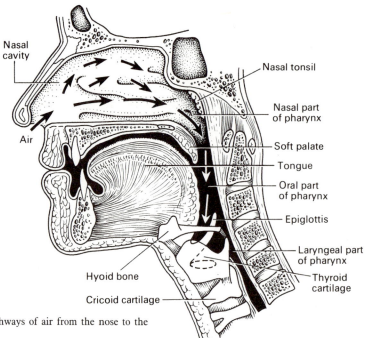

Figure 7:6 Arrows showing the pathways of air from the nose to the larynx.

roof of the nose in the area of the cribriform plates of the ethnoid bones and the superior conchae. These nerves and fibres are stimulated by chemical substances given off by odorous materials. The resultant nerve impulses are conveyed by the *olfactory nerves* to the brain where the sensation of smell is perceived (see p. 267).

PHARYNX

POSITION

The pharynx is a tube 12 to 14 cm long that extends from the base of the skull to the level of the sixth cervical vertebra. It lies behind the nose, mouth and larynx and is wider at its upper end.

Structures associated with the pharynx

Superiorly — the inferior surface of the base of the skull
Inferiorly — it is continuous with the oesophagus
Anteriorly — the wall is incomplete because of the openings into the nose, mouth and larynx
Posteriorly — areolar tissue, involuntary muscle and the bodies of the first six cervical vertebrae

For descriptive purposes the pharynx is divided into three parts: *nasal, oral* and *laryngeal*.

The nasal part of the pharynx lies behind the nose above the level of the soft palate. On its lateral walls are the two openings of the *auditory tubes*, one leading to each middle ear. On the posterior wall there is the *pharyngeal tonsil* (adenoid), consisting of lymphoid tissue. It is most

prominent in children up to approximately the age of 7 years and thereafter it gradually atrophies.

The oral part of the pharynx lies behind the mouth, extending from below the level of the soft palate to the level of the upper part of the body of the 3rd cervical vertebra. The walls of the pharynx blend with the soft palate to form two folds on each side. Between each pair of folds there is a collection of lymphoid tissue called the *palatine tonsil*.

During swallowing, the nasal and oral parts are separated by the soft palate and the *uvula*.

The laryngeal part of the pharynx extends from the oral part above and continues as the oesophagus below, i.e., from the level of the 3rd to the 6th cervical vertebrae.

STRUCTURE

The pharynx is composed of three layers of tissue:

1. *Mucous membrane lining*. The type varies slightly in the different parts. In the nasal part it is continuous with the lining of the nose and consists of *ciliated columnar epithelium*; in the oral and laryngeal parts it is *stratified squamous epithelium* which is continuous with the lining of the mouth and oesophagus.

2. *Fibrous tissue*. This forms the intermediate layer. It is thicker in the nasal part, where there is little muscle, and becomes thinner towards the lower end, where the muscle layer is thicker.

3. *Muscle tissue*. This consists of several *constrictor muscles*. The muscle fibres are unstriped and play an important part in the mechanism of swallowing (deglutition) which, in the pharynx, is not under voluntary control. The upper end of the oesophagus is closed by the lower constrictor muscle, except during swallowing.

BLOOD AND NERVE SUPPLY

Blood is supplied to the pharynx by several branches of the facial artery. The venous return is into the facial and internal jugular veins.

The nerve supply is from the pharyngeal plexus, formed by parasympathetic and sympathetic nerves. Parasympathetic supply is mainly through the *vagus* and *glossopharyngeal* nerves. Sympathetic supply is by nerves from the *superior cervical ganglion* (see p. 234).

FUNCTIONS

The pharynx is an organ involved in both the respiratory and the digestive systems: air passes through the nasal and oral parts, and food through the oral and laryngeal parts. By the same methods as in the nose the air is further warmed and moistened as it passes through the pharynx.

There are nerve endings of the sense of taste in the epithelium of the oral and pharyngeal parts.

The auditory tube extending from the nasal part to each middle ear allows air to enter the middle ear. Satisfactory hearing depends on the presence of air at atmospheric pressure on each side of the *tympanic membrane* (ear drum) (see p. 252).

The lymphatic tissue of the pharyngeal and laryngeal tonsils produces antibodies in response to antigens, e.g., microbes. The tonsils are larger in children and tend to atrophy in adults.

LARYNX

POSITION

The larynx or 'voice box' extends from the root of the tongue and the hyoid bone to the trachea. It lies in front of the laryngeal part of the pharynx at the level of the 3rd, 4th, 5th and 6th cervical vertebrae. Until puberty there is little difference in the size of the larynx between the sexes. Thereafter it grows larger in the male, which explains the prominence of the 'Adam's apple' and the generally deeper voice.

Structures associated with the larynx

Superiorly — the hyoid bone and the root of the tongue
Inferiorly — it is continuous with the trachea
Anteriorly — the muscles attached to the hyoid bone and the muscles of the neck
Posteriorly — laryngeal part of the pharynx and cervical vertebrae
Laterally — the lobes of the thyroid gland

STRUCTURE

The larynx is composed of several irregularly shaped cartilages attached to each other by ligaments and membranes. The main cartilages are:

1 thyroid cartilage
1 cricoid cartilage   } hyaline cartilage
2 arythenoid cartilages
1 epiglottis   — elastic fibrocartilage

The thyroid cartilage is the most prominent and consists of two flat pieces of cartilage, or *laminae*, fused anteriorly, forming the *laryngeal prominence* (Adam's apple). Immediately above the laryngeal prominence the laminae are separated by a V-shaped notch known as the *thyroid notch*. The thyroid cartilage is incomplete posteriorly and the posterior border of each lamina is extended to form two processes called the *superior* and *inferior cornu* (Fig. 7:7).

The upper part of the thyroid cartilage is lined with stratified squamous epithelium like the larynx, and the lower part with ciliated columnar epithelium like the trachea. There are many muscles attached to its outer surface.

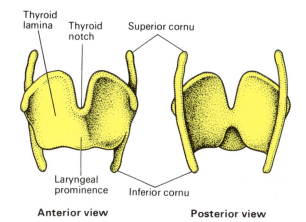

Thyroid lamina   Thyroid notch   Superior cornu

Laryngeal prominence   Inferior cornu

Anterior view   **Posterior view**

Figure 7:7 Thyroid cartilage.

The cricoid cartilage lies below the thyroid cartilage. It is shaped like a signet ring, completely encircling the larynx with the narrow part anteriorly and the broad part posteriorly. The broad posterior part articulates with the arytenoid cartilages above and with the inferior cornu of the thyroid cartilage below. It is lined with ciliated columnar epithelium and there are muscles and ligaments attached to its outer surface (Fig. 7:8).

The arytenoid cartilages are two roughly pyramid-shaped cartilages situated on top of the broad part of the cricoid cartilage forming part of the posterior wall of the larynx. They give attachment to the vocal cords and to muscles and are lined with ciliated columnar epithelium.

The epiglottis is a leaf-shaped cartilage attached to the inner surface of the anterior wall of the thyroid cartilage

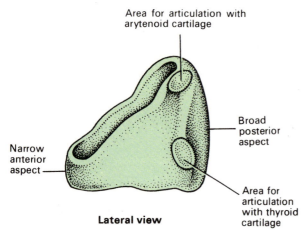

Figure 7:8   Cricoid cartilage.

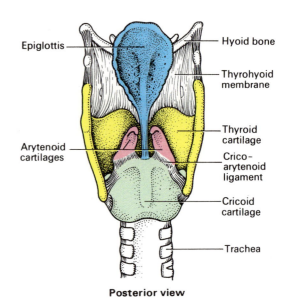

Figure 7:9   Larynx—viewed from behind.

immediately below the thyroid notch. It rises obliquely upwards behind the tongue and the body of the hyoid bone. It is covered with stratified squamous epithelium.

LIGAMENTS AND MEMBRANES

There are several ligaments that attach the cartilages to each other and to the hyoid bone (see Figs 7:9, 7:10 and 7:11).

BLOOD AND NERVE SUPPLY

Blood is supplied to the larynx through the superior and inferior laryngeal arteries and drained by the thyroid vein, which joins the internal jugular vein.

The parasympathetic nerve supply is from the superior laryngeal and recurrent laryngeal nerves, which are branches of the vagus nerves, and the sympathetic nerves are from the superior cervical ganglion (see p. 234). These provide the motor nerve supply to the muscles of the larynx and sensory fibres to the lining membrane.

INTERIOR

The vocal cords are two pale folds of mucous membrane with cord-like free edges which extend from the inner wall of the thyroid prominence anteriorly to the arytenoid cartilages posteriorly.

When the muscles of the arytenoid cartilages contract, the cartilages adduct and rotate medially, pulling the vocal cords together and narrowing the gap between them, thus forming the *chink of the glottis*. If air is forced through this chink it causes vibration of the cords and sound is produced. When the muscles relax the cartilages rotate laterally and abduct, separating the cords, and no sound is produced (Fig. 7:12).

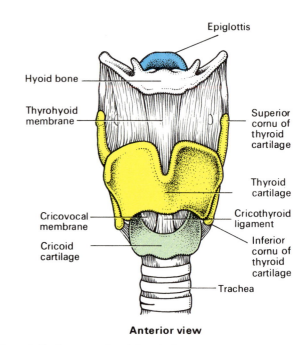

Figure 7:10   Larynx—viewed from in front.

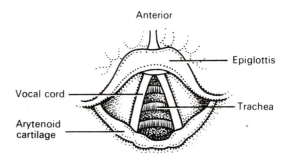

Figure 7:11   Interior of the larynx viewed from above.

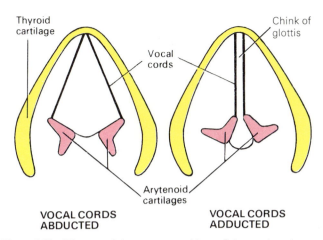

Figure 7:12 Diagram of the extreme positions of the vocal cords.

VOCAL CORDS ABDUCTED

VOCAL CORDS ADDUCTED

Sound has the properties of *pitch, loudness* and *quality*.

The pitch of the voice depends upon the *length* and *tightness* of the cords. In adults the vocal cords are longer in the male than in the female, thus the male voice has the lower pitch.

The loudness of the voice depends upon the *force* with which the cords vibrate. The greater the force of expired air the more vibration of the cords and the louder the sound.

The quality and resonance of the voice depend upon the shape of the mouth, the position of the tongue and the lips, the facial muscles and the air sinuses in the bones of the face and skull.

FUNCTIONS

1. The larynx provides a passageway for air between the pharynx and trachea. As air from the outside passes through, it is further moistened, filtered and warmed thus continuing the process started in the nose.

2. The vocal cords produce sounds of varying loudness and pitch.

3. During swallowing (deglutition) the larynx moves upwards, occluding the opening into it from the pharynx. This ensures that food passes into the oesophagus and not into the lower respiratory passages.

TRACHEA

POSITION

The trachea or windpipe is a continuation of the larynx and extends to about the level of the 5th thoracic vertebra where it divides (bifurcates) into the right and left bronchi, one bronchus going to each lung. It is approximately 10 to 11 cm long and lies mainly in the median plane in front of the oesophagus.

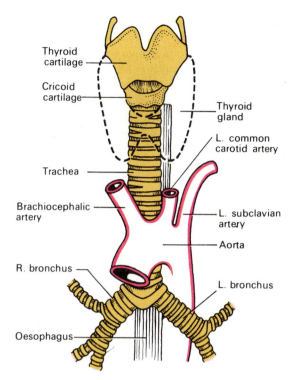

Figure 7:13 The trachea and some of its associated structures.

Structures associated with the trachea (Fig. 7:13)

Superiorly — the larynx

Inferiorly — the right and left bronchi

Anteriorly — upper part: the isthmus of the thyroid gland

lower part: the arch of the aorta and the sternum

Posteriorly — the oesophagus separates the trachea from the vertebral column

Laterally — the lungs and the lobes of the thyroid gland

STRUCTURE

The trachea is composed of from 16 to 20 incomplete (C-shaped) rings of hyaline cartilages situated one above the other. The cartilages are incomplete posteriorly. Connective tissue and involuntary muscle join the cartilages and form the posterior wall where they are incomplete. The soft tissue posterior wall is in contact with the oesophagus.

There are three layers of tissue which 'clothe' the cartilages of the trachea.

The outer layer consists of fibrous and elastic tissue and encloses the cartilages.

The middle layer consists of cartilages and bands of smooth muscle that wind round the trachea in a helical arrangement. There is some areolar tissue, containing blood and lymph vessels and autonomic nerves.

The inner lining consists of ciliated columnar epithelium, containing mucus-secreting goblet cells.

The trachea is easily distended and the cartilages prevent collapse of the tube when the internal pressure is less than intrathoracic pressure, e.g., at the end of forced expiration. The arrangement of cartilage and elastic tissue prevents kinking and obstruction of the airway as the head and neck move. The absence of cartilages posteriorly allows the trachea to dilate and constrict in response to nerve stimulation, and for indentation as the oesophagus dilates during swallowing.

BLOOD AND NERVE SUPPLY, LYMPH DRAINAGE

The arterial blood supply is mainly through the inferior thyroid and bronchial arteries and the *venous return* is through the inferior thyroid veins into the brachiocephalic veins.

The *nerve supply* is by parasympathetic and sympathetic fibres. Parasympathetic supply is by the recurrent laryngeal nerves and other branches of the vagi. Sympathetic supply is by nerves from the sympathetic ganglia.

Lymph from the respiratory passages passes through lymph nodes situated round the trachea and in the area where it divides into two bronchi.

DISORDERS OF THE UPPER RESPIRATORY PASSAGES

INFLAMMATION

Upper respiratory tract infections are usually caused by viruses which lower the resistance of the mucous membrane and allow bacteria dormant in the tract to invade the tissues. Such infections are not usually a threat to life unless:

1. They spread to the lungs or other organs
2. Inflammatory swelling and exudate block the airway

ACUTE INFLAMMATION

Microbes are usually spread by droplet infection, in dust or by contaminated equipment and dressings. If not completely resolved, acute infection may become chronic.

VIRAL INFECTIONS

Viruses are the initial cause of most upper respiratory tract infections and are small enough to pass through cell walls. They multiply inside the cells and cannot be killed by antibacterial or antifungal agents because the molecules of these are too large to penetrate cell walls. Viral infections cause acute inflammation of mucous membrane, leading to tissue congestion and profuse exudate of watery fluid. Secondary infection by bacteria usually results in purulent discharge. Viral respiratory diseases include:

Common cold	Influenza
Pharyngitis	Laryngotracheobronchitis
Tracheitis	(croup in young children)
Tonsillitis	Laryngitis

Viral infections commonly cause severe illness and sometimes death in infants and young children. In adults influenza is an incapacitating condition but is rarely fatal unless infection spreads to the lungs.

HAY FEVER

In this condition, *atopic* ('immediate') hypersensitivity develops to foreign proteins (antigens), e.g., pollen, mites in pillow feathers, animal dander. The acute non-microbial inflammation of nasal mucosa and conjunctiva causes excessive watery exudate from the nose, redness of the eyes and excessive secretion of tears. The first time the antigen is encountered and absorbed, IgE antibodies (reagins) are produced by B-lymphocytes which adhere to the surface of mast cells and basophils in the mucosa. Subsequent contact with the same antigen causes an immediate antigen/antibody reaction, stimulating the release of hista-

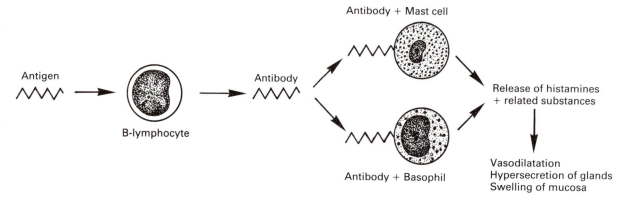

Figure 7:14   The sequence of events in the development of hay fever and other allergies.

mine and related substances by mast cells and basophils. These substances cause vasodilatation, increased permeability of capillary walls, hypersecretion by glands and swelling of tissues (see Fig. 7:14). Atopic hypersensitivity tends to run in families, but no genetic factor has yet been identified. Other forms of atopic hypersensitivity include extrinsic asthma, eczema in infants and young children, food allergies.

ACUTE SINUSITIS

This is usually caused by spread of microbes from the nose and pharynx to the mucous membrane lining the air sinuses in maxillary, sphenoidal, ethmoidal and frontal bones. The primary virus infection is usually followed by bacterial infection, e.g., *Streptococcus pyogenes*, *Streptococcus pneumoniae*, *Staphylococcus aureus*. The congested mucosa may block the openings between the nose and the sinuses, preventing drainage of mucopurulent discharge. If there are repeated attacks or if recovery is not complete, the infection may become chronic.

ACUTE PHARYNGITIS

This usually accompanies a common cold or tonsillitis. Viruses, with superimposed bacterial infection, cause acute inflammation of the mucous membrane of the pharynx, nose and sinuses.

ACUTE LARYNGITIS AND TRACHEITIS

The larynx and trachea are subject to the same viral and bacterial infections as the nose and pharynx. The infection may become chronic, especially in tobacco smokers or people who live or work in a polluted atmosphere. Acute inflammation may follow damage caused by an endotracheal tube, especially if it is used for long periods or has an inflated cuff.

NASAL POLYPI

These are small pear-shaped masses, covered with mucous membrane, that protrude into the nasal cavity. They consist of oedematous connective tissue, lymphocytes and plasma cells. Each polypus is suspended by a mucous membrane pedicle, usually from the lateral wall of the cavity. Chronic inflammation of the nasal mucous membrane is the main predisposing factor.

ENLARGED NASOPHARYNGEAL TONSILS
(Adenoids)

Adenoids consist of lymphoid tissue that reacts to foreign material and produces appropriate antibodies.

Temporary enlargement is a protective reaction to infection of nasal and pharyngeal mucosa. After repeated inflammatory episodes the adenoids may remain enlarged due to fibrosis, causing partial or complete obstruction of the airway and preventing nasal breathing, especially in children.

TONSILLITIS

Viruses and *Streptococcus pyogenes* are common causes of inflammation of the palatine tonsils, palatine arches and walls of the pharynx. Severe infection may lead to suppuration and abscess formation (*quinsy*). Following acute tonsillitis, swelling subsides and the tonsil returns to normal but repeated infection may lead to chronic inflammation, fibrosis and permanent enlargement. Toxins from tonsillitis caused by *Streptococcus pyogenes* are associated with the development of rheumatic fever and glomerulonephritis.

TUMOURS

BENIGN (Haemangioma)

These occur in the nasal septum. They consist of abnormal proliferations of blood vessels interspersed with collagen fibres of irregular size and arrangement. The blood vessels tend to rupture and cause persistent bleeding.

MALIGNANT TUMOURS

Carcinoma of the nose, sinuses, nasopharynx and larynx are relatively rare.

BRONCHI AND SMALLER AIR PASSAGES

The two bronchi are formed when the trachea divides, i.e., about the level of the 5th thoracic vertebra (Fig. 7:15).

The right bronchus is a wider, shorter tube than the left bronchus and it lies in a more vertical position. It is approximately 2.5 cm long. After entering the right lung at the hilum it divides into three branches, one of which passes to each lobe. Each branch then subdivides into numerous smaller branches.

The left bronchus is about 5 cm long and is narrower than the right. After entering the lung it divides into two branches, one of which goes to each lobe. Each branch then subdivides into progressively smaller tubes within the lung substance.

STRUCTURE

The bronchi are composed of the same tissues as the trachea. They are lined with ciliated columnar epithelium.

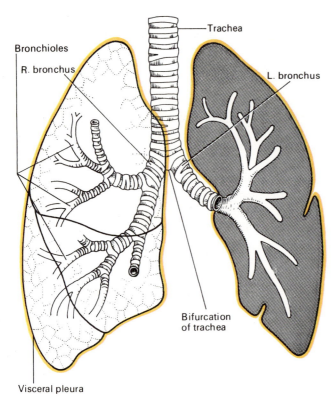

Figure 7:15   The bronchial tree and the lungs. Anterior part of the left lung removed. Visceral pleura — yellow.

The bronchi subdivide into smaller air passages and the smallest that retain bronchial structure are the *bronchioles*. The smallest bronchioles progressively divide into *terminal bronchioles, respiratory bronchioles, alveolar ducts* and finally *alveoli* (see Fig. 7:16). At the level of terminal bronchioles the cartilages become irregular in size and shape and the walls become thinner as muscle and areolar tissue disappear until, at the alveolar duct level, the walls consist of a single layer of flattened epithelial cells. The interchange

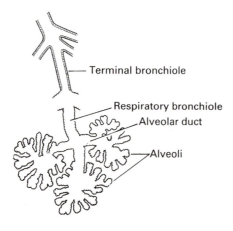

Figure 7:16   Diagram of small air passages and alveoli.

of gases takes place across two membranes: the walls of alveolar ducts and alveoli and those of pulmonary artery capillaries.

BLOOD AND NERVE SUPPLY, LYMPH DRAINAGE

The arterial blood supply to the walls of the bronchi and smaller air passages is through branches of the *right and left bronchial arteries* and the *venous return* is mainly through the bronchial veins. On the right side they empty into the azygos vein and on the left into the superior intercostal vein (see Figs 5:33 and 5:34).

The nerve supply is by parasympathetic and sympathetic nerves. The vagus nerves (parasympathetic) stimulate constriction of the bronchial tree and sympathetic stimulation causes dilatation (see Ch. 12).

The lymphatic vessels and lymph nodes. Lymph is drained from the walls of the air passages in a network of lymph vessels. It passes through lymph nodes situated around the trachea and bronchial tubes then into the thoracic duct on the left side and the right lymphatic duct on the other.

FUNCTIONS OF AIR PASSAGES NOT INVOLVED IN INTERCHANGE OF GASES

1. The cartilages in the larger air passages keep them open, allowing for the unobstructed passage of air between the outside atmosphere and the alveoli of the lungs.

2. The mucus which coats the lining membrane is of a sticky consistency to which particles in the air adhere, preventing them from reaching the alveoli, e.g., dust.

3. The wave motion of the cilia of the lining membrane wafts mucus and adherent particles towards the throat. When it reaches the pharynx it is usually swallowed but it may be expectorated. The process may be aided by coughing.

4. The diameter of the respiratory passages may be altered by contraction or relaxation of the involuntary muscle in their walls, thus regulating the volume of air entering the lungs. These changes are controlled by the autonomic nerve supply: parasympathetic stimulation constricts the air passages and sympathetic stimulation dilates them (see p. 234).

5. Cells in connective tissue protect against infection and inhaled foreign particles not trapped by mucus. Lymphocytes and plasma cells produce antibodies in the presence of antigens, and macrophages and polymorphonuclear leukocytes are phagocytic. These cells are most active in the distal air passages where ciliated epithelium has been replaced by flattened epithelial cells.

6. Inspired air is warmed (or cooled) to body temperature by contact with the walls of the air passages.

7. Inhaled air not already saturated with water vapour takes up moisture from surface mucus. Inhalation of dry air over a period of time causes irritation of the mucosa and facilitates the entry of pathogenic microbes.

DISEASES OF THE BRONCHI

ACUTE BRONCHITIS

This is a bacterial infection of the bronchi. It is usually preceded by a common cold or influenza and it may also complicate measles and whooping cough in children. The viruses depress normal defence mechanisms, allowing bacteria already present in the respiratory tract to multiply, e.g., *Streptococcus pneumoniae, Haemophilus influenzae, Streptococcus pyogenes, Staphylococcus aureus.* Downward spread of infection may lead to bronchiolitis, and bronchopneumonia may develop, especially in children and in debilitated, often elderly, adults.

CHRONIC BRONCHITIS

This is a progressive inflammatory disease caused by exposure to irritants, including tobacco smoke and atmospheric pollutants, e.g., motor vehicle exhaust fumes, industrial chemicals, sulphur dioxide, urban fog. It develops mostly in middle-aged men who are chronic heavy smokers and may have a familial predisposition. The changes occurring in the mucous membrane of the bronchi include:

1. Thickening
2. Increase in the number and size of mucous glands
3. Oedema
4. Reduction in the number of ciliated cells
5. Narrowing of bronchioles due to fibrosis following repeated inflammatory episodes

Reduced ciliary activity causes stagnation in the bronchi of the excessive amount of mucus secreted. If the mucus and bronchial lining are colonised by bacteria, pus is formed. Stagnant mucus may partially or completely obstruct small bronchioles. The severe difficulty in breathing (*dyspnoea*) with active expiratory effort raises the air pressure in the alveoli and may rupture the alveolar walls, causing *emphysema*. Damp cold conditions tend to exacerbate the disease, and spread of infected mucus to the alveoli may cause pneumonia. As the disease progresses, ventilation of the lungs is severely impaired, leading to hypoxia, pulmonary hypertension and right-sided heart failure.

BRONCHIAL ASTHMA

There are two types of bronchial asthma, *extrinsic* and *intrinsic*. In both types the mucous membrane and muscle layers of the bronchi become thickened and the mucous glands enlarge. During an asthmatic attack spasmodic contraction of bronchial muscle constricts the airway and there is excessive secretion of thick sticky mucus which further reduces the airway. Inspiration is normal but only partial expiration is achieved. The lungs become hyperinflated and there is severe dyspnoea. The duration of attacks usually varies from a few minutes to hours, and very occasionally, days (*status asthmaticus*). In severe attacks the bronchi may be obstructed by mucus plugs, leading to acute respiratory failure, hypoxia and possibly death.

EXTRINSIC ASTHMA (Allergic, Atopic)

This type occurs in children and young adults who have atopic (Type 1) hypersensitivity to foreign protein, e.g., pollen, dust containing mites from feather pillows, animal dander, fungi. A history of infantile eczema or food allergies is common.

The same disease process occurs as in hay fever. Antigens (allergens) are inhaled and absorbed by the bronchial mucosa. This stimulates the production of IgE antibodies that bind to the surface of mast cells and basophils round the bronchial blood vessels. When the allergen is encountered again, the antigen/antibody reaction results in the release of histamine and other related substances. Attacks tend to become less frequent and less severe with age. (See Fig. 7:14.)

INTRINSIC (Chronic) ASTHMA

This type occurs later in adult life and there is no history of childhood allergic reactions. It is associated with chronic inflammation of the upper respiratory tract, e.g., chronic bronchitis, nasal polypi. Antigens are rarely identified but drug hypersensitivity may develop later, especially to aspirin and penicillin. Attacks tend to increase in severity and there may be irreversible damage to the lungs.

BRONCHIECTASIS

This is permanent abnormal dilatation of small bronchi. It is associated with chronic bacterial infection and in some cases there is a history of childhood bronchiolitis and bronchopneumonia. The bronchi become obstructed by mucus, pus and inflammatory exudate and the alveoli *distal* to the blockage collapse as trapped air is absorbed. Interstitial elastic tissue degenerates and is replaced by fibrous adhesions that attach the bronchi to the pleura. The pressure of inspired air in these damaged bronchi leads to dilatation *proximal* to the blockage. The persistent severe coughing to remove copious purulent sputum causes intermittent increases in pressure in the blocked bronchi, leading to further dilatation.

The lower lobe of the lung is usually affected. Suppuration is common. If a blood vessel is eroded, haempotysis may occur or pyaemia, leading to abscess formation elsewhere in the body, commonly in the brain. Progressive fibrosis in the lung leads to hypoxia, pulmonary hypertension and right-sided heart failure.

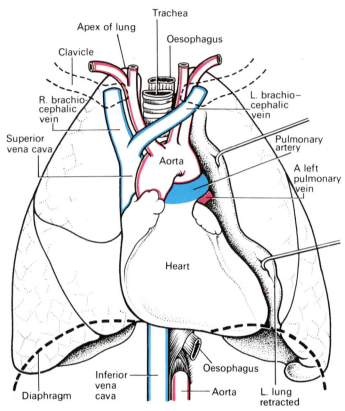

Figure 7:17 Organs associated with the lungs.

LUNGS

POSITION AND ASSOCIATED STRUCTURES
(Fig. 7:17)

There are two lungs, one lying on each side of the midline in the thoracic cavity. They are cone-shaped and are described as having an *apex*, a *base*, a *costal surface* and a *medial surface*.

The apex is rounded and rises into the root of the neck, about 25 mm (1 inch) above the level of the middle third of the clavicle. The structures associated with it are the first rib and the blood vessels and nerves in the root of the neck.

The base is concave and semilunar in shape and is closely associated with the thoracic surface of the diaphragm.

The costal surface is convex and is closely associated with the costal cartilages, the ribs and the intercostal muscles.

The medial surface is concave and has a roughly triangular-shaped area known as the *hilus* at the level of the 5th, 6th and 7th thoracic vertebrae. Structures which form the *root of the lung* enter and leave at the hilus. These include the following (see Fig. 7:18):

1 bronchus
1 pulmonary artery
2 pulmonary veins
1 bronchial artery
Bronchial veins
Lymph vessels
Parasympathetic and sympathetic nerves

The area between the lungs is the *mediastinum*, occupied by the heart, great vessels, trachea, right and left bronchus, oesophagus, lymph nodes, lymph vessels and nerves.

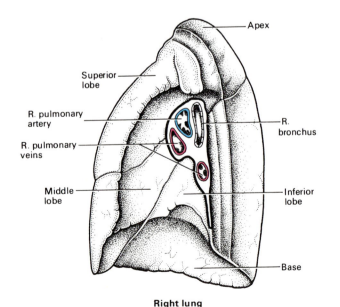

Right lung

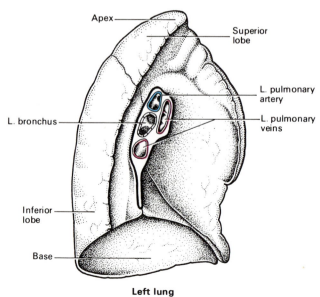

Left lung

Figure 7:18 The parts of the lungs and some structures entering at the hilum.

STRUCTURE

LOBES

The *right lung* is divided into three distinct lobes: superior, middle and inferior.

The *left lung* is divided into only two lobes: superior and inferior.

PLEURA

The pleura consists of a closed sac of serous membrane (one for each lung) which contains a small amount of serous fluid. The lung is invaginated into this sac so that it forms two layers: one adheres to the lung and the other lines the thoracic cavity (see Fig. 7:1).

The visceral pleura is the adherent layer, covering each lobe and passing into the fissures which separate them.

The parietal pleura is adherent to the inside of the chest wall and the thoracic surface of the diaphragm. It remains detached from the adjacent structures in the mediastinum and is continuous with the visceral pleura round the edges of the hilus.

The pleural cavity is only a potential space because in healthy people the two layers of pleura are separated by only a thin film of serous fluid, sufficient to prevent friction between them during breathing. The serous fluid is secreted by the epithelial cells of the membrane.

INTERIOR OF THE LUNGS

The lungs are composed of the bronchi and smaller air passages, alveoli, connective tissue, blood vessels, lymph vessels and nerves. The left lung is divided into two lobes and the right, into three. Each lobe is made up of a large number of lobules (see Fig. 7:16).

Lobules are composed of *terminal bronchioles, respiratory bronchioles, alveolar ducts* and *alveoli*. As these air passages subdivide, their structure undergoes a gradual change. The cartilages become irregular in size and shape and eventually disappear. The walls become thinner until muscle and connective tissue are replaced by a single layer of flattened epithelial cells in the alveolar ducts and alveoli.

Drying of the inside surface of the alveolar ducts and alveoli is prevented by a small amount of fluid containing a phospholipid 'surfactant' secreted by cells in the connective tissue between the alveoli. This fluid also helps to prevent the movement of water from capillary blood into the alveoli as the pressure falls when the lungs expand during inspiration. Surfactant begins to be secreted about the 35th week of fetal life and is an important aid to expansion of the lungs and to the establishment of respiration after birth.

Pulmonary blood supply (Fig. 7:19)
The *pulmonary artery* divides into two, one branch conveying *deoxygenated* blood to each lung. Within the

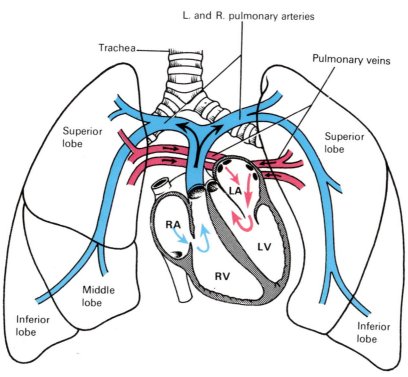

Figure 7:19   Diagram of the flow of blood between the heart and lungs.

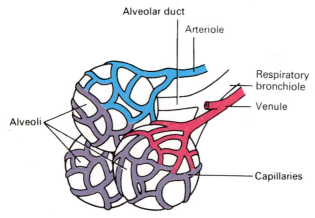

Figure 7:20 Diagram of the capillary network surrounding the alveoli.

lungs each pulmonary artery divides into many branches which eventually end in a dense capillary network in the walls of the alveoli (Fig. 7.20). The walls of the alveoli and those of the capillaries consist of only one layer of flattened epithelial cells. The exchange of gases between the air in the alveoli and the blood in the capillaries takes place across these two very fine membranes. The pulmonary capillaries join up, eventually becoming *two pulmonary veins* in each lung. They leave the lungs at the hilus and convey *oxygenated blood* to the left atrium of the heart. The innumerable blood capillaries and blood vessels in the lungs are supported by connective tissue.

The blood supply to the respiratory passages, lymphatic drainage and nerve supply have already been described.

EXTERNAL RESPIRATION

Expansion and contraction of the lungs ensure that a regular exchange of gases takes place between the alveoli and the external air. This is dependent upon the arrangement of the pleura and the contraction and relaxation of the muscles of respiration.

MECHANISM OF RESPIRATION

This is the process by which the lungs expand to take in air then contract to expel it. The cycle of respiration, which occurs about 15 times per minute, consists of three phases:

    Inspiration
    Expiration
    Pause

The expansion of the chest during inspiration occurs as a result of muscular activity, partly voluntary and partly involuntary. The main muscles of respiration in normal quiet breathing are the *intercostal muscles* and the *diaphragm*. During difficult or deep breathing they are assisted by the muscles of the neck, shoulders and abdomen.

Intercostal muscles

There are 11 pairs of intercostal muscles that occupy the spaces between the 12 pairs of ribs. They are arranged in two layers, the external an internal intercostal muscles (Fig. 7.21)

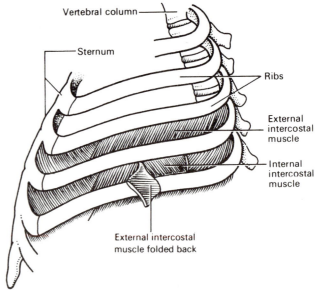

Figure 7:21 The intercostal muscles and the bones of the thorax.

The external intercostal muscle fibres extend in a downwards and forwards direction from the lower border of the rib above to the upper border of the rib below.

The internal intercostal muscle fibres extend in a downwards and backwards direction from the lower border of the rib above to the upper border of the rib below, crossing the external intercostal muscle fibres at right angles.

The first rib is fixed. Therefore, when the intercostal muscles contract they pull all the other ribs towards the first rib. Because of the shape of the ribs they move outwards when pulled upwards. In this way the thoracic cavity is enlarged anteroposteriorly and laterally. The intercostal muscles are stimulated to contract by the *intercostal nerves*.

Diaphragm

The diaphragm is a dome-shaped structure which separates the thoracic cavity from the abdominal cavity. It forms the floor of the thoracic cavity and the roof of the abdominal cavity and consists of a central tendon from which muscle fibres radiate to be attached to the lower ribs and sternum and to the vertebral column by two crura. When the muscle of the diaphragm is relaxed the central tendon is

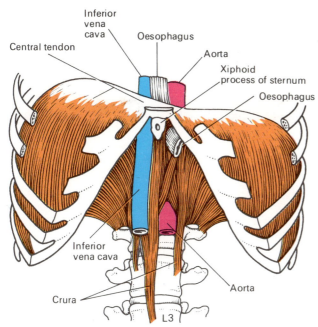

Figure 7:22 Diaphragm.

at the level of the 8th thoracic vertebra (Fig. 7.22). When it contracts, its muscle fibres shorten and the central tendon is pulled downwards, enlarging the thoracic cavity in length. This increases the pressure in the abdominal and pelvic cavities. The diaphragm is supplied by the *phrenic nerves*.

The intercostal muscles and the diaphragm contract *simultaneously* ensuring the enlargement of the thoracic cavity in all directions, that is from back to front, side to side and from top to bottom (Fig. 7.23).

CYCLE OF RESPIRATION

As described previously, the visceral pleural is adherent to the lungs and the parietal pleura to the inner wall of the thorax and to the diaphragm. There is a potential space between these two layers of serous membrane called the pleural cavity which contains a very thin layer of serous fluid.

When the capacity of the thoracic cavity is increased by simultaneous contraction of the intercostal muscles and the diaphragm, the parietal pleura moves with the walls of the thorax and the diaphragm. This reduces the pressure in the pleural cavity to a level considerably lower than atmospheric pressure. The visceral pleura tends to follow the parietal pleura. During this process the lungs are stretched and the pressure within the alveoli and in the air passages is reduced. This results in air being drawn into the lungs in an attempt to equalise the atmospheric and alveolar air pressures. This is the process of *inspiration* which is described as *active* because it is the result of muscle contraction.

When the diaphragm and intercostal muscles *relax*, the inspiratory processes are reversed and there is elastic recoil of the lungs, resulting in *expiration*. This is a passive process. After expiration there is a pause before the next cycle begins.

In normal quiet breathing there are about 15 complete respiratory cycles per minute. The lungs and the air passages are never empty and, as the exchange of gases takes place across only the walls of the alveolar ducts and alveoli, the remaining capacity of the respiratory passages is called the *anatomical dead space* (about 150 ml).

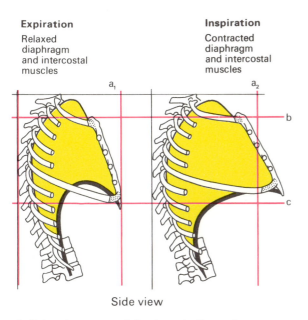

1. Outward movement of ribs shown by lines a_1 & a_2.
2. Upward movement of ribs & sternum shown by lines b & c.
3. Lowering of diaphragm shown by line c.

1. Outward movement of ribs shown by lines a_3 & a_4.
2. Upward movement of ribs shown by lines b_1 & c_1.
3. Lowering of diaphragm shown by line c_1.

Figure 7:23 Diagram of the changes in capacity of the thoracic cavity (and the lungs) during breathing.

Tidal volume (about 500 ml) is the amount of air which passes into and out of the lungs during each cycle of quiet breathing.

Inspiratory capacity is the amount of air that can be inspired with the maximum effort. In consists of the tidal volume (500 ml) plus the *inspiratory reserve volume*.

Functional residual volume is the amount of air remaining in the air passages and alveoli at the end of quiet expiration. Tidal air mixes with this air, causing relatively small changes in the composition of alveolar air. As blood flows continuously through the pulmonary capillaries this means that the interchange of gases is not interrupted between breaths, preventing marked changes in the concentration of blood gases. The functional residual volume also prevents collapse of the alveoli on expiration.

Vital capacity is the tidal volume plus the inspiratory and expiratory reserve volumes.

Alveolar ventilation is the amount of air which moves into and out of the alveoli each minute.

Alveolar ventilation = respiratory rate (tidal volume − dead space
 volume)
 = 15 (500 − 150) ml
 = 5.25 litres per minute

INTERCHANGE OF GASES

The interchange of gases in the lungs occurs between the blood in the capillary network surrounding the alveoli and the air in the alveoli. Some *properties of gases* include:
1. The molecules of gases are always in motion
2. Gases always tend to diffuse from an area of higher concentration to one of lower concentration, i.e., down the concentration gradient
3. Gases always exert pressure upon all the walls of their container. Unlike liquids, gases always fill their container. If not enclosed on all sides a gas will escape

The atmospheric pressure at sea level is 101.3 kilopascals (kPa) or 760 millimetres of mercury (mmHg).* This pressure is exerted by the mixture of gases which make up the inspired air in the following proportions:

Oxygen	21%
Carbon dioxide	0.04%
Nitrogen and other inert gases	78%
Water vapour	variable

During respiration, the lungs and the respiratory passages are never empty of air. The tidal ebb and flow of air results in inspired air being mixed with the air already in the lungs. When it reaches the alveoli the air is saturated with water vapour and because of the tidal movement the *concentration of gases* in the alveoli remains fairly constant.

The total pressure exerted on the walls of the alveoli by the mixture of gases is the same as atmospheric pressure: 101.3 kPa (760 mmHg). Each gas in the mixture exerts a part of the total pressure proportional to its concentration, i.e., the *partial pressure* (see Table 7:1).

Table 7:1 Partial pressures of gases

Gas	Alveolar air kPa	mmHg	Deoxygenated blood kPa	mmHg	Oxygenated blood kPa	mmHg
Oxygen	13.3	100	5.3	40	13.3	100
Carbon dioxide	5.3	40	5.8	44	5.3	40
Nitrogen and other inert gases	76.4	573	76.4	573	76.4	573
Water vapour	6.3	47				
	101.3	760				

*1 mmHg = 133.3 Pa = 0.1333 kPa
 1 kPa = 7.5 mmHg

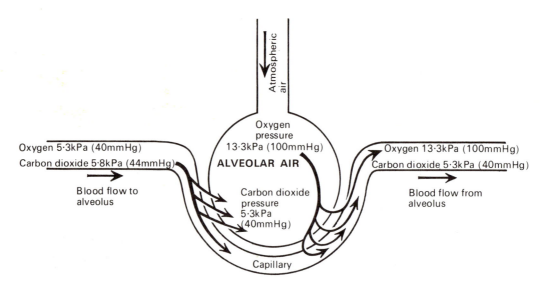

Figure 7:24 Diagram of the interchange of gases between air in the alveoli and the blood capillaries.

The partial pressure of nitrogen (P_{N_2}) is the same in the alveoli as it is in the blood. This stable state is maintained because nitrogen as a gas is not used by the body but it can diffuse across the walls of the alveoli and the capillaries.

The partial pressure of oxygen (P_{O_2}) in the alveoli is higher than that in the deoxygenated blood in the capillaries of the pulmonary arteries (see Table 7:1). As gases diffuse from an area of higher to one of lower concentration, the movement of oxygen is from the alveoli to the blood.

The reverse is true of carbon dioxide. The P_{CO_2} is higher in deoxygenated blood than in alveolar air, so carbon dioxide passes across the walls of the capillaries and the alveoli into the alveolar air (see Table 7:1 and Fig. 7:24).

The partial pressure of each gas in the blood when leaving the lungs via the pulmonary veins is the same as in the alveolar air.

The slow movement of blood through the capillaries surrounding the alveoli allows time for the interchange of gases to take place and for the uptake of oxygen by the erythrocytes in the blood. Oxygen is transported round the body in solution in the blood water and in combination with haemoglobin in the erythrocytes.

CONTROL OF RESPIRATION

This control is partly *chemical* and partly *nervous*. Respiration is controlled by nerve cells in the brain stem: the *respiratory centre* in the *medulla oblongata* and the *pneumotaxic centre* in the *pons varolii* (see Ch. 12). The cells in the respiratory centre are concerned with inspiration and those in the pneumotaxic centre with the inhibition of inspiration, which results in expiration. Nerve impulses which originate in the respiratory centre pass to the diaphragm in the phrenic nerves and to the intercostal muscles in the intercostal nerves. This results in contraction of these muscles, and inspiration occurs (Fig. 7:25).

There are nerve endings in the lungs sensitive to stretch which are stimulated when the lungs are inflated. The nerve impulses produced are passed to the pneumotaxic centre in the afferent fibres of the vagus nerves, and expiration occurs.

There are *chemoreceptors* in the walls of the aorta and carotid arteries. They are the *aortic* and *carotid bodies*, consisting of cells that are sensitive to changes in P_{CO_2} and P_{O_2} in the blood. The nerve impulses which originate in these cells are transmitted to the respiratory centre in the *glossopharyngeal* and the *vagus nerves*. The chemoreceptors and the respiratory centre are stimulated by an increase in the P_{CO_2} in the blood which results in increased ventilation of the lungs. A small reduction in the P_{O_2} has the same effect but a substantial reduction tends to have a depressing effect.

Normally, quiet breathing is sufficient to maintain a

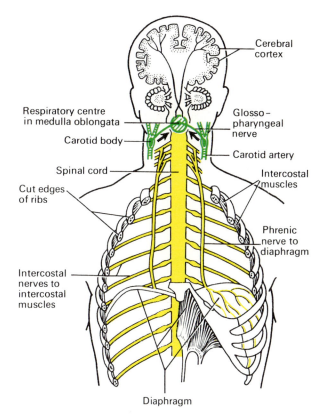

Figure 7:25   Diagram of some of the nerves involved in the control of respiration.

balance between the blood P_{CO_2} and P_{O_2} while the individual is at rest or taking light exercise. During strenuous exercise breathing becomes deeper and more rapid in response to the needs of the muscles for more oxygen and in order to excrete the excess carbon dioxide produced.

In normal quiet breathing the intercostal muscles and the diaphragm are the only muscles involved, but in deep or forced breathing other muscles come into play. They are the *accessory muscles of respiration* and the most important are the sternocleidomastoid (Figs 17:6 and 18:1). The contraction of these muscles in addition to the diaphragm and intercostal muscles ensures the maximum increase in the capacity of the thoracic cavity.

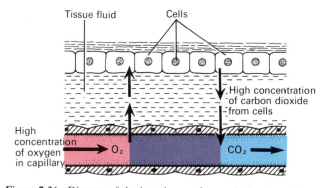

Figure 7:26   Diagram of the interchange of gases during internal respiration.

INTERNAL OR CELL RESPIRATION

This is the name given to the interchange of gases which takes place between the blood and the cells of the body.

Oxygen is carried from the lungs to the tissues dissolved in plasma and in chemical combination with haemoglobin as *oxyhaemoglobin*. The exchange in the tissues takes place between the arterial end of the capillaries and the tissue fluid. The process involved is the same as that which occurs in the lungs, that is, *diffusion from a higher concentration to a lower concentration* (Fig. 7:26). In this case the higher concentration of oxygen is in the blood and the lower concentration is in the tissue fluid. The cells therefore obtain their oxygen from the tissue fluid by diffusion.

Oxyhaemoglobin is an unstable compound which breaks up easily to liberate oxygen. As the cells of the body require a constant supply of oxygen, the process of diffusion of oxygen from the blood across the capillary wall to the tissue fluid and then into the cells is continuous. The rate at which this process is carried on is increased in the presence of a high concentration of carbon dioxide, which occurs when cells in a particular area are more than usually active. The higher P_{CO_2} assists the release of oxygen from oxyhaemoglobin. In this way cells receive a supply of oxygen consistent with their activity and the supply changes as the amount of activity changes.

Carbon dioxide is one of the waste products of carbohydrate and fat metabolism in the cells. The method of transfer of carbon dioxide from the cells into capillary blood is also by *diffusion*. Blood transports carbon dioxide by three different mechanisms:

1. Some is dissolved in the water of the blood plasma
2. Some is transported in chemical combination with sodium in the form of sodium bicarbonate
3. The remainder is transported in combination with haemoglobin

DISEASES OF THE LUNGS

EMPHYSEMA (Fig. 7:27)

PULMONARY EMPHYSEMA

In this form of the disease there is irreversible distension of the respiratory bronchioles, alveolar ducts and alveoli. There are two main types and both are usually present.

ALVEOLAR EMPHYSEMA

The walls between *adjacent alveoli* break down, the *alveolar ducts* dilate and there is loss of interstitial elastic tissue. The lungs become distended and their capacity is increased. Because the volume of air in each breath remains unchanged it constitutes a smaller proportion of the total volume of air in the distended alveoli, reducing the partial pressure of oxygen. The merging of alveoli reduces the surface area for diffusion of gases. As the disease progresses the combined effect of these changes may lead to hypoxia, pulmonary hypertension and eventually right-sided heart failure. Predisposing factors include:

1. Tobacco smoking, believed to promote the release of proteolytic enzymes from mast cells and basophils in the lungs
2. Acute inflammation of bronchi and lungs
3. Increased pressure caused by coughing which stretches the already damaged structures
4. Congenital deficiency of an antiproteolytic enzyme, α_1-antitrypsin, which causes deficiency of supporting elastic tissue in the lungs

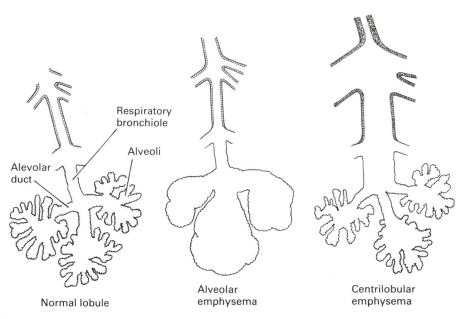

Normal lobule   Alveolar emphysema   Centrilobular emphysema

Figure 7:27 Emphysema.

CENTRILOBULAR EMPHYSEMA

In this form there is irreversible dilatation of the *respiratory bronchioles* in the centre of lobules. When inspired air reaches the dilated area the pressure falls, leading to a reduction in alveolar air pressure, reduced ventilation efficiency and reduced partial pressure of oxygen. As the disease progresses the resultant hypoxia leads to pulmonary hypertension and right-sided heart failure. Predisposing conditions, exacerbated by persistent severe coughing, include recurrent bronchiolitis, pneumoconiosis and chronic bronchitis.

INTERSTITIAL EMPHYSEMA

Air may gain access to thoracic interstitial tissues when a lung is ruptured in one of the following ways:
1. From the outside by injury, e.g., fractured rib, stab wound
2. From the inside when an alveolus ruptures through the pleura, e.g., during an asthmatic attack, in bronchiolitis, coughing as in whooping cough

The air in the tissues usually tracks upwards to the soft tissues of the neck where it is gradually absorbed, causing no damage. A large quantity in the mediastinum may limit heart movement.

ATELECTASIS

In atelectasis, expansion of the lungs at birth is defective. There are two types, distinguished by the time of onset.

IMMEDIATELY AFTER BIRTH

There may be partial or complete lack of expansion of the lungs due to:
1. Obstruction of the airways by secretions
2. Incomplete development of the respiratory passages
3. Lack of stimulation by the respiratory centre in the medulla oblongata due to congenital abnormality or prematurity

WITHIN MINUTES OR HOURS OF BIRTH

Breathing appears to be normal at first but acute respiratory distress and cyanosis develop later. The condition is associated with prematurity. Normally fluid containing phospholipid surfactant is secreted between the surfaces of the unexpanded alveoli after about the 35th week of gestation. Its function is to reduce the tension between the alveolar surfaces, and so reduce the effort needed to expand the lungs at birth. Surfactant deficiency, possibly accompanied by immaturity of the respiratory centre, causes atelectasis. Reasons for its occurrence following caesarian section and in maternal diabetes are not known.

PNEUMONIA

This occurs when protective processes fail to prevent inhaled or blood-borne microbes reaching and *colonising* the lungs. The following are some predisposing factors.

Impaired coughing. The effectiveness or coughing as an aid to the removal of infected mucus may be reduced by damage to:
1. Sensory nerve endings in the walls of the respiratory passages
2. The cough reflex centre in the medulla oblongata
3. Nerves to the respiratory passages, lungs and muscles of respiration
4. The diaphragm and respiratory muscles

Voluntary inhibition may occur if coughing causes pain, e.g., following abdominal surgery.

Damage to the epithelial lining of the tract. Ciliary action may be impaired or the epithelium destroyed by, e.g., tobacco smoking, inhaling noxious gases, infection.

Defective alveolar phagocytosis. Depressed macrophage activity may be caused by tobacco smoking, alcohol, anoxia, oxygen intoxication.

Pulmonary oedema and congestion. Bronchopneumonia frequently occurs in patients with hypostatic pulmonary oedema and congestive heart failure. The relationship is not clear.

General lowering of resistance to infection. Factors involved include:
1. Leukopenia
2. Chronic diseases
3. Impaired immune response by, e.g., X-rays, corticosteroid drugs
4. Unusually virulent infections
5. Hypothermia

LOBAR PNEUMONIA

This is infection of one or more lobes by *Streptococcus pneumoniae*, usually Type 1 or 3. The infection leads to production of watery inflammatory exudate in the alveoli. This accumulates and fills the lobule then overflows into adjacent lobules, spreading the microbes. If not treated by antibacterial drugs the disease goes through a series of stages followed by resolution and reinflation of the lobes in 2 to 3 weeks.

COMPLICATIONS OF LOBAR PNEUMONIA

Incomplete resolution. If the fibrous exudate, formed as the infection subsides, is not completely cleared it may become permanently solid, tough, leathery and airless, reducing the surface area for gaseous exchange.

Pleural effusion and empyema. When infection spreads to the pleura, inflammatory exudate accumulates (*effusion*) and pus is formed (*empyema*). Healing may lead to formation of fibrous adhesions which prevent normal expansion of the lung.

Spread of infection may be to local tissues or cause:

Pericarditis   Meningitis   Arthritis
Acute endocarditis   Acute otitis media

ACUTE BRONCHOPNEUMONIA

Infection is spread from the bronchi to terminal bronchioles and alveoli. As these become inflamed, fibrous exudate accumulates and there is an influx of leukocytes. Small foci of consolidation develop. There is frequently incomplete resolution with fibrosis. Bronchiectasis is a common complication leading to further acute attacks, lung fibrosis and progressive destruction of lung substance. Bronchopneumonia occurs most commonly in infancy and old age. Predisposing factors include:

1. Debility due to, e.g., cancer, uraemia, cerebral haemorrhage, congestive heart failure, malnutrition, hypothermia
2. Chronic bronchitis
3. Bronchiectasis
4. Cystic fibrosis
5. General anaesthetics which depress respiratory and ciliary activity
6. Acute virus infections
7. Inhalation of gastric contents in, e.g., unconsciousness, drunkenness
8. Inhalation of infected material from air sinuses or upper respiratory tract

MICROBES CAUSING BRONCHOPNEUMONIA

Staphylococcus aureus. Infection is usually preceded by influenza, measles, whooping cough or chronic lung disease. Incomplete resolution may cause abscess formation, with rupture into the pleural cavity, empyema and possibly pneumothorax. Pleural adhesions may form during healing and limit lung expansion.

Friedlander's bacillus (*Klebsiella pneumoniae*). This commensal is sometimes present in the upper respiratory tract, especially where there is advanced dental caries. It commonly causes pneumonia in men over 50.

Legionella pneumonophila. These microbes are widely distributed in water tanks, shower heads and air conditioning systems, and are therefore commonly found in institutions such as hospitals and hotels. They may cause a severe form of pneumonia (*Legionnaires' disease*), complicated by gastrointestinal disturbances, headache, mental confusion and renal failure.

Streptococcus pyogenes. Infection is usually preceded by influenza or measles. In very severe cases death may occur within a few days.

Pseudomonas pyocyanea. This is a commensal in the bowel that may cause a type of pneumonia acquired by cross-infection in hospitals, especially in patients with mechanically assisted ventilation or tracheostomy. It is resistant to many antibacterial agents and multiplies at room temperature in, e.g., water, soap solutions, eye drops, ointments, weak antiseptics.

Streptococcus pneumoniae (pneumococcus). This is a commensal in the respiratory tract which may cause lobar pneumonia or bronchopneumonia, usually preceded by virus infection. It affects mainly debilitated bed-ridden patients when there is stagnation of mucus in the respiratory passages.

Other organisms. Some viruses, protozoa and fungi may cause pneumonia in people whose general resistance is lowered or whose immune systems are depressed by, e.g., drugs.

LUNG ABSCESS

Local suppuration and necrosis within the lung substance is most commonly caused by: *Streptococcus viridans, Streptococcus pyogenes, Staphylococcus aureus, Streptococcus pneumoniae.*

SOURCES OF INFECTION

Inhalation. Infected matter from air sinuses, gums and the upper respiratory passages may be inhaled. An important predisposing factor is inhalation of regurgitated gastric contents that may occur during coma, emergency anaesthesia or drunkenness. The hydrochloric acid and food particles cause severe irritation of the upper respiratory tract, allowing bacteria already present to invade the tissues and spread to the lungs.

Pneumonia. Lung abscess may complicate pneumonia, especially when the latter is caused by *Staphylococcus aureus, Streptococcus pneumoniae* or Freidlander's bacillus.

Septic embolism. Thrombophlebitis and right-sided endocarditis are the main sources of septic emboli that cause lung abscess.

Traumatic penetration of the lung. Pathogenic microbes may enter the lung when both skin and lung are penetrated. e.g., in compound fracture of rib, stab wound, gunshot wound or during surgery.

Local spread of infection. Microbes may spread from the pleural cavity, oesophagus, spine or a subphrenic abscess. Recovery from lung abscess may be complete or chronic suppuration may develop. Septic emboli may spread to other parts of the body, causing, e.g., meningitis.

COLLAPSE OF LUNG (Fig. 7:28)

This may be caused by airway obstruction or by positive pressure in the pleural cavity.

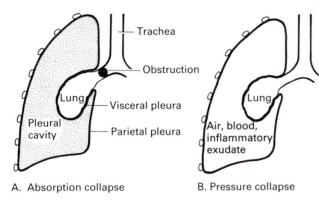

A. Absorption collapse   B. Pressure collapse

Figure 7:28 Collapsed lung.

ABSORPTION COLLAPSE (Airway obstruction)

The amount of lung affected depends on the size of the obstructed air passage. Distal to the obstruction air is trapped and absorbed, the lung collapses and secretions collect. These may become infected, causing abscess formation. Short-term obstruction is usually followed by reinflation of the lung without lasting ill-effects. Prolonged obstruction leads to progressive fibrosis and permanent collapse. Sudden obstruction may be due to inhalation of a foreign body or a mucous plug formed during an asthmatic attack. Gradual obstruction may be due to a bronchial tumour or pressure on a bronchus by, e.g., enlarged mediastinal lymph nodes, aortic aneurysm.

PRESSURE COLLAPSE

When air or fluid enters the pleural cavity the negative pressure becomes positive, preventing lung expansion. The collapse usually affects only one lung and may be partial or complete. There is no obstruction of the airway.

Pneumothorax (air in the pleural cavity) may be caused by:
1. Penetrating injury, e.g., compound fracture of rib, stab or gun-shot wound, surgery
2. Rupture of the lung over a diseased area, e.g., emphysema
3. Therapeutic introduction of air to rest the lung

Haemothorax (blood in the pleural cavity) may be caused by:
1. Penetrating chest injury involving blood vessels
2. Ruptured aortic aneurysm
3. Erosion of a blood vessel by a malignant tumour

Pleural effusion (excess fluid in the pleural cavity may be caused by:
1. Inflammation, usually due to infection
2. Malignant tumour involving the pleura

Fibrous adhesions which limit reinflation may form between the layers of pleura, following haemothorax and pleural effusion.

TUBERCULOSIS

This infection is caused by one of two forms of mycobacteria.

Mycobacterium tuberculosis. Man is the main host. The microbes cause pulmonary tuberculosis and are spread either by droplet infection from an individual with active tuberculosis, or in dust contaminated by infected sputum.

Mycobacterium bovis. Animals are the main host. The microbes are usually spread to man by untreated milk from infected cows, causing infection of the alimentary tract. In Britain the incidence has been greatly reduced by the elimination of bovine tuberculosis and the heat treatment of milk. It is still a significant infection in many countries.

PHASES OF PULMONARY TUBERCULOSIS

PRIMARY TUBERCULOSIS

When microbes are inhaled they colonise a lung bronchiole, usually towards the apex of the lung. There may be no evidence of clinical disease during the initial stage of non-specific inflammation. Later, cell-mediated T-lymphocytes respond to the microbes (antigens) and the individual becomes *sensitised*. Macrophages surround the microbes at the site of infection, forming *Ghon foci* (tubercles). Some macrophages containing live microbes are spread in lymph and infect hilar lymph nodes. *Primary complexes* are formed, consisting of Ghon foci plus infected hilar lymph nodes. Necrosis (caseation) may reduce the core of foci to a cheesy substance consisting of dead macrophages, dead lung tissue and live and dead microbes. At this stage:
1. The disease may be permanently arrested, the foci becoming fibrosed and calcified
2. Microbes may survive in the foci and become the source of postprimary infection months or years later
3. The disease may spread:
   a. throughout the lung, forming multiple small foci; to the respiratory passages, causing bronchopneumonia or bronchiectasis; to the pleura causing pleurisy, with or without effusion
   b. to other parts of the body via lymph and blood

SECONDARY (POSTPRIMARY) TUBERCULOSIS

This phase occurs only in people *previously sensitised* by a primary lesion. It may be caused by a new infection or by reactivation of infection by microbes surviving in Ghon foci. As sensitisation has already occurred, T-lymphocytes stimulate an immediate immune reaction, followed by phagocytosis. The subsequent course of the disease is variable, e.g.:
1. It may be arrested, healing occurring with fibrosis and calcification of the foci

2. Foci, containing live microbes, may be walled off by fibrous tissue, becoming a potential source of future infection
3. Live microbes in foci may break out through the fibrous walls and cause further infection
4. Pleurisy with or without effusion or empyema may develop
5. Haemorrhage (haemoptysis) may occur if a blood vessel is eroded
6. Caseous material containing live microbes may be coughed up, leaving cavities in the lung, and become a source of infection of:
   a. the other lung, bronchi, trachea and larynx
   b. the alimentary tract when microbes are swallowed
7. Microbes may spread in blood and lymph, leading to widespread infection and the development of numerous small foci throughout the body (miliary tuberculosis)

Bovine tuberculosis follows the same course but primary complexes develop in the intestines.

PNEUMOCONIOSIS

This group of lung diseases is caused by inhaling organic or inorganic atmospheric pollutants. To cause pneumoconiosis, particles must be so small that they are carried in inspired air to the level of the respiratory bronchioles and alveoli where they can only be cleared by phagocytosis. Larger particles are trapped by mucus higher up the tract and expelled by ciliary action and coughing. Other contributory factors include:

1. High concentration of pollutants in the air
2. Long exposure to pollutants
3. Reduced numbers of macrophages and ineffectual phagocytosis
4. Tobacco smoking

COAL WORKERS' PNEUMOCONIOSIS

This is caused by inhaling dust from soft bituminous coal. It occurs in two forms.

SIMPLE PNEUMOCONIOSIS

The particles of coal dust lodge mainly in the upper two-thirds of the lungs and are ingested by macrophages inside the alveoli. Some macrophages remain in the alveoli and some move out into the surrounding tissues and adhere to the outside of the alveolar walls, respiratory bronchioles, blood vessels and to the visceral pleura. When macrophages fail to digest the inorganic particles they die and are surrounded by fibrous tissue. Fibrosis is progressive during exposure to coal dust but tends to stop when

exposure stops. Early in the disease there may be few clinical signs unless emphysema develops or there is concurrent chronic bronchitis.

PNEUMOCONIOSIS WITH PROGRESSIVE MASSIVE FIBROSIS

This develops in a small number of cases, sometimes after the worker is no longer exposed to coal dust. Masses of fibrous tissue develop and progressively encroach on the blood vessels and bronchioles. Large parts of the lung are destroyed and emphysema is extensive, leading to pulmonary oedema, pulmonary hypertension and right-sided cardiac failure. The reasons for the severity of the disease are not clear. One factor may be hypersensitivity to antigens released by the large number of dead macrophages. About 40% of the patients have tuberculosis.

SILICOSIS

This may be caused by long-term exposure to dust containing silicon compounds. High risk industries are:

1. Quarrying granite, slate, sandstone
2. Mining hard coal, gold, tin, copper
3. Stone masonry and sand blasting
4. Glass and pottery work

When silica particles are inhaled they accumulate in the alveoli. The particles are ingested by macrophages, some of which remain in the alveoli and some move out into the connective tissues around respiratory bronchioles and blood vessels and close to the pleura. Progressive fibrosis is stimulated which eventually obliterates the blood vessels and respiratory bronchioles. Fibrous adhesions form between the two layers of pleura, and eventually fix the lung to the chest wall. The relationship between the presence of silicon and the production of excess fibrous tissue is not clear. It may be that:

1. Inflammation is caused by silicic acid which gradually forms when silicon compounds dissolve
2. There is an immune reaction in which silicon is the antigen
3. Fibrosis is stimulated by enzymes released when macrophages containing silicon die

Silicosis appears to predispose to the development of tuberculosis which rapidly progresses to tubercular bronchopneumonia and possibly to miliary tuberculosis. Gradual destruction of lung tissue leads to pulmonary hypertension and right-sided heart failure.

ASBESTOS-RELATED DISEASES

Diseases caused by inhaling asbestos fibres containing silicon usually develop after 10 to 20 years exposure but sometimes after only 2 years. Asbestos miners and workers involved in making and using some products containing asbestos are at risk. The types associated with disease are crocidolite (blue asbestos), chrysotile (white asbestos) and amosite (brown asbestos).

ASBESTOSIS

This occurs when asbestos fibres are inhaled in dust. In spite of their large size the particles penetrate to the level of respiratory bronchioles and alveoli. Macrophages accumulate in the alveoli and the shorter fibres are ingested. The larger fibres form *asbestos bodies*, consisting of fibres surrounded by macrophages, protein material and iron deposits. Their presence in sputum indicates exposure to asbestos but not necessarily that there is asbestosis. The macrophages that have engulfed fibres move out of the alveoli and accumulate round respiratory bronchioles and blood vessels, stimulating the formation of fibrous tissue. There is progressive destruction of lung tissue, with the development of dyspnoea, chronic hypoxia, pulmonary hypertension and right-sided cardiac failure. The link between inhaled asbestos and fibrosis is not clear. It may be that asbestos stimulates the macrophages to secrete enzymes that promote fibrosis or that it stimulates an immune reaction, causing fibrosis.

PLEURAL MESOTHELIOMA (Malignant tumours)

The majority of cases of carcinoma of the pleura are linked with previous exposure to asbestos dust, e.g., asbestos workers and people living near asbestos mines and factories. Mesothelioma may develop after widely varying duration of exposure to asbestos, e.g., 3 months to 60 years. The latent period between exposure and the appearance of symptoms may range from 10 to 40 years. The tumour involves both layers of pleura and as it grows it obliterates the pleural cavity, compressing the lung. Lymph- and blood-spread metastases are commonly found in the hilar and mesenteric lymph nodes, the other lung, liver, thyroid and adrenal glands, bone, skeletal muscle and brain.

BYSSINOSIS

This is caused by the inhalation of fibres of cotton, flax and hemp over a period of years. The fibres cause bronchial irritation and possibly the release of histamine-like substances. At first, breathless attacks similar to asthma occur only when the individual is at work. Later, they become more persistent and chronic bronchitis and emphysema may develop, leading to chronic hypoxia, pulmonary hypertension and right-sided heart failure.

EXTRINSIC ALLERGIC ALVEOLITIS

This group of conditions is caused by inhaling materials contaminated by moulds and fungi, e.g.:

Disease	Contaminant
Farmer's lung	mouldy hay
Bagassosis	mouldy sugar waste
Bird handler's lung	moulds in bird droppings
Malt worker's lung	mouldy barley

The contaminants act as antigens causing antigen/antibody reactions in the walls of the alveoli. There is excess fluid exudate and the accumulation of platelets, lymphocytes and plasma cells. The alveolar walls become thick and there is progressive fibrosis, leading to pulmonary hypertension and right-sided heart failure.

CHEMICALLY INDUCED LUNG DISEASES
PARAQUAT (1,1-dimethyl-4,4 bipyridlium chloride)

Within hours of ingestion of this weedkiller it is blood-borne to the lungs and begins to cause irreversible damage. The alveolar membrane becomes swollen, pulmonary oedema develops and alveolar epithelium is destroyed. The kidneys are also damaged and death may be due to combined respiratory and renal failure or cardiac failure.

CYTOTOXIC DRUGS (Bisulphan, Bleomycin, Methotrexate)

These and other drugs used in cancer treatment may cause fibrosis of interstitial tissue in the lungs followed by alveolar fibrosis.

OXYGEN POISONING

The lungs may be damaged by a high concentration of oxygen administered for several days, e.g., in intensive care. The permeability of the alveoli is increased, alveolar and capillary walls break down and fluid and blood accumulate. In severe cases pneumonia may develop followed by fibrosis, pulmonary hypertension and right-sided heart failure.

BRONCHIAL CARCINOMA

Primary bronchial carcinoma is a common form of malignancy. The tumour usually develops in a main bronchus, forming a large friable mass that projects into the lumen.

As the tumour grows it may erode a blood vessel, causing haemoptysis. The cause is not known but there is a strong positive association with cigarette smoking.

SPREAD OF BRONCHIAL CANCER

This does not follow any particular pattern or sequence. The modes of spread are infiltration of local tissues and the transport of tumour fragments in blood and lymph. If blood or lymph vessels are eroded fragments may spread while the tumour is quite small. A metastatic tumour may, therefore, cause symptoms before the primary in the lung has been detected.

Local spread may be within the lung, to the other lung or to mediastinal structures, e.g., blood vessels, nerves, oesophagus.

Lymphatic spread. Tumour fragments spread along lymph vessels to successive lymph nodes in which they may cause metastatic tumours. Fragments may enter lymph draining from a tumour or gain access to a larger vessel when its walls have been eroded by a growing tumour. Early symptoms may be due to pressure caused by enlarged lymph nodes in the thorax.

Blood spread. Tumour cells usually enter the blood when a blood vessel is eroded by local spread of the tumour. The most common sites of blood-borne metastases are the liver, brain, adrenal glands, bones and kidneys.

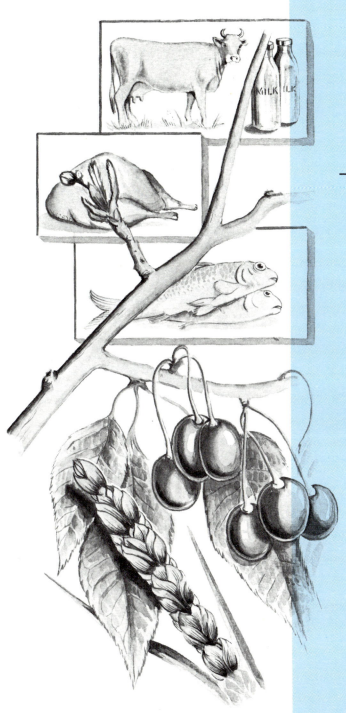

Nutrition

Constituents of the diet: sources, chemical
   composition, functions
Carbohydrates: monosaccharides,
   disaccharides, polysaccharides
Proteins: first class, second class
   Constituent amino acids: essential, non-
      essential
Fats: animal, vegetable
Vitamins: fat-soluble, water-soluble
   Chemical stability
   Effects of deficiency in diet
Mineral salts (inorganic compounds)
Roughage
Water

8. Nutrition

Before discussing the digestive system it is necessary to have an understanding of the needs of the body regarding diet. The essentials include:

Carbohydrates
Proteins
Fats
Vitamins
Mineral salts
Water
Roughage

If the cells of the body are to be able to function efficiently, these nutritional substances must be available in the *correct proportions*. Many foods contain a number of the essential dietary components, e.g., potatoes and bread are mainly carbohydrate but both contain protein and some vitamins. Foods are described as carbohydrate or protein because they are composed mainly of one or other of these nutrients.

CARBOHYDRATES

These are found in sugar, jam, cereals, bread, biscuits, potatoes, fruit and vegetables. They consist of carbon, hydrogen and oxygen, the hydrogen and oxygen being in the same proportion as in water. Carbohydrates are classified according to the complexity of the chemical substances of which they are formed.

MONOSACCHARIDES

Carbohydrates are digested in the alimentary canal and when absorbed they are in the form of monosaccharides. Examples include: glucose, fructose and galactose. These are, chemically, the simplest form in which a carbohydrate can exist. They are made up of single units or molecules which, if they were broken down further, would cease to be monosaccharides.

DISACCHARIDES

These consist of two monosaccharide molecules chemically combined to form sugars, e.g., sucrose, maltose and lactose.

POLYSACCHARIDES

These consist of complex molecules made up of large numbers of monosaccharide molecules in chemical combination, e.g., starches, glycogen, cellulose and dextrins.

Not all polysaccharides can be digested by human beings, e.g., fibre and cellulose present in vegetables, fruit and some cereals pass through the alimentary canal almost unchanged.

Functions

1. To provide energy and heat
2. To save protein, i.e., when there is an adequate supply of carbohydrate in the diet, protein does not need to be used to provide energy and heat
3. If carbohydrate is eaten in excess of the body's needs it is converted to fat and deposited in the fat depots, e.g., under the skin

PROTEINS OR NITROGENOUS FOODS

Proteins are made up of a large number of units called amino acids linked together chemically. Each protein consists of a specific number of different amino acids arranged in a way which is characteristic of that protein.

Proteins are broken down into their constituent amino acids by digestion and it is in this form that they are absorbed through the intestinal wall.

AMINO ACIDS

These are composed of the elements carbon, hydrogen, oxygen, nitrogen, sulphur and phosphorus. They are divided into two categories, *essential* and *non-essential*.

Essential amino acids cannot be synthesised in the body, therefore they must be included in the diet. They are:

isoleucine	methionine	tryptophan
leucine	phenylalanine	valine
lysine	threonine	

Non-essential amino acids are those which can be synthesised in the body. They are:

alanine	cystine	hydroxyproline
arginine	glutamic acid	proline
asparagine	glutamine	serine
aspartic acid	glycine	tyrosine
cysteine	histidine	

FIRST-CLASS PROTEINS

This is the name given to protein foods which contain all the essential amino acids in the correct proportions. They are derived almost entirely from animal sources and include:

meat	fish	soya beans
milk	eggs	

SECOND-CLASS PROTEINS

These do not contain all the essential amino acids in the correct proportions; they are mainly of vegetable origin and can be found in peas, beans and lentils, called the *pulses*. A variable proportion of protein is to be found in other vegetables and in some of the mainly carbohydrate foods, such as bread and potatoes.

Functions

Amino acids are used for:

1. Growth and repair of body cells and tissues
2. Synthesis of hormones, enzymes, plasma proteins and antibodies (immunoglobulins)
3. Provision of energy, normally a secondary function, becomes important only when there is not enough carbohydrate in the diet and fat stores are depleted

When protein is eaten in excess of the body's needs, the nitrogenous part is detached and excreted by the kidneys and the remainder is converted to fat for storage in the fat depots, e.g., in the fat cells of adipose tissue.

FATS

Fats consist of carbon, hydrogen and oxygen, but they differ from carbohydrates in that the hydrogen and oxygen are not in the same proportions as in water. Fats are divided into two groups, *animal* and *vegetable*.

Animal fat, containing mainly saturated fatty acids and glycerol, is found in milk, cheese, butter, eggs, meat and oily fish such as herring, cod and halibut. All the animal sources of protein contain some animal fat.

Vegetable fat, containing mainly unsaturated fatty acids and glycerol, is found in margarine and in vegetable oils.

Functions

1. To produce energy and heat
2. To support certain organs of the body, for example, the kidneys and the eyes
3. To transport the fat-soluble vitamins A, D, E and K
4. Fat is present in the nerve sheaths and in the secretions of the sebaceous glands in the skin
5. It is used in the formation of cholesterol and steroid hormones
6. When eaten in excess of that required by the body it is stored in the fat depots

VITAMINS

Vitamins are chemical compounds which are essential for health. They are found widely distributed in food and are divided into two main groups:

Fat-soluble vitamins: A, D, E and K
Water-soluble vitamins: B complex, C

FAT-SOLUBLE VITAMINS

VITAMIN A (RETINOL)

This vitamin is found in such foods as cream, egg yolk, fish oil, milk, cheese and butter. It is absent from vegetable fats and oils but is added to margarine during manufacture. It can be formed in the body from certain carotenes the main dietary sources of which are green vegetables, fruit and carrots.

Vitamin A and carotene are only absorbed from the small intestine satisfactorily if fat absorption is normal.

Functions

1. It is necessary for the regeneration of the visual purple in the retina of the eye which is bleached by bright light. If there is insufficient vitamin A adaptation to seeing in dim light is delayed
2. It influences the nutrition of epithelial cells and tends to reduce the severity of microbial infection. Because of this, it is sometimes known as the *anti-infective vitamin*
3. It is necessary to maintain the cornea of the eye in a healthy state

VITAMIN D

Vitamin D_3, the *antirachitic vitamin*, is found mainly in animal fats such as eggs, butter, cheese, fish liver oils.

Humans and other animals can synthesise cholicalciferol (vitamin D) by the action of the ultraviolet rays of the sun on a form of cholesterol in the skin (7-dehydrocholesterol).

Calciferol (vitamin D_2) is formed in plants and is used widely in therapeutics.

Functions

Vitamin D regulates calcium and phosphorus metabolism and is therefore associated with the calcification of bones and teeth. Deficiency causes rickets in children and osteomalacia in adults.

VITAMIN E

This is a group of substances, the most active of which are *tocopherols*. The sources of this vitamin are peanuts, lettuce, egg yolk, wheat germ, whole cereal, milk and butter.

Functions

Lack of this vitamin in animals causes muscle wasting and failure in reproduction, but it is not quite certain whether it has the same functions in humans.

VITAMIN K

The sources of vitamin K are fish liver, leafy green vegetables and fruit. It is synthesised in the intestine by microbes and significant amounts and absorbed. Absorption is dependent upon the presence of bile salts in the small intestine.

Functions

It is necessary for the formation by the liver of prothrombin and factors VII, IX and X, all essential for the clotting of blood.

WATER-SOLUBLE VITAMINS

VITAMIN B COMPLEX

This consists of a group of water-soluble vitamins which are more or less closely associated.

VITAMIN B₁ (THIAMINE)

This vitamin is present in nuts, yeast, egg yolk, liver, legumes, meat and the germ of cereals.

Functions

1. It is essential for normal carbohydrate metabolism
2. It stimulates appetite
3. It helps to regulate the functioning of the nervous system
4. It is associated with the control of water balance in the body

VITAMIN B₂ (RIBOFLAVIN)

This is found in yeast, leafy vegetables, milk, liver, eggs, kidney, cheese and roe.

Functions

1. It is concerned with the oxidation of all foods
2. It is associated in some way with the physiology of the skin and eyes

FOLIC ACID

This is found in liver, kidney, fresh leafy green vegetables and yeast. It is synthesised by bacteria in the large intestine, and significant amounts derived from this source are believed to be absorbed.

Functions

It is essential for the maturation of erythrocytes in the red bone marrow.

NICOTINIC ACID (NIACIN)

This is found in liver, cheese, yeast, whole cereal, eggs, fish, peanuts and Bemax. Deficiency causes pellagra.

Functions

It is necessary for:
1. The metabolism of carbohydrates
2. The normal functioning of the gastrointestinal tract
3. The satisfactory functioning of the nervous system

VITAMIN B₆ (PYRIDOXINE)

This is found in egg yolk, peas, beans, soya bean, yeast, meat and liver.

Functions

It is believed to be necessary for satisfactory protein and fat metabolism and for the synthesis of haem in haemoglobin.

VITAMIN B₁₂

This name is used to describe a number of *cobalamin compounds* (containing cobalt). It is found in liver, meat, eggs, milk and fermented liquors.

Functions

It is essential for the maturation of erythrocytes in the red bone marrow.

PANTOTHENIC ACID

This is found in many foods and is associated with the metabolism of fats and carbohydrates.

BIOTIN

This is found in egg yolk, liver and tomatoes and is synthesised by microbes in the intestine. It is associated with the metabolism of carbohydrates.

VITAMIN C (ASCORBIC ACID)

This is found in fresh fruit, especially blackcurrants, oranges, grapefruit and lemons, and also in rose-hips and green vegetables.

Functions

It is necessary for:
1. The maintenance of the strength of the walls of the blood capillaries
2. The development and maintenance of healthy bones
3. The formation of red blood cells
4. The production of antibodies
5. The formation of connective tissue

MINERAL SALTS

Mineral salts (inorganic compounds) are necessary within the body for all body processes. They are usually required in small quantities. They include many compounds some of which are in very small quantities. The main elements involved are:

calcium   phosphorus   sodium
iron   iodine   potassium

CALCIUM

This is found in milk, cheese, eggs, green vegetables and some fish. An adequate supply should be obtained in a normal, well-balanced diet.

Functions

In association with vitamin D and phosphorus it is essential for the hardening of bones and teeth. Therefore an adequate supply in young people is important.

It plays an important part in the coagulation of blood and is associated with the mechanism of muscle contraction.

PHOSPHORUS

Sources of phosphorus include cheese, oatmeal, liver and kidney. If there is sufficient calcium in the diet it is unlikely that there will be a deficiency of phosphorus.

Functions

It is associated with calcium and vitamin D in the hardening of bones and teeth and helps to maintain the constant composition of the body fluids. Phosphates are an essential part of systems of energy transport inside cells.

SODIUM

Sodium is found in most foods, especially fish, meat, eggs, milk, artificially enriched bread and as cooking and table salt. The normal intake of sodium chloride per day varies from 5 to 20 g and the daily requirement is between 2 and 5 g. Excess is excreted in the urine.

Functions

It is the most commonly occurring *extracellular cation* and is associated with:
1. The contraction of muscle
2. The transmission of nerve impulses in nerve fibres
3. The maintenance of the electrolyte balance in the body

POTASSIUM

This substance is to be found widely distributed in all foods. The normal intake of potassium chloride varies from 5 to 7 g per day and this meets the potassium requirements.

Functions

It is the most commonly occurring *intracellular cation*.
It is involved in:
1. Many chemical activities inside cells
2. The contraction of muscles
3. The transmission of nerve impulses
4. The maintenance of the electrolyte balance in the body

IRON

Iron, as a soluble compound, is found in liver, kidney, beef, egg-yolk, wholemeal bread and green vegetables. In normal adults about 1 mg of iron is lost from the body daily. The normal daily diet contains more, but the amount absorbed is only equal to the amount lost. Higher intake is needed by women especially during pregnancy.

Functions

Iron is essential for the formation of *haemoglobin* in the red blood cells. It is necessary for tissue oxidation.

IODINE

Iodine is found in salt-water fish and in vegetables grown in soil containing iodine. In some parts of the world where iodine is deficient in soil very small quantities are added to table salt. The daily requirement of iodine depends upon the individual's metabolic rate. Some people have a higher normal metabolic rate than others and their iodine requirements are greater. The minimum daily requirement is about 20 μg. In favourable conditions the normal daily intake is 100–200 μg.

Function

It is essential for the formation of *thyroxine* and *triiodo-thyronine*, the hormones secreted by the thyroid gland.

ROUGHAGE (Fibre)

Fibre is the undigestible part of the diet, consisting of bran, cellulose and the polysaccharides. It is widely distributed in wholemeal flour, the husks of cereals and vegetables.

Functions

1. It gives bulk to the diet and helps to satisfy the appetite
2. It stimulates peristalsis (muscular activity) of the alimentary tract
3. It stimulates bowel movement
4. An adequate supply is believed to prevent some gastrointestinal disorders

WATER

Water is a liquid compound of hydrogen and oxygen, formed by the chemical combination of two parts hydrogen and one part oxygen (H_2O). Water makes up about 70% of the body weight in men and about 60% in women.

A large amount of water is lost from the body each day. Under normal circumstances this is balanced by intake. Dehydration with serious consequences may occur if the balance is not maintained.

Functions

1. It provides the moist environment which is required by all living cells in the body, i.e., all the cells of the body except the superficial layers of the skin, the nails, the hair and the outer hard layer of the teeth
2. It participates in all the chemical reactions which occur inside and outside the body cells
3. It dilutes and moistens food
4. It assists in the regulation of body temperature as a constituent of sweat, which is secreted onto the skin. The evaporation of sweat cools the body
5. As a major component of blood and tissue fluid it transports some substances in solution and some in suspension round the body
6. It dilutes waste products and poisonous substances in the body
7. It contributes to the formation of urine and faeces

Tables 8:1 and 8:2 summarise the vitamins — their chemical names, sources, stability, functions, deficiency diseases and daily adult requirements.

Table 8:1 Summary — fat-soluble vitamins

Vitamin	Chemical name	Source	Stability	Functions	Deficiency diseases	Daily requirement (adults)
A	Retinol (carotene provitamin in plants)	Milk, butter, cheese, egg yolk, fish liver oils, green and yellow vegetables	Some loss at high temperatures and long exposure to light and air	Maintains healthy epithelial tissues and cornea. Formation of visual purple	Keratinisation Xerophthalmia Stunted growth Night blindness	750 μg
D	Calciferol	Fish liver oils, milk, cheese, egg-yolk, irradiated 7-dehydrocholesterol in human skin	Very stable	Facilitates the absorption and utilisation of calcium and phosphorus = healthy bones and teeth	Rickets Osteomalacia	2.5 μg
E	Tocopherols	Egg-yolk, milk, butter, green vegetables, nuts	Destroyed by rancid fat and iron salts	Maintains healthy muscular system		
K	Phylloquinone	Leafy vegetables, fish liver, fruit	Destroyed by light, strong acids and alkalis	Formation of prothrombin and Factors VII, IX and X in the liver	Slow blood clotting Haemorrhages in the newborn	

Bile is necessary for the absorption of these vitamins.
Mineral oils interfere with absorption.

Table 8:2   Summary — water-soluble vitamins

Vitamin	Chemical name	Source	Stability	Functions	Deficiency diseases	Daily requirement (adults)
B_1	Thiamin	Yeast, liver, germ of cereals, nuts, pulses rice polishings, egg-yolk, liver, legumes	Stable	Metabolism of carbo-hydrates and nutri-tion of nerve cells Efficient water ex-change in the body	General fatigue and loss of muscle tone Ultimately leads to beriberi	1–1.5 mg
B_2	Riboflavin	Liver, yeast, milk, eggs, green vegetables, kidney	Destroyed by light and alkalis	Necessary for tissue oxidation and growth	Angular stomatitis Cheilosis Dermatitis Eye lesions	1.5–2 mg
B_6	Pyridoxine	Meat, liver, vegetables, bran of cereals, egg-yolk, beans, soya beans	Stable	Protein metabolism Formation of RBCs and WBCs	Very rare	1–2 mg
B_{12}	Cobalamins	Liver, milk, moulds, fermenting liquors, egg	Destroyed by heat	Maturation of RBCs	Pernicious anaemia Degeneration of nerve fibres of the spinal cord	2–3 µg
B	Folic acid	Dark green vegetables, liver, kidney, eggs Synthesised in colon	Destroyed by heat and moisture	Formation of RBCs	Anaemia	200 µg
B	Nicotinic acid (Niacin)	Yeast, offal, fish, pulses, wholemeal cereals. Synthesised in the body from tryptophan	Fairly stable	Necessary for tissue oxidation	Prolonged deficiency causes pellagra, i.e., dermatitis, diarrhoea, dementia	15–20 mg
B	Pantothenic acid	Liver, yeast, egg-yolk, fresh vegetables	Destroyed by ex-cessive heat and freezing	Probably required for formation of RBCs	Dermatitis, adrenal insufficiency	Unknown
B	Biotin	Yeasts, liver, kidney, pulses, nuts	Stable	Carbohydrates and fat metabolism Growth of bacteria	Dermatitis, con-junctivitis	Unknown
C	Ascorbic acid	Citrus fruits, currants, berries, green vege-tables, potatoes, liver and glandular tissue in animals	Destroyed by heat, ageing, acids, alkalis, chopping, salting and drying	Formation of inter-cellular matrix Maturation of RBCs	Multiple haemorrhages Slow wound healing Anaemia Gross deficiency causes scurvy	30–50 mg

Para-aminobenzoic acid and inositol are two vitamins of the B group about which little is known.

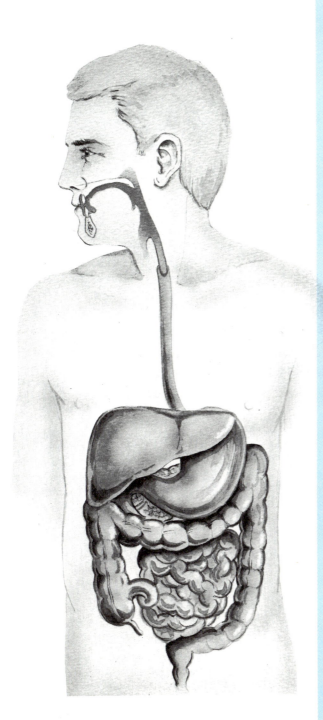

9 The Digestive System

9. The Digestive System

The digestive system is the collective name used to describe the *alimentary canal*, some *accessory organs* and a variety of *digestive processes* which take place at different levels in the canal to prepare food eaten in the diet for absorption. The alimentary canal begins at the mouth, passed through the thorax, abdomen and pelvis and ends at the anus. It has a general structure which is modified at different levels to provide for the processes occurring at each level. The complex of digestive processes gradually simplify the foods eaten until they are in a form suitable for absorption. For example, meat, even when cooked, is chemically too complex to be absorbed from the alimentary canal. It therefore goes through a series of changes which release its constituent nutrients: amino acids, mineral salts, fat and vitamins. Chemical substances or *enzymes** which effect these changes are secreted into the canal by special glands, some of which are in the walls of the canal and some outside the canal but with ducts, leading into it.

After they are absorbed, the nutrient materials are used in the synthesis of the constituents of the body. They provide the raw materials for the manufacture of new cells, hormones and enzymes, and the energy needed for these and other processes and for the disposal of waste materials.

The activities in the alimentary canal can be grouped under four main headings.

Ingestion, taking food into the alimentary tract.

Digestion, which can be divided into the *mechanical* breakdown of food by *mastication* (chewing) and *chemical* breakdown by *enzymes* present in secretions produced by glands of the digestive system. The secretions include:

Saliva from the salivary glands
Gastric juice from the stomach
Intestinal juice from the small intestine
Pancreatic juice from the pancreas
Bile from the liver

Absorption is the process by which digested food substances pass through the walls of some organs of the alimentary canal into the blood and lymph capillaries for circulation round the body.

Elimination. Food substances which have been eaten but cannot be digested and absorbed are excreted by the bowel as faeces.

* An enzyme is a chemical substance which causes, or speeds up, a chemical change in other substances without itself being changed.

ORGANS OF THE DIGESTIVE SYSTEM (Fig. 9:1)

Alimentary tract
This is a long tube through which food passes. It commences at the mouth and terminates at the anus, and the various parts are given separate names, although structurally they are remarkably similar. The parts are:

Mouth
Pharynx
Oesophagus
Stomach
Small intestine
Large intestine
Rectum and anal canal

Accessory organs
Various secretions are poured into the alimentary tract, some by glands in the lining membrane of the organs, e.g., gastric juice by the lining of the stomach, and some by glands situated outside the tract. The latter are the accessory organs of digestion and their secretions pass through ducts to enter the tract. They consist of:

3 pairs of salivary glands
Pancreas
Liver and the biliary tract

From these lists it can be seen that this is a large system involving a considerable number of organs and glands. They are linked physiologically as well as anatomically in that digestion and absorption occur in stages, each stage being dependent upon the previous stage or stages.

For descriptive purposes the system will be dealt with in six sections:

1. The general plan of the alimentary tract
2. Mouth, pharynx, oesophagus, salivary glands
3. Stomach
4. Pancreas, liver, biliary tract
5. Small intestine, large intestine, rectum and anal canal
6. Metabolism

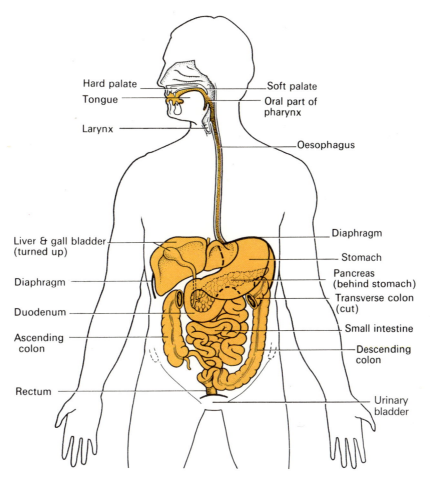

Figure 9:1   The organs of the digestive system.

GENERAL PLAN OF THE ALIMENTARY TRACT (Fig. 9:2)

The structure of the alimentary canal follows a consistent pattern from the level of the oesophagus onwards. This *general plan* does not apply so obviously to the mouth and the pharynx so these parts of the tract have been excluded at this stage.

In the different organs from the oesophagus onward, modifications of structure are found which are associated with special functions. The general plan is described here and the modification in structure and function are described in the appropriate section.

The walls of the alimentary tract are formed by four layers of tissue:

Adventitia or outer covering
Muscle layer
Submucous layer
Mucous membrane lining

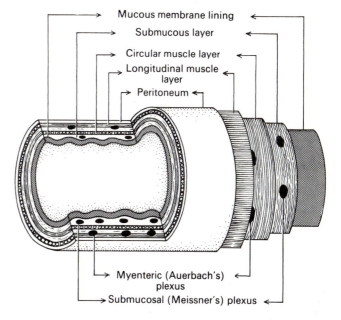

Figure 9:2   General plan of the alimentary canal.

ADVENTITIA (Outer covering)

In the thorax this consists of *loose fibrous tissue* and in the abdomen the organs are covered by a serous membrane called *peritoneum.*

PERITONEUM (Fig. 9:3)

The peritoneum is the largest serous membrane of the body. It consists of a closed sac within the abdominal cavity and has two layers:

The parietal layer, which lines the abdominal wall

The visceral layer, which covers the organs (viscera) within the abdominal and pelvic cavities

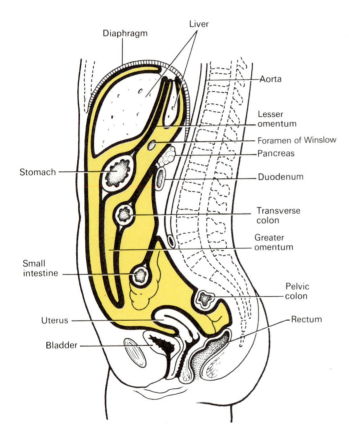

Figure 9:3   The peritoneum (coloured) and its association with the abdominal organs of the digestive system and the pelvic organs, viewed from the side.

The arrangement of the peritoneum is complicated; it is as though the organs had been invaginated into the closed sac from below, behind and above, taking with them a covering of visceral peritoneum. This means that:

1. The pelvic organs, invaginated from below, are covered only on their superior surfaces
2. The stomach and intestines, deeply invaginated from behind, are almost completely surrounded by peritoneum and have a double fold that attaches them to the posterior abdominal wall
3. The pancreas, spleen, kidneys and adrenal glands are invaginated from behind but only their anterior surfaces are covered
4. The liver is invaginated from above and is almost competely covered by peritoneum which attaches it to the inferior surface of the diaphragm
5. The main blood vessels and nerves pass close to the posterior abdominal wall and send branches to the organs between folds of peritoneum

The parietal peritoneum lines the anterior abdominal wall.

The two layers of peritoneum are actually in contact and friction between them is prevented by the presence of serous fluid secreted by the peritoneal cells, thus the *peritoneal cavity* is only a *potential cavity*. In the male it is completely closed but in the female the uterine tubes open into it and the ovaries are the only structures inside (see Ch. 15).

MUSCLE LAYER

With some exceptions this consists of two layers of *smooth muscle*. The muscle fibres of the outer layer are arranged longitudinally, and those of the inner layer encircle the wall of the tube. Between these two muscle layers there are blood vessels, lymph vessels and a plexus of sympathetic and parasympathetic nerves, called the *myenteric* or *Auerbach's plexus*. These nerves supply the adjacent smooth muscle and blood vessels.

The contraction of these muscle layers occurs in waves which push the contents of the tract onwards. This type of contraction of smooth muscle is called *peristalsis*. The other main effect of muscle contraction is to mix the contents of the tract with the digestive juices.

Onward movement of the contents is prevented by *sphincters* situated at various points on the tract. These consist of an increased number of circular muscle fibres. They delay onward movement allowing time for digestion and absorption to take place.

SUBMUCOUS LAYER

This layer consists of loose connective tissue with some elastic fibres. Within this layer there are plexuses of blood vessels and nerves, lymph vessels and varying amounts of lymphoid tissues. The blood vessels consist of arterioles, venules and capillaries. The nerve plexus is the *submucosal* or *Meissner's plexus* and it contains sympathetic and parasympathetic nerves which supply the mucous membrane lining.

MUCOUS MEMBRANE

This layer has three main functions: protective, secretory and absorptive. In parts of the tract which are subject to mechanical injury this layer consists of *stratified squamous epithelium* with mucus-secreting glands just below the surface. In areas where the food is already soft and moist

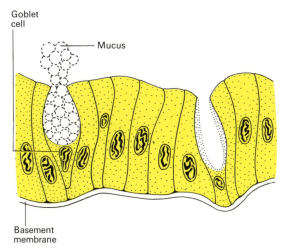

Goblet cell

Mucus

Basement membrane

Figure 9:4   Columnar epithelium with goblet cells.

and where the secretion of digestive juices and absorption occur, the mucous membrane consists of *columnar epithelial cells* interspersed with goblet cells which secrete mucus (Fig. 9:4). Below the surface in the part lined by columnar epithelium there are collections of specialised cells, or glands, which pour their secretions into the lumen of the tract. These are *digestive juices* and they contain the enzymes which chemically simplify food. Under the epithelial lining there are varying amounts of lymphoid tissue.

NERVE SUPPLY

The alimentary tract is supplied by nerves from both parts of the autonomic nervous system, i.e., parasympathetic and sympathetic, and in the main their actions are antagonistic (Fig. 9:5). In the normal healthy state one influence may outweigh the other according to the needs of the body as a whole at a particular time.

The parasympathetic supply to most of the alimentary tract is provided by one pair of cranial nerves, the *vagus nerves*. Stimulation causes muscle contraction and the secretion of digestive juices. The most distal part of the tract is supplied by sacral nerves.

The sympathetic supply is provided by numerous nerves which emerge from the spinal cord in the thoracic and lumbar regions. These form *plexuses* in the thorax, abdomen and pelvis, and from them nerves pass to the organs of the alimentary tract. Their action is to reduce muscle contraction and grandular secretion.

Within the walls of the canal there are two nerve plexuses from which both sympathetic and parasympathetic fibres are distributed. The *myenteric* or *Auerbach's plexus* lies between the two layers of muscle tissue and supplies the muscle, and the *submucosal* or *Meissner's plexus* lies in the submucosa and supplies the mucous membrane and secretory glands.

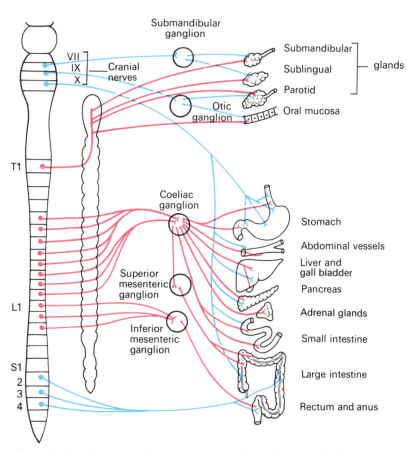

Figure 9:5   Autonomic nerve supply to the digestive system. Blue — parasympathetic, red — sympathetic.

BLOOD SUPPLY

ARTERIAL BLOOD SUPPLY

In the thorax

The oesophagus is supplied by paired *oesophageal arteries*, branches from the thoracic aorta.

In the abdomen and pelvis

The alimentary tract, pancreas, liver and biliary tract are supplied by the unpaired branches from the aorta: the *coeliac artery* and the *superior* and *inferior mesenteric arteries* (Figs 9:6 and 9:7).

The *coeliac artery* divides into three branches which supply to stomach, duodenum, pancreas, spleen, liver, gall bladder and bile ducts. They are:

Left gastric artery
Splenic artery
Hepatic artery

The *superior mesenteric artery* supplies the whole of the small intestine, the caecum, ascending colon and most of the transverse colon.

The *inferior mesenteric artery* supplies a small part of the transverse colon, the descending colon, pelvic colon and most of the rectum.

The distal part of the rectum and the anus are supplied by the *middle* and *inferior rectal arteries*, branches of the internal iliac arteries.

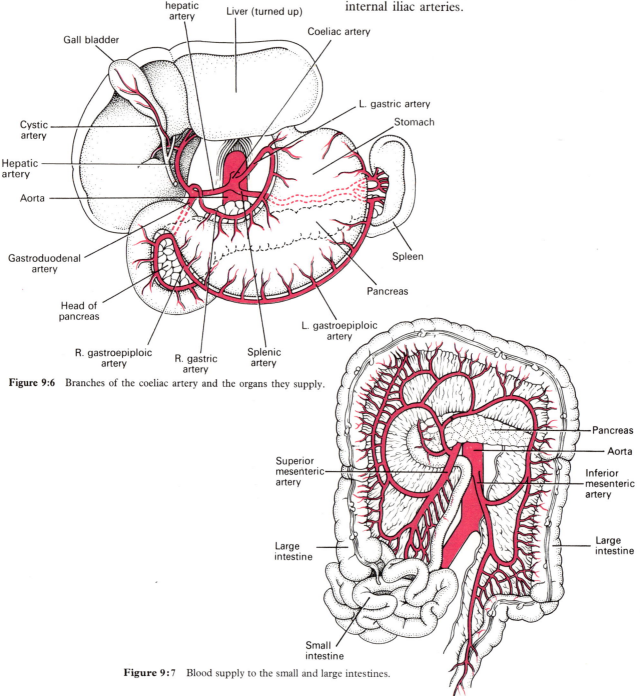

Figure 9:6   Branches of the coeliac artery and the organs they supply.

Figure 9:7   Blood supply to the small and large intestines.

VENOUS DRAINAGE

In the thorax
Venous blood from the oesophagus enters the *azygos* and *hemiazygos veins*. The azygos vein joins the superior vena cava near the heart, and the hemiazygos joins the left brachiocephalic vein.

Some blood from the lower part of the oesophagus drains into the *left gastric vein*. There are anastomotic vessels between the azygos, hemiazygos and left gastric veins.

In the abdomen and pelvis
The veins that drain blood from the lower part of the oesophagus, the stomach, pancreas, small intestine, large intestine and most of the rectum join to form the *portal vein* (Fig. 9:8). This blood, containing a high concentration of absorbed nutritional materials, is conveyed first to the liver then to the inferior vena cave. The circulation of blood in the liver is described later (see p. 160).

Blood from the lower part of the rectum and the anal canal drains into the *internal iliac veins*.

MOUTH (Fig. 9:9)

The mouth or oral cavity is bounded by muscles and bones:

Anteriorly — by the lips
Posteriorly — it is continuous with the oral part of the pharynx
Laterally — by the muscles of the cheeks
Superiorly — by the bony hard palate and muscular soft palate
Inferiorly — by the muscular tongue and the soft tissues of the floor of the mouth

The oral cavity is lined throughout with *mucous membrane*, consisting of *stratified squamous epithelium* containing small mucus-secreting glands.

The part of the mouth outside the gums and teeth is the *vestibule* and the remainder of the cavity, the *mouth proper*. The mucous membrane lining of the cheeks and the lips is reflected on to the gums or *alveolar ridges*.

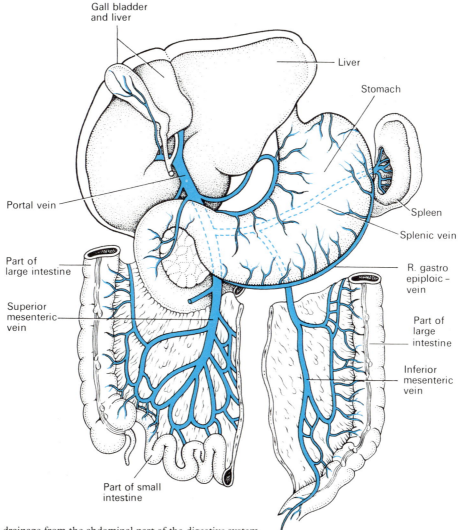

Figure 9:8   Venous drainage from the abdominal part of the digestive system.

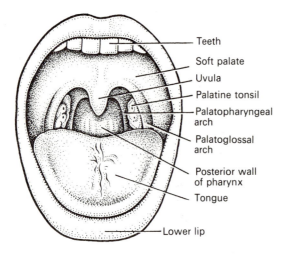

Figure 9:9   Structures seen in the widely open mouth.

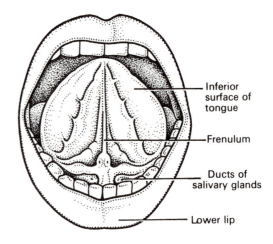

Figure 9:10   The inferior surface of the tongue.

The *palate* is divided into the anterior *hard palate* and the posterior *soft palate*. The bones forming the hard palate are the maxilla and the palatine bones. The soft palate is muscular, curves downwards from the posterior end of the hard palate and blends with the walls of the pharynx at the sides.

The *uvula* is a curved fold of muscle covered with mucous membrane, hanging down from the middle of the free border of the soft palate. Originating from the upper end of the uvula there are four folds of mucous membrane, two passing downwards at each side to form membranous arches. The posterior folds, one on each side, are the *palatopharynegeal arches* and the two anterior folds are the *palatoglossal arches*. On each side, between the arches, there is a collection of lymphoid tissue called the *palatine tonsil*.

TONGUE

The tongue is a voluntary muscular structure which occupies the floor of the mouth. It is attached by its base to the *hyoid bone* and by a fold of its mucous membrane covering, called the *frenulum*, to the floor of the mouth (Fig. 9:10). The superior surface consists of stratified squamous epithelium, with numerous *papillae* (little projections), containing nerve endings of the sense of taste; these are sometimes called the *taste buds*. There are three varieties of papillae (Fig. 9:11).

Vallate papillae are usually about 8 to 12 in number and are arranged in an inverted V shape towards the base of the tongue. These are the largest of the papillae and are the most easily seen.

Fungiform papillae are situated mainly at the tip and the edges of the tongue and are more numerous than the vallate papillae.

Filiform papillae are the smallest of the three types. They are most numerous on the surface of the anterior two-thirds of the tongue.

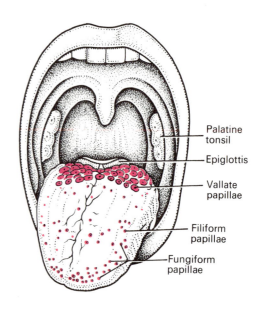

Figure 9:11   Diagram of the papillae of the tongue and related structures.

Blood supply
The main arterial blood supply to the tongue is by the *lingual branch* of the *external carotid artery*. Venous drainage is by the *lingual vein* which joins the internal jugular vein.

Nerve supply
The nerves involved are:

The *hypoglossal nerves* (12th cranial nerves) which supply the voluntary muscle tissue

The *lingual branch of the mandibular nerves* which are the nerves of somatic (ordinary) sensation, i.e., pain, temperature and touch

The *facial* and *glossopharyngeal nerves* (7th and 9th cranial nerves) which are the nerves of the special sensation of taste

Functions of the tongue

The tongue plays an important part in mastication (chewing), deglutition (swallowing) and speech. It is the main organ of taste. Nerve endings of the sense of taste are present in the papillae and widely distributed in the epithelium of the tongue, soft palate, pharynx and epigloltis.

TEETH

The teeth are embedded in the alveoli or sockets of the alveolar ridges of the mandible and the maxilla (Fig. 9:12).

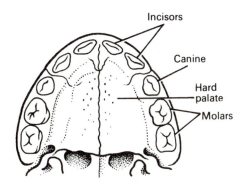

Figure 9:13 The roof of the mouth and the deciduous teeth. Viewed from below.

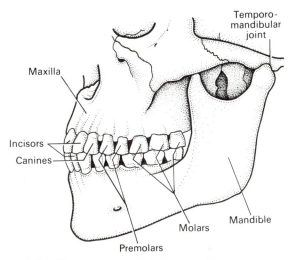

Figure 9:12 The permanent teeth and the jaw bones.

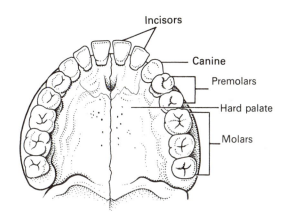

Figure 9:14 The roof of the mouth and the permanent teeth. Viewed from below.

Each individual has two sets, the *temporary* or *deciduous teeth* and the *permanent teeth* (Figs 9:13 and 9:14). At birth the teeth of both dentitions are present in immature form in the mandible and maxilla.

The *temporary teeth* are 20 in number, 10 in each jaw.

Deciduous teeth						
Jaw	Molars	Canine	Incisors	Incisors	Canine	Molars
Upper	2	1	2	2	1	2
Lower	2	1	2	2	1	2

They begin to erupt when the child is about *6 months* old, and should all be present by the end of *24 months*.

The *permanent teeth* begin to replace the deciduous teeth in the *6th year* of age and this dentition, consisting of 32 teeth, is usually complete by the *24th year*.

Permanent teeth							
Jaw	Molars	Premolars	Canine	Incisors	Incisors	Canine	Premolars	Molars
Upper	3	2	1	2	2	1	2	3
Lower	3	2	1	2	2	1	2	3

The *incisor* and *canine* teeth are the cutting teeth and are used for biting off pieces of food, whereas the *premolar*

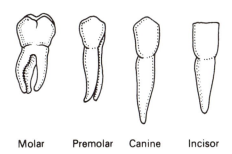

Figure 9:15 The shapes of the permanent teeth.

and *molar* teeth, with broad, flat surfaces, are used for grinding or chewing food (Fig. 9:15).

Structure of a tooth (Fig. 9:16)

Although the shapes of the different teeth vary, the structure is the same and consists of:

The crown — the part which protrudes from the gum
The root — the part embedded in the bone
The neck — the slightly constricted part where the crown merges with the root

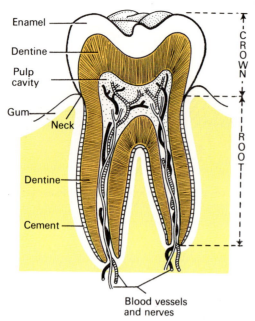

Figure 9:16 A section of a tooth.

In the centre of the tooth there is the *pulp cavity* containing blood vessels, lymph vessels and nerves, and surrounding this there is a hard ivory-like substances called *dentine*. Outside the dentine of the crown of the tooth there is a thin layer of very hard substance, the *enamel*. The root of the tooth, on the other hand, is covered with a substance resembling bone, called *cement*, which fixes the tooth in its socket. Blood vessels and nerves pass to the tooth through a small foramen at the apex of each tooth.

Blood supply
Most of the arterial blood supply to the teeth is by branches of the *maxillary arteries*. The venous drainage is by a number of veins which empty into the *internal jugular veins*.

Nerve supply
The nerve supply to the upper teeth is by branches of the *maxillary nerves* and to the lower teeth by branches of the *mandibular nerves*. These are both branches of the *trigeminal nerves* (5th cranial nerves) (see p. 233).

DISEASES OF THE MOUTH

INFLAMMATION

This may be caused by solids or liquids taken into the mouth, or by infection.

INJURIOUS SUBSTANCES TAKEN INTO THE MOUTH

The injury may be due to:
1. Excessive heat or cold
2. The rough texture of the materials
3. Corrosive substances

Corrosive chemicals are the most likely to cause serious tissue demage and acute inflammation. The outcome depends on the extent and depth of the injury.

INFECTIONS

THRUSH (CANDIDIASIS)

This acute infection of the lining epithelium of the mouth is caused by *Candida albicans*. In adults it causes infection mainly in debilitated people and in those whose immunity is suppressed by steroid or cytotoxic drugs. In babies it may be a severe infection, sometimes causing epidemics in nurseries by cross infection. It occurs most commonly in bottle-fed babies. *Chronic thrush* may develop, affecting the roof of the mouth in people who wear dentures. The fungus survives in the fine grooves on the upper surface of the denture and repeatedly reinfects the epithelium.

ANGULAR CHEILITIS

Painful cracks develop in folds of tissue at the corners of the mouth, usually occurring in elderly debilitated people, especially if they do not wear their dentures and the folds remain moist. The usual causal organisms are *Candida albicans* and *Staphylococcus aureus*.

VINCENT'S ANGINA

This is an acute infection with severe ulceration of the lips, mouth, throat and the palatine tonsil. It is caused by two organisms acting together, *Bordetella vincenti* and a fusiform bacillus. Both organisms may be present in the mouth and only cause the disease in the presence of:
1. Malnutrition
2. Debilitating disease
3. Poor mouth hygiene
4. Injury caused by previous infection

VIRUS INFECTIONS

Aphthous stomatitis (Herpetic gingivostomatitis)
Ulcers, caused by *Herpes simplex* virus, occur singly or in crops inside the mouth. They are often found in association with iron and vitamin B group deficiency but a link has not been established.

Secondary or recurrent herpes lesions (cold sores)
Lesions, caused by *Herpes simplex* virus, occur round the nose and on the lips. After an outbreak the viruses remain alive in the cells, but inactive. Later outbreaks, usually at the same site, are believed to be precipitated by a variety of stimuli not yet clearly identified but failing immune response is a major factor in old age.

TUMOURS OF THE MOUTH
SQUAMOUS CELL CARCINOMA

This is the most common type of malignant tumour in the mouth. The usual sites are the lower lip and the edge of the tongue. Ulceration occurs frequently and there is early spread to cervical lymph nodes .

DEVELOPMENTAL DEFECTS
CLEFT PALATE AND CLEFT LIP (Harelip)

Before birth the right and left halves of the palate develop separately then fuse with the midline nasal septum. One or both lines of fusion may be incomplete, leaving an opening between the mouth and the nose. The membranous soft palate also develops in two parts and fuses in the midline. Thus the lines of fusion extend from the uvula posteriorly to the upper lip anteriorly. The extent of defective fusion varies from bilateral cleft palate and cleft lip to cleft uvula. The causes may be:
1. Genetic abnormality
2. Defective development between the 8th and 12th week of gestation caused by, e.g., nutritional disturbances or hypoxia

PHARYNX

As has already been described (see p. 108) the pharynx is divided for descriptive purpose into three parts, the *nasal*, *pharyngeal* and *laryngeal* parts. Of these only the pharyngeal and laryngeal parts are associated with the alimentary tract. Food passes from the oral cavity to the pharynx then to the oesophagus below, with which it is continuous.

The lining membrane is stratified squamous epithelium, continuous with the lining of the mouth at one end and with the oesophagus at the other.

The middle layer consists of fibrous tissue which becomes thinner towards the lower end.

The outer layer consists of a number of involuntary *constrictor* muscles which are involved in swallowing. When food reaches the pharynx swallowing is no longer under voluntary control.

Blood supply
The blood supply to the pharynx is by several branches of the *facial arteries*. Venous drainage is into the *facial veins* and the *internal jugular veins*.

Nerve supply
This is from the *pharyngeal plexus* and consists of parasympathetic and sympathetic nerves. Parasympathetic supply is mainly by the *glossopharyngeal* and *vagus nerves* and sympathetic from the *cervical ganglia* (see p. 233).

DISEASES OF THE PHARYNX

INFECTIONS
TONSILLITIS

Viruses and *Streptococcus pyogenes* are common causes of inflammation of the palatine tonsils, palatine arches and walls of the pharynx. Severe infection may lead to suppuration and abscess formation (*quinsy*). Following acute tonsillitis, swelling subsides and the tonsil returns to normal but repeated infection may lead to chronic inflammation, fibrosis and permanent enlargement. Toxins from tonsillitis caused by *Streptococcus pyogenes* are associated with the development of rheumatic fever and glomerulonephritis.

DIPHTHERIA

This is an infection of the pharynx, caused by *Corynebacterium diphtheriae*, that may extend to the larynx. A thick fibrous membrane forms over the area and may obstruct the airway. Powerful exotoxins may severely damage cardiac and skeletal muscle, the liver, kidneys and adrenal glands. Where immunisation is widespread diphtheria is now comparatively rare.

SALIVARY GLANDS (Fig. 9:17)

There are three pairs of *compound racemose glands* which pour their secretions into the mouth. They are:
2 parotid
2 submandibular
2 sublingual

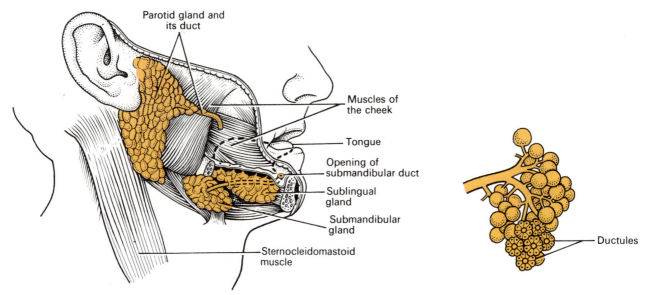

Figure 9:17 The positions of the salivary glands. *Right*: an enlargement of part of a gland.

PAROTID GLANDS

These are situated one on each side of the face just below the external acoustic meatus. Each gland has a *parotid duct* opening into the mouth at the level of the second upper molar tooth.

SUBMANDIBULAR GLANDS

These lie one on each side of the face under the angle of the jaw. The two *submandibular ducts* open on the floor of the mouth, one on each side of the frenulum of the tongue.

SUBLINGUAL GLANDS

These glands lie under the mucous membrane of the floor of the mouth in front of the submandibular glands. They have numerous small ducts that pierce the mucous membrane of the floor of the mouth.

Structure of the salivary glands
The glands are all surrounded by a *fibrous capsule*. They consist of a number of *lobules* made up of small alveoli lined with *secretory cells*. The secretions are poured into small ducts which join up to form larger ducts leading into the mouth.

Nerve supply
All the glands are supplied by parasympathetic and sympathetic nerve fibres.

    Parasympathetic supply — stimulates secretion
    Sympathetic supply   — depresses secretion

Blood supply
Arterial supply is by various branches from the *external carotid arteries* and venous drainage is into the *external jugular veins*.

Saliva
This is the combined secretions from the salivary glands and the small mucus-secreting glands of the lining of the oral cavity. It consists of:
    Water
    Mineral salts
    Enzyme; salivary amylase (ptyalin)
    Mucus

DISEASES OF THE SALIVARY GLANDS

MUMPS

This is an acute inflammatory condition of the salivary glands, especially the parotids. It is caused by the mumps virus, one of the para-influenza group. The virus is inhaled in infected droplets and during the 18 to 21 day incubation period viruses multiply in the lungs before spreading to the salivary glands. They may also spread to:

1. The pancreas, causing pancreatitis
2. The testes, causing orchitis after puberty and sometimes atrophy of the glands and sterility

CALCULUS FORMATION

Calculi are formed in the salivary glands by the crystalisation of mineral salts in saliva. They may partially or completely block the ducts, leading to swelling of the gland, a predisposition to infection and, in time, atrophy. The causes is not known.

TUMOURS

MIXED TUMOURS (Pleomorphic salivary adenoma)

This tumour consists of epithelial and connective tissue cells and occurs mainly in the parotid gland. A second tumour may develop in the same gland several years after the first has been removed. It rarely undergoes malignant change.

CARCINOMA

Malignant tumours occur in any salivary gland or duct. Some forms have a tendency to infiltrate nerves in the surrounding tissues, causing severe pain. Lymph spread is to the cervical nodes.

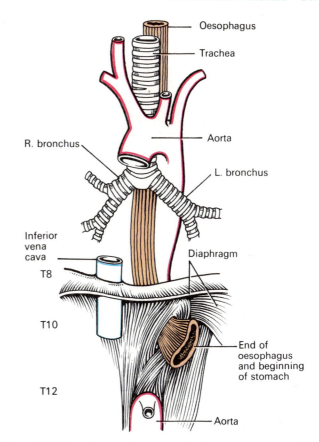

Figure 9:18   Oesophagus and some associated structures.

OESOPHAGUS (Fig. 9:18)

The oesophagus or gullet is the first part of the alimentary tract to which the general plan described previously applies (see p. 138). It is about 25 cm long and about 2 cm in diameter. It lies in the median plane in the thorax in front of the vertebral column behind the trachea and the heart. It is continuous with the pharynx above and just below the diaphragm it joins the stomach. It passes between muscle fibres of the diaphragm behind the central tendon at the level of the 10th thoracic vertebra. Immediately the oesophagus passes through the diaphragm it curves upwards before becoming the stomach. This sharp angle is believed to be one of the factors which prevents the regurgitation (backward flow) of gastric contents into the oesophagus.

STRUCTURE

There are four layers of tissue as described in the general plan. As the oesophagus is almost entirely in the thorax the outer covering consists of *elastic fibrous tissue*. The proximal third is lined by stratified squamous epithelium and the distal third by columnar epithelium. The middle third is lined by a mixture of the two.

Blood supply

Arterial. The thoracic part of the oesophagus is supplied mainly by the oesophageal arteries, branches from the aorta. The abdominal part is supplied by branches from the inferior phrenic arteries and the left gastric branch of the coeliac artery.

Venous drainage. From the thoracic part venous drainage is into the azygos and hemiazygos veins. The abdominal part drains into the left gastric vein. There is a venous plexus at the distal end that links the upward and downward venous drainage, i.e., the general and portal circulations.

Nerve supply

Sympathetic and parasympathetic nerves terminate in the myenteric and submucosal plexuses. Parasympathetic fibres are branches of the vagus nerves (see Fig. 9:5).

FUNCTIONS OF THE MOUTH, PHARYNX, OESOPHAGUS AND SALIVARY GLANDS

Digestion in the mouth

When food is taken into the mouth it is masticated or chewed by the teeth and moved round the mouth by the

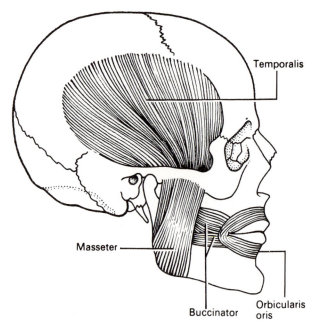

Figure 9:19 The muscles used when chewing.

tongue and muscles of the cheeks (Fig. 9:19). It is mixed with saliva and formed into a soft mass or *bolus* ready for *deglutition* or swallowing. The length of time that food remains in the mouth depends, to a large extent, on the consistency of the food. Some foods need to be chewed longer than others before the individual feels that the mass is ready for swallowing.

Functions of saliva

Digestion. The enzyme salivary amylase acts on cooked starches (polysaccharides), changing them to disaccharide *maltose*. The pH is between 6.2 and 7.4 depending on the rate of flow. The more rapid the flow of saliva the higher the pH. The optimum pH for salivary amylase action is 6.8 (slightly acid). Enzyme action continues after the bolus is swallowed, until it is finally inhibited by the strongly acid reaction of gastric juices—pH 1.5 to 1.8.

Lubrication of food. Dry food entering the mouth is moistened and lubricated by saliva before it can be made into a bolus ready for swallowing.

Cleansing and lubricating. An adequate flow of saliva is necessary to cleanse the mouth and keep its tissues soft and pliable.

Taste. The taste buds are stimulated by particles present in the food which are dissolved in water. Dry foods stimulate the sense of taste only after thorough mixing with saliva.

Secretion of saliva

The flow of saliva is controlled by sympathetic and parasympathetic nerve supply. Parasympathetic stimulation causes an increase in the secretion of saliva and sympathetic stimulation has an inhibitory effect. These are two types of reflex action involved in salivation.

Unconditioned reflex. This is the automatic response to the presence of an object in the mouth. It may be demonstrated by placing something in a child's mouth before he or she has learned what substances satisfy hunger.

Conditioned reflex. This too is an automatic response but it is one that the individual has *learned* from previous experience. The sight, smell and even the thought of appetising food results in salivation or 'mouth watering'. This type of salivation occurs on anticipation of food.

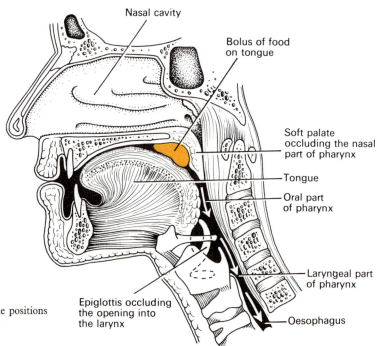

Figure 9:20 Section of the face and neck showing the positions of structures during swallowing.

Deglutition or swallowing (Fig. 9:20)

This occurs in three stages after mastication is complete and the bolus has been formed. It is initiated voluntarily but later it is under autonomic nerve control.

1. The mouth is closed and the voluntary muscles of the tongue and cheeks push the bolus backwards into the pharynx.

2. The muscles of the pharynx propel the bolus down into the oesophagus. All other routes that the bolus could possibly take are closed. The soft palate rises up and occludes the nasal part of the pharynx; the tongue and the pharyngeal folds close the way back into the mouth; and the larynx is lifted up and forward so that its opening is occluded by the overhanging epiglottis coming into contact with the base of the tongue.

3. The presence of the bolus in the pharynx stimulates a wave of peristalsis which propels the bolus through the oesophagus to the stomach (Fig. 9:21).

DISEASES OF THE OESOPHAGUS

OESOPHAGEAL VARICES (Fig. 9:22)

Varicosities of the venous plexus at the distal end of the oesophagus occur when there is impairment of the flow of portal vein blood through the liver. This causes venous congestion in its tributaries, including the gastric veins draining blood from the oesophageal plexus. The anastomotic veins between the venous plexus and the azygos vein become distended, causing valvular incompetence. The distended thin-walled veins may be ruptured by increased venous pressure or by bulky food passing through the oesophagus. This may result in steady slow blood loss causing iron-deficiency anaemia, or in severe acute haemorrhage.

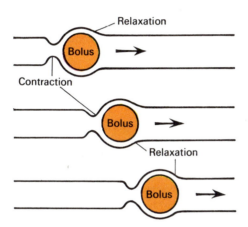

Figure 9:21 Illustration of the movement of the bolus through the oesophagus by peristalsis.

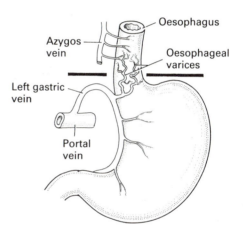

Figure 9:22 Oesophageal varices.

Peristaltic waves of contraction only pass along the oesophagus after swallowing. Otherwise the walls are relaxed. It is closed at its proximal end by a band of pharyngeal muscle and at its distal end by a 'physiological sphincter', the lower oesophageal sphincter (LOS). On X-ray the LOS can be seen to relax ahead of a peristaltic wave but the mechanism is not clearly understood. It prevents regurgitation of acid gastric contents into the oesophagus which would damage the columnar epithelial lining. Other factors that help to prevent reflux of gastric juice include:

1. The attachment of the stomach to the diaphragm by peritoneum
2. The maintenance of an acute angle between the oesophagus and the fundus of the stomach, i.e., an acute cardio-oesophageal angle

The walls of the oesophagus are lubricated by mucus which assists the passage of the bolus during the peristaltic contraction of the muscular wall.

INFLAMMATORY CONDITIONS

PEPTIC REFLUX OESOPHAGITIS

This condition is caused by regurgitation of acid gastric juice into the oesophagus, causing irritation and ulceration. Haemorrhage occurs when blood vessels are eroded. Persistent reflux leads to chronic inflammation and if damage is extensive, secondary healing with fibrosis occurs. Shrinkage of ageing fibrous tissue may cause stricture of the oesophagus.

The reflux of gastric contents is associated with:

1. Increase in the intra-abdominal pressure, e.g., in pregnancy and obesity
2. High acid content of gastric juice
3. Low levels of secretion of the hormone gastrin, leading to reduced sphincter action at the lower end of the oesophagus
4. The presence of hiatus hernia (see p. 176).

SWALLOWING CAUSTIC MATERIALS

When swallowed, caustic materials burn the walls of the oesophagus causing an inflammatory reaction. The extent of the damage depends on the concentration and amount swallowed. Following severe injury, healing causes fibrosis, and there is a risk of oesophageal stricture developing later, as the fibrous tissue shrinks.

MICROBIAL INFECTIONS

THRUSH

This is caused by *Candida albicans*, usually spread to the oesophagus from the mouth and pharynx. It occurs most commonly in bottle-fed babies (see p. 146).

ACHALASIA

In this disease the cardiac end of the oesophagus is constricted and there is dilatation and muscle hypertrophy proximal to the stricture. This is caused by peristaltic waves of contraction that are not preceded by the normal waves of relaxation. There is an abnormal autonomic nerve supply to the oesophageal muscle but the cause is not known.

SPONTANEOUS RUPTURE OF OESOPHAGUS

This may occur, usually at the distal end, if the oesophagus is suddenly distended during an acute vomiting attack. Gastric contents pass into the mediastinum, causing acute inflammation. The cause of weakness in the wall of the oesophagus is not known.

TUMOURS

Benign tumours rarely occur.

MALIGNANT TUMOURS

These occur more often in males than females. The most common sites are the distal end of the oesophagus and at the levels of the larynx and bifurcation of the trachea. The tumours are mainly of two types, both of which may eventually lead to oesophageal obstruction.

Scirrhous (fibrous) tumours. These spread round the circumference and along the oesophagus. They cause thickening of the wall and loss of elasticity.

Soft tissue tumours. These grow into the lumen and spread along the wall.

The causes of malignant change are not known but may be associated with diet and the temperature of swallowed food.

CONGENITAL ABNORMALITIES

In most common congenital abnormalities of the oesophagus:
1. The lumen is narrow or completely blocked
2. There is an opening between the oesophagus and the trachea through which milk or regurgitated gastric contents are aspirated

One or both abnormalities may be present. The causes of these developmental deficiencies are not known.

STOMACH AND GASTRIC JUICE

The stomach is a J-shaped dilated portion of the alimentary tract situated in the epigastric, umbilical and left hypochondriac regions of the abdominal cavity.

Organs associated with the stomach (Fig. 9:23)

Anteriorly — left lobe of liver and anterior abdominal wall

Posteriorly — abdominal aorta, pancreas, spleen, left kidney and adrenal gland

Superiorly — diaphragm, oesophagus and left lobe of liver

Inferiorly — transverse colon and small intestine

To the left — diaphragm and spleen

To the right — liver and duodenum

STRUCTURE (Fig. 9:24)

The stomach is continuous with the oesophagus at the *cardiac orifice*, and with the duodenum at the *pyloric orifice*. It is described as having two curvatures. The *lesser curvature* is short, lies on the posterior surface of the stomach and is the downwards continuation of the posterior wall of the oesophagus. Just before the pyloric sphincter it curves upwards to complete the J shape. Where the oesophagus joins the stomach the anterior part angles acutely upwards, curves downwards forming the greater curvature then slightly upwards towards the pyloric orifice.

The part of the stomach above the cardiac orifice is the *fundus*, the main part is the *body* and the lower part, *pyloric antrum*. At the distal end of the pyloric antrum there is a sphincter, the *pyloric sphincter*, guarding the opening between the stomach and the duodenum. When the stomach is inactive the pyloric sphincter is relaxed and open.

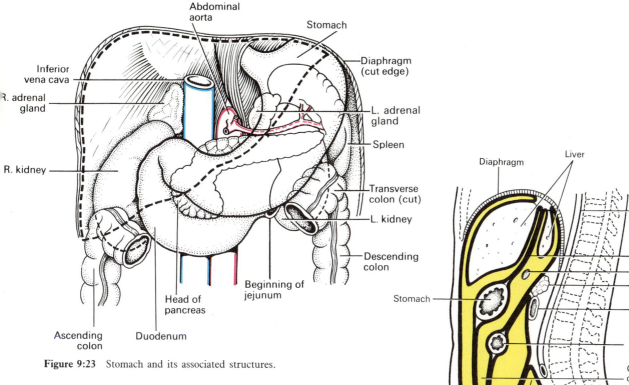

Figure 9:23   Stomach and its associated structures.

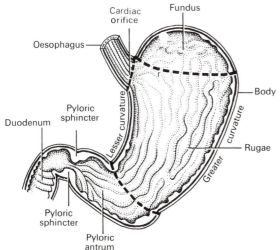

Figure 9:24   Longitudinal section of the stomach.

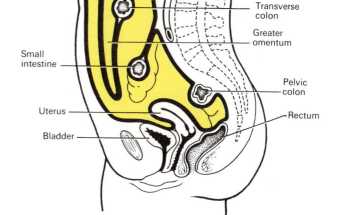

Figure 9:25A   The peritoneum viewed from the side.

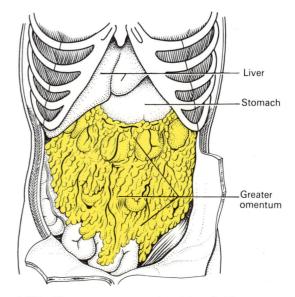

Figure 9:25B   The greater omentum viewed from the front.

WALLS OF THE STOMACH

The four layers of tissue described in the *general plan* of the alimentary canal are to be found in the stomach but with some modifications.

Peritoneum

From Figure 9:25 it will be seen that the fold of peritoneum which attaches the stomach to the posterior abdominal wall extends beyond the greater curvature of the stomach. This is the *greater omentum*. It is free at its distal end and hangs down in front of the abdominal organs like an apron.

The greater omentum stores fat, is richly supplied with blood and lymph vessels and contains a considerable number of lymph nodes. It has the ability to isolate an area of slowly developing inflammation, such as chronic appendicitis, preventing the spread of infection to the peritoneal cavity as a whole.

Muscle layer (Fig. 9:26)

This consists of *three layers* of smooth muscle fibres. The outer layer has *longitudinal fibres*, the middle layer has *circular fibres* and the inner layer, *oblique fibres*. This arrangement allows for the churning motion characteristic of gastric activity, as well as the peristaltic movement. Circular muscle is strongest in the pyloric antrum and sphincter.

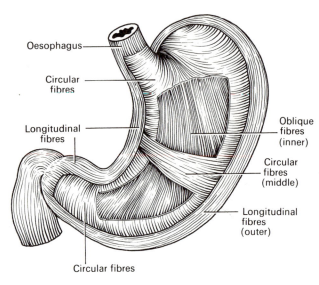

Oesophagus

Circular fibres

Longitudinal fibres

Oblique fibres (inner)

Circular fibres (middle)

Longitudinal fibres (outer)

Circular fibres

Figure 9:26   The muscle fibres of the stomach wall. Sections have been removed to show the three layers.

Mucous membrane lining

When the stomach is empty the mucous membrane lining is thrown into longitudinal folds or *rugae*, and when full the rugae are 'ironed out' and the surface has a smooth, velvety appearance. There are numerous *gastric glands* situated below the surface in the mucous membrane. They consist of specialised cells that secrete *gastric juice* into the stomach.

BLOOD SUPPLY

Arterial blood is supplied to the stomach by branches of the coeliac artery and venous drainage is into the portal vein. Figures 9:7 and 9:8 give details of the names of these vessels.

NERVE SUPPLY

The sympathetic supply to the stomach is mainly from the coeliac plexus and the parasympathetic supply is from the *vagus nerves*. Sympathetic stimulation reduces the motility of the stomach and the secretion of gastric juice; vagal stimulation has the opposite effect (see Fig. 9:5, p. 141).

GASTRIC JUICE AND FUNCTIONS OF THE STOMACH

The size of the stomach varies with the amount of food it contains. When a meal has been eaten the food accumulates in the stomach in layers, the last part of the meal remaining in the fundus for some time. Mixing with the gastric juice takes place gradually and it may be some time before the food is sufficiently acidified to stop the action of ptyalin.

Gastric muscle contraction consists of a churning movement that breaks down the bolus and mixes it with gastric juice, and peristaltic waves that propel the stomach contents towards the pylorus. When the stomach is active the pyloric sphincter closes. Strong peristaltic contraction of the pyloric antrum forces gastric contents, after they are sufficiently liquefied, through the pylorus into the duodenum is small spurts.

GASTRIC JUICE

This is secreted by special secretory glands in the mucosa and consists of:

Water

Mineral salts } secreted by gastric glands

Mucus secreted by cells in the glands and on the stomach surface

Hydrochloric acid

Intrinsic factor } secreted by *parietal cells* in the gastric glands

Enzymes; pepsinogens secreted by *chief (peptic or zymogen) cells* in the glands

Functions of gastric juice

1. *Water* further liquefies the food swallowed.
2. *Hydrochloric acid*:
   a. acidifies the food and stops the action of ptyalin
   b. kills many microbes which may be harmful to the body.
   c. provides the acid environment needed for effective digestion by pepsin
3. *Pepsinogen* is activated to *pepsin* by hydrochloric acid and pepsin already present in the stomach lumen. It begins the digestion of proteins, breaking them up into smaller molecules. Pepsin acts most effectively at pH 1.5 to 3.5.
4. *Intrinsic factor* (a protein compound) is neccessary for the absorption of vitamin B_{12}, the *anti-anaemic factor* (see p. 45).
5. *Mucus* prevents mechanical injury to the stomach wall by lubricating the contents. It prevents chemical injury by acting as a barrier between the stomach wall and

the other constituents of gastric juice. Hydrochloric acid is present in potential damaging concentrations and pepsin digests protein.

Secretion of gastric juice

There is always a small quantity of gastric juice present in the stomach, even when it contains no food. This is known as *fasting juice*.

There are three phases of secretion of gastric juice (Fig. 9:27).

Cephalic phase. This flow of juice occurs *before* food reaches the stomach and is due to reflex stimulation of the vagus nerves initiated by the sight, smell or taste of food. When the vagus nerves have been cut this phase of gastric secretion stops.

Gastric phase. When stimulated by the presence of food the stomach produces *gastrin*, a hormone which passes directly into the circulating blood. Gastrin, circulating in the blood which supplies the stomach, stimulates the gastric glands to produce more gastric juice. In this way the secretion of digestive juice is continued after the completion of the meal and the end of the cephalic phase.

Intestinal phase. When the partially digested contents of the stomach reach the small intestine, a hormone complex *enterogastrone** is produced which slows down the secretion of gastric juice and reduces gastric motility. By slowing the

---

* Enterogastrone has been described as any hormone or combination of hormones released by the intestine that inhibits gastric secretion.

emptying rate of the stomach, the contents of the duodenum become more thoroughly mixed with bile and pancreatic juice. This phase of gastric secretion is most marked when the meal has had a high fat content.

FUNCTIONS OF THE STOMACH

1. The stomach acts as a temporary reservoir for food, allowing the digestive enzymes time to act.

2. It produces *gastric juice* which begins the chemical digestion of proteins.

3. Muscular action mixes the food with gastric juice then moves it on to the small intestine. When the contents of the pyloric end of the stomach have reached a suitable degree of acidity and liquefaction the pyloric antrum forces small jets of gastric contents through the pyloric sphincter into the duodenum. The rate at which the stomach empties depends to a large extent on the type of food eaten. A carbohydrate meal leaves the stomach in 2 to 3 hours, a protein meal remains longer and a fatty meal remains in the stomach longest. The stomach contents entering the duodenum are called *chyme*.

4. Absorption takes place in the stomach to a limited extent. Water, alcohol and some drugs are absorbed through the walls of the stomach into the venous circulation.

5. Although iron absorption takes place in the small intestine it is dissolved out of foods most effectively in the presence of hydrochloric acid in the stomach.

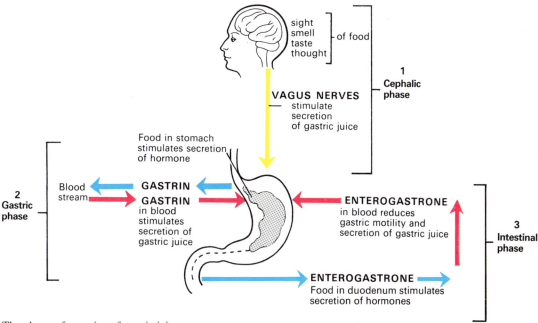

Figure 9:27 The phases of secretion of gastric juices.

DISEASES OF THE STOMACH

GASTRITIS

This is a common condition which occurs when the amount of mucus in the stomach is insufficient to protect the surface epithelium from the destructive effects of hydrochloric acid. It may be acute or chronic.

ACUTE GASTRITIS

Gastritis occurs with varying degrees of severity. The most severe form is *acute haemorrhagic gastritis*. When the surface epithelium of the stomach is exposed to acid gastric juice the cells absorb hydrogen ions which increase their internal acidity, disrupt their metabolic processes and trigger the inflammatory reaction. The causes of acute gastritis include:

1. Regular prolonged use of aspirin and other anti-inflammatory drugs, especially the non-steroids
2. Regular excessive alcohol consumption
3. Food poisoning caused by, e.g., *Staphylococcus aureus*, *Salmonella paratyphi*, viruses
4. Heavy cigarette smoking
5. Treatment with cytotoxic drugs and X-radiation
6. Ingestion of corrosive poisons, acids and alkalis
7. Regurgitation of bile into the stomach

The outcome depends on the extent of the damage. In many cases recovery is uneventful after the cause is removed. In the most severe forms there is ulceration of the mucosa, haemorrhage, perforation of the stomach wall and peritonitis. Where there has been extensive tissue damage, healing is by fibrosis causing reduced elasticity and peristalsis.

CHRONIC GASTRITIS

Chronic gastritis is a milder longer lasting form. It may follow repeated acute attacks or be an autoimmune disease.

AUTOIMMUNE CHRONIC GASTRITIS

This is a progressive form of the disease. There are destructive inflammatory changes that begin on the surface of the mucous membrane and may extend to affect its whole thickness, including the gastric glands. When this stage is reached the secretion of digestive enzymes, hydrochloric acid and intrinsic factor are markedly reduced. The antigens are the *acid-secreting cells* and the *intrinsic factor* they secrete. When these cells are destroyed the inflammation subsides. The initial causes of the autoimmunity are not known but there is a familial predisposition and an association with chronic thyroiditis, thyrotoxicosis and atrophy of the adrenal glands. Secondary effects include:

1. Pernicious anaemia due to lack of intrinsic factor
2. Impairment of digestion due to lack of enzymes
3. Microbial infection due to lack of hydrochloric acid

PEPTIC ULCERATION

Ulceration of the gastrointestinal mucosa is caused by the action of acid gastric juice and may be viewed as an extension of the cell damage found in acute gastritis. The most common sites for ulcers are the stomach and the first few centimetres of the duodenum. More rarely they occur in the oesophagus, following reflux of gastric juice, and round the anastomosis of the stomach and small intestine, following gastrectomy. The underlying causes are not known but if factors associated with the maintenance of healthy mucous membrane are defective, acid gastric juice gains access to the epithelium, causing the initial cell damage that leads to ulceration. The main factors are: normal blood supply, mucus secretion and cell replacement.

Blood supply
Reduced blood flow and ischaemia may be caused by excessive cigarette smoking and stress, physical or mental. In a stressful situation there is an increase in the secretion of the hormones noradrenaline and adrenaline and these cause constriction of the blood vessels supplying the alimentary tract.

Secretion of mucus
The composition and the amount of mucus may be altered, e.g.:

1. By regular and prolonged use of aspirin and other anti-inflammatory drugs
2. By the reflux of bile acids and salts
3. In chronic gastritis

Epithelial cell replacement
There is normally a rapid turnover of gastric and intestinal epithelial cells. This may be reduced:

1. By raised levels of steroid hormones, e.g., in response to stress or when they are used as drugs
2. In chronic gastritis
3. By irradiation and the use of cytotoxic drugs

ACUTE PEPTIC ULCERS

The tissues involved are the mucous and submucous layers and the ulcers may be single or multiple. They are found in many sites in the stomach and in the first few centimetres of the duodenum. The underlying causes are unknown but their development is often associated with severe stress, e.g., severe illness, shock, burns, severe emotional disturbance and following surgery. Healing without the formation of fibrous tissue usually occurs when the cause of the stress is removed.

CHRONIC PEPTIC ULCERS

These ulcers penetrate through the epithelial and muscle layers of the stomach wall and may include the adjacent pancreas or liver. In the majority of cases they occur singly in the pyloric antrum of the stomach and in the duodenum. Occasionally there are two ulcers facing each other in the duodenum, called kissing ulcers. Healing occurs with the formation of fibrous tissue and subsequent shrinkage may cause:

1. Stricture of the lumen of the stomach
2. Stenosis of the pyloric sphincter
3. Adhesions to adjacent structures, e.g., pancreas, liver, transverse colon

HAEMORRHAGIC ULCERS

In some cases acid gastric juice causes widespread damage to the superficial cells of the mucosa. Many tiny ulcers form, leading to multiple capillary bleeding points.

PERFORATION OF PEPTIC ULCERS

When an ulcer erodes through the full thickness of the wall of the stomach or duodenum their contents enter the peritoneal cavity, causing acute peritonitis. Infected inflammatory material may collect under the diaphragm, forming a subphrenic abscess and the infection may spread through the diaphragm to the pleural cavity.

TUMOURS OF THE STOMACH

Benign tumours of the stomach rarely occur.

MALIGNANT TUMOURS

This is a relatively common form of malignancy and it occurs more frequently in men than women. The local growth of the tumour gradually destroys the normal tissue so that achlorhydria and pernicious anaemia are frequently secondary features. The causes have not been established but there appears to be:

1. A familial predisposition
2. An association with environmental factors, e.g., foods, methods of cooking
3. The presence of other diseases, e.g., chronic gastritis, chronic ulceration, pernicious anaemia

Spread of gastric carcinoma

Local spread. These tumours spread locally to the remainder of the stomach, the omentum, liver and pancreas. The spleen is seldom affected.

As the tumour grows, the surface may ulcerate and become infected, especially when achlorhydria develops.

Lymphatic spread. This occurs early in the disease. At first the spread is within the lymph channels in the stomach wall then to lymph nodes round the stomach, in the mesentery, omentum and walls of the small intestine and colon.

Blood spread. The common sites for blood-spread metastases are the liver, lungs, brain and bones.

Peritoneal spread. When a tumour includes the full thickness of the stomach wall, small groups of cells may break off and spread throughout the peritoneal cavity. Metastases may develop in any tissue in the abdominal or pelvic cavity where the fragments settle.

CONGENITAL PYLORIC STENOSIS

In this condition there is spasmodic constriction of the pyloric sphincter. In an attempt to overcome the spasms, hypertrophy of the muscle of the pyloric antrum develops, causing obstruction of the pylorus 2 to 3 weeks after birth. The reason for the excess stimulation or neuromuscular abnormality of the pylorus is not known but there is a familiar tendency and it is more common in males.

PANCREAS (Fig. 9:28)

The pancreas is a pale grey gland weighing about 60 grams. It is about 12 to 15 cm long and is situated in the *epigastric* and *left hypochondriac* regions of the abdominal cavity. It consists of a broad head, a body and a narrow tail. The head lies in the curve of the duodenum,

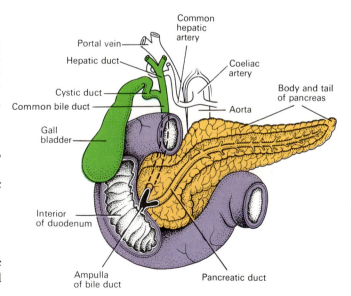

Figure 9:28   The pancreas in relation to the duodenum and biliary tract. Part of the anterior wall of the duodenum removed.

the body behind the stomach and the tail lies in front of the left kidney and just reaches the spleen. The abdominal aorta and the inferior vena cava lie behind the gland.

STRUCTURE

The pancreas consists of a large number of *lobules* made up of small *alveoli*, the walls of which consist of *secretory cells*. Each lobule is drained by a tiny duct and these unite eventually to form the *pancreatic duct* that extends the whole length of the gland and opens into the duodenum at its midpoint. Just before entering the duodenum the pancreatic duct joins the *common bile duct* to form the *ampulla of the bile duct*. The duodenal opening of the ampulla is controlled by the *sphincter of Oddi*.

Islets of Langerhans

Distributed throughout the substances of the pancreas there are little collections of the different type of cell. These cells form what are known as the *islets of Langerhans*. They secrete the hormones *insulin* and *glucagon* that pass directly into the blood (Ch. 14).

FUNCTIONS

The function of the alveoli of the pancreas is to produce *pancreatic juice* containing enzymes that digest carbohydrates, proteins and fats (see p. 168).

Pancreatic juice

Pancreatic juice enters the duodenum at its midpoint. It is strongly alkaline (pH 8) and consists of:

> Water
> Mineral salts
> Enzymes: amylase, lipase and precursors of the peptidases, e.g., trypsinogen, chymotrypsinogen

Secretion is stimulated by the hormones *secretin* and *cholecystokinin-pancreozymin* secreted by cells in the intestinal wall when acid chyme enters the duodenum.

BLOOD SUPPLY

The splenic and mesenteric arteries supply arterial blood to the pancreas and the venous drainage is by the veins of the same names that join other veins to form the portal vein.

NERVE SUPPLY

As in the alimentary tract, parasympathetic stimulation increases the secretion of pancreatic juice and sympathetic stimulation depresses it.

DISEASES OF THE PANCREAS

ACUTE PANCREATITIS

Protein-splitting enzymes are activated while still in the pancreas and digest the parenchymal cells. The severity of the disease is directly related to the amount of pancreatic tissue destroyed.

Mild forms may damage only those cells near the ducts.

Severe forms cause widespread damage with necrosis and haemorrhage. Common complications include infection, suppuration, and local venous thrombosis. Pancreatic enzymes enter and circulate in the blood, causing similar damage to other structures.

The causes of acute pancreatitis are not clear but known predisposing factors are gallstones and alcoholism. When a gallstone obstructs the ampulla there is reflux of bile into the pancreas and the spread of infection from cholangitis. Alcoholism is known to be present in a large number of cases but the mechanisms by which it causes pancreatitis are not clear. Other associated conditions include:

1. Cancer of the ampulla or head of pancreas
2. Virus infections
3. Chronic renal failure
4. Renal transplantation
5. Hyperthyroidism
6. Hypothermia
7. Drugs, e.g., corticosteroids, cytotoxic agents
8. Mumps

CHRONIC RELAPSING PANCREATITIS

This is due to repeated attacks of acute pancreatitis and is frequently associated with fibrosis and distortion of the main pancreatic duct.

TRUE CHRONIC PANCREATITIS

This condition develops gradually and is independent of the acute disease. There is obstruction of the tiny acinar ducts by protein material secreted by the acinar cells. This eventually leads to the formation of cysts which may rupture into the peritoneal cavity. Intact cysts may cause obstruction of the:

1. Common bile duct, causing jaundice
2. Portal vein, causing venous congestion in the organs drained by its tributaries

The causes of these changes are not known but they are associated mainly with heavy wine drinking.

CYSTIC FIBROSIS (Fibrocystic disease)

This is one of the most common genetic diseases. It is estimated that almost 20% of people carry the abnormal recessive gene which must be present in *both parents* to cause the disease.

The secretions of all exocrine glands have abnormally high viscosity but the most severely affected are those of the pancreas, intestines, biliary tract, lungs and the reproductive system in the male. Sweat glands secrete abnormally large amounts of salt during excessive sweating. In the pancreas highly viscous mucus is secreted by the walls of the ducts and causes obstruction, parenchmal cell damage, the formation of cysts and defective enzyme secretion. In the newborn, intestinal obstruction may be caused by a plug of meconium and viscid mucus, leading to perforation and meconium peritonitis which is often fatal. In older children:

1. Digestion of food and absorption of nutrients is impaired

2. There may be obstruction of bile ducts in the liver, causing cirrhosis

3. Bronchitis, bronchietasis and pneumonia may develop

TUMOURS OF THE PANCREAS

Benign tumours of the pancreas are very rare.

Malignant tumours are relatively common. They occur most frequently in the head of the pancreas, obstructing the flow of bile and pancreatic juice into the duodenum.

Jaundice and acute pancreatitis usually develop. Tumours is the body and tail of the gland rarely cause symptoms until the disease is advanced. Metastatases are often recognised before the primary tumour. The causes of the malignant changes are not known but it is believed that there may be an associated with:

1. Cigarette smoking
2. Alcohol consumptiom
3. Diet high in fats and carbohydrates

LIVER

The liver is the largest gland in the body, weighing between 1 and 2.3 kg. It is situated in the upper part of the abdominal cavity occupying the greater part of the *right hypochondriac region*, part of the *epigastric region* and extending into the *left hypochondriac region*. Its upper and anterior surfaces are smooth and curved to fit the under surface of the diaphragm (Fig. 9:29); its posterior surface is irregular in outline (Fig. 9:30).

Organs associated with the liver
Superiorly and
   anteriorly — diaphragm and anterior abdominal wall
Inferiorly — stomach, bile ducts, duodenum, right colic flexure of the colon, right kidney and adrenal gland

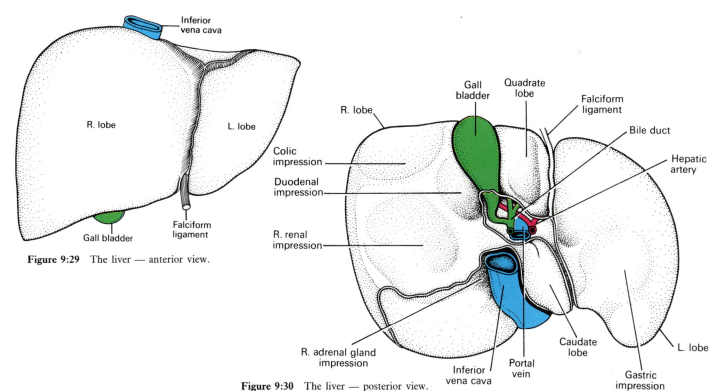

Figure 9:29 The liver — anterior view.

Figure 9:30 The liver — posterior view.

Posteriorly — oesophagus, inferior vena cava, aorta, gall bladder, vertebral column and diaphragm

Laterally — lower ribs and diaphragm

The liver is enclosed in a thin capsule and incompletely covered by a layer of peritoneum. Folds of peritoneum form supporting ligaments attaching the liver to the inferior surface of the diaphragm. It is held in position partly by these ligaments and partly by the pressure of the organs in the abdominal cavity.

The liver is described as having four lobes. The two most obvious are the large *right lobe* and the smaller, wedge-shaped, *left lobe*. The other two, the *caudate* and *quadrate* lobes, are areas on the posterior surface (Fig. 9:30).

The portal fissure

This is the name given to the part on the posterior surface of the liver where various structures enter and leave the gland.

The *portal vein* enters, carrying blood from the stomach, spleen, pancreas and the small and large intestines.

The *hepatic artery* enters, carrying arterial blood. It is a branch from the coeliac artery which is a branch from the abdominal aorta.

Nerve fibres, sympathetic and parasympathetic.

The *right* and *left hepatic ducts* leave, carrying bile from the liver to the gall bladder.

Lymph vessels leave the liver, draining some lymph to abdominal and some to thoracic nodes.

BLOOD SUPPLY

The hepatic artery and the portal vein take blood to the liver. Hepatic veins, varying in number, leave the posterior surface and immediately enter the inferior vena cava just below the diaphragm.

STRUCTURE

The lobes of the liver are made up of tiny lobules just visible to the naked eye. These lobules are *hexagonal* in outline and are formed by cubical-shaped cells, the *hepatocytes*, arranged in *pairs of columns* radiating from a *central vein*. Between two pairs of columns of cells there are *sinusoids* (blood vessels with incomplete walls) containing *a mixture of blood* from the tiny branches of the *portal vein* and *hepatic artery* (Fig. 9:31). This arrangement allows the arterial blood and venous blood (with a high concentration of nutritional materials) to mix and come into close contact with the liver cells. Some cells, lining the sinusoids, are *hepatic macrophages* (Kupffer cells of the reticuloendothelial system).

Blood drains from the sinusoids into *central* or *centrilobular veins*. These then join with veins from other lobules, forming larger veins, until eventually they become the

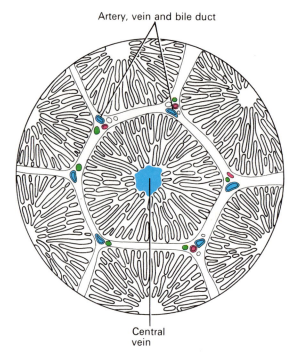

Figure 9:31A   Diagram of a magnified transverse section of liver lobules.

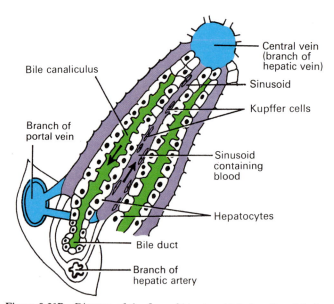

Figure 9:31B   Diagram of the flow of blood and bile in a liver lobule.

hepatic veins which leave the liver and empty blood into the inferior vena cave just below the diaphragm. Figure 9:32 shows the system of blood flow through the liver. One of the functions of the liver is to secrete *bile*. In Figure 9:31 it will be seen that *bile canaliculi* run between the columns of liver cells. This means that each column has a blood sinusoid on one side and a bile canaliculus on the other. The canaliculi join up to form larger bile canals until eventually they form the *right and left hepatic ducts* which drain bile from the liver.

Lymphoid tissue and a system of lymph vessels are present in each lobule.

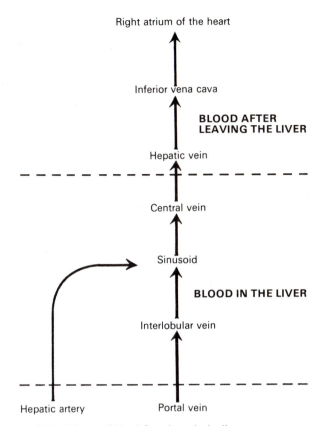

Right atrium of the heart

Inferior vena cava

**BLOOD AFTER
LEAVING THE LIVER**

Hepatic vein

Central vein

Sinusoid

BLOOD IN THE LIVER

Interlobular vein

Hepatic artery   Portal vein

Figure 9:32 Scheme of blood flow through the liver.

FUNCTIONS

The liver is an extremely active organ. Some of its functions have already been described, therefore they will only be mentioned here.

1. *Deaminates amino acids*
   a. Removes the nitrogenous portion from the amino acids, not required for the formation of new protein, and forms *urea* from this nitrogenous portion which is excreted in urine.
   b. Breaks down the nucleoprotein of worn-out cells of the body to form *uric acid* which is excreted in the urine
2. *Converts glucose to glycogen* in the presence of *insulin*, and changes liver glycogen to glucose in the presence of *glucagon*.
3. *Desaturates fat*, i.e., converts stored fat to a form in which it can be used by the tissues to provide energy.
4. *Produces heat*. The liver uses a considerable amount of energy, has a high metabolic rate and produces a great deal of heat. It is the main heat producing organ of the body.
5. *Secretes bile*. The liver cells synthesise the constituents of bile from the mixed arterial and venous blood in the sinusoids. These include bile salts, bile pigments, cholesterol.

6. *Stores*
   a. vitamin B_{12} (anti-anaemic factor)
   b. fat-soluble vitamins A, D, E, K
   c. iron
7. *Synthesises vitamin A* from carotene, the provitamin found in some plants, e.g., carrots and green leaves of vegetables.
8. *Forms the plasma proteins* and most of the *blood clotting factors* from the available amino acids.
9. *Detoxicates drugs and noxious substances*, such as toxins produced by microbes.
10. *Metabolises ethanol* in alcoholic drinks.
11. *Inactivates hormones*, including insulin, glucagon, cortisol, aldosterone, thyroid and sex hormones.

DISEASES OF THE LIVER

New liver cells develop only when needed to replace damaged cells. Capacity for regeneration is considerable and damage is usually extensive before it is evident. The effects of disease or toxic agents are seen when:

1. Regeneration of hepatocytes (liver cells) does not keep pace with damage, leading to hepatocelluar failure
2. There is a gradual replacement of damaged cells by fibrous tissue, leading to portal hypertension

In most liver disease both conditions are present.

ACUTE PARENCHYMAL DISEASE (Acute hepatitis)

Area of necrosis develop as groups of hepatocytes die and the eventual outcome depends on the size and number of the these areas. Causes of the damage may be a variety of conditions, including:

1. Infections
2. Toxic substances
3. Circulatory disturbances

INFECTIONS

ACUTE VIRAL HEPATITIS

The viruses involved are described as Type A, Type B and Type non-A and non-B. The types are distinguished serologically, i.e., by the antibodies produced to combat the infection. The severity of the ensuing disease caused by the different virus types varies considerably but the pattern is similar. The viruses enter the liver cells, causing degenerative changes by mechanisms not yet understood. An inflammatory reaction ensues, accompanied by production of an exudate containing lymphocytes, plasma cells and granulocytes. There is reactive hyperplasia of the reticuloendothelial Kupffer cells in the walls of the sinusoids.

As groups of cells die, necrotic areas of varying sizes develop, phagocytes remove the necrotic material and the lobules collapse. The basic lobule framework (see Fig. 9:31) becomes distorted and blood vessels develop kinks. These changes interfere with the circulation of blood to the remaining heptocytes and the resultant hypoxia causes further damage. Fibrous tissue develops in the damaged area, and adjacent hepatocytes proliferate. The effect of these changes on the overall functioning of the liver depends on the size of the necrotic areas, the amount of fibrous tissue formed and the extent to which the blood and bile channels are distorted.

Type A virus (Infectious hepatitis)
This virus has only one known serological type. It occurs endemically, affecting mainly children, causing a mild illness. Infection is spread by hands, food, water and fomites contaminated by infected faeces. The incubation period in 15 to 40 days and the viruses are excreted in the faeces for 7 to 14 days before clinical symptoms appear and for about 7 days after. Antibodies develop and immunity persists after recovery. Subclinical disease may occur but carriers do not develop.

Type B virus (Serum hepatitis)
This virus has a number of serological types. Infection occurs at any age, but mostly in adults. The incubation period is 50 to 90 days. The virus enters the blood and is spread by blood and blood products. Antibodies are formed and immunity persists after recovery. Infection usually leads to severe illness lasting 2 to 6 weeks, often followed by a protracted convalescence. Carriers may, or may not, have had clinical disease. Type B virus may cause massive liver necrosis and death. In less severe cases recovery may be complete. In chronic hepatitis which may develop, live viruses continue to circulate in the blood.

Type non-A and non-B hepatitis
These viruses are known to exist because of the occurrence of virus hepatitis that cannot be attributed to Type A or Type B viruses. Serological identity has not yet been established. One know method of spread is transfusion of infected blood. The incubation period is 10 to 90 days. Clinical illness of varying severity occurs and it may be followed by complete recovery or become chronic. Carriers may, or may not, have had clinical disease.

TOXIC SUBSTANCES

Many drugs undergo chemical change in the liver before excretion by it or other organs. They may damage the liver cells in their original form or while in various intermediate stages. Some substances always cause liver damage (predictably toxic) while others only do so when hypersensitivity develops (unpredictably toxic). In both types the extent of the damage depends on the size of the dose and/or the duration of intake. Examples include:

Predictable group (dose related)	Unpredictable group (individual idiosyncrasy)
Chloroform	Phenothiazine compounds
Tetracyclines	Halothane
Cytotoxic drugs	Methyldopa
Anabolic steroids	Phenylbutazone
Alcohol	Indomethazine
Aspirin	Chlorpropamide
Some hydrocarbons	Thiouracil
Some fungi	Sulphonamides

CIRCULATORY DISTURBANCES

The intensely active hepatocytes are particularly vulnerable to damage by hypoxia which is usually due to deficient blood supply caused by:
1. Fibrosis in the liver following inflammation
2. Compression of the portal vein, hepatic artery or vein by a tumour
3. Acute general circulatory failure and shock
4. Venous congestion caused by acute or chronic right-sided heart failure.

CHRONIC HEPATITIS

This is defined as any form of hepatitis which persists for more than 6 months. It may be caused by viruses or drugs, but in some cases the cause is unknown.

Chronic persistent hepatitis
This is a mild, persistent inflammation following acute virus hepatitis. There is usually little or no fibrosis.

Chronic active hepatitis
This is a continuing progressive inflammation with cell necrosis and the formation of fibrous tissue that may lead to cirrhosis of the liver. There is distortion of the liver blood vessels and hypoxia, leading to further hepatocyte damage. This condition is commonly associated with Type B virus hepatitis and with some forms of autoimmunity.

NON-VIRAL INFECTIONS OF THE LIVER
PYOGENIC

Ascending cholangitis
Infection, usually by *Escherichia coli*, may spread from the

biliary tract. The most common predisposing factor is obstruction of the common bile duct by gallstones.

Suppurative phlebitis

when fragments break off from infected thrombi in tributaries of the portal vein. Common sources of this type of infection are acute appendicitis, diverticulitis and inflamed haemorrhoids.

CIRRHOSIS OF LIVER

This is the result of long-term inflammation caused by a wide variety of agents. The commonest cause is alcohol abuse but it may follow acute or chronic liver cell disease of any origin. As the inflammation subsides:

1. Destroyed liver tissue is replaced by fibrous tissue
2. There is hyperplasia of hepatocytes adjacent to the damaged area, in an attempt to compensate for the destroyed cells. This leads to the formation of nodules consisting of hepatocytes confined within sheets of fibrous tissue

As the condition progresses there is the development of:

1. Portal hypertension, leading to congestion in the organs drained by the tributaries of the portal vein, to ascites and possibly to the development of varicose veins at the lower end of the oesophagus
2. Liver failure when hyperplasia is unable to keep pace with cell destruction

LIVER FAILURE

This occurs when liver function is reduced to such an extent that other body activities are impaired. It may be acute or chronic and may be the outcome of a wide variety of disorders, e.g.:

1. Acute viral hepatitis
2. Extensive necrosis due to poisoning, e.g., some drug overdoses, hepatotoxic chemicals
3. Cirrhosis of the liver
4. Following some medical procedures, e.g., abdominal paracentesis, porta-caval shunt operations

EFFECTS OF LIVER FAILURE ON OTHER PARTS OF THE BODY

Hepatic encaphalopathy

The cells affected are the astrocytes in the brain. The condition is characterised by apathy, disorientation, muscular rigidity, delirium and coma. Several factors may be involved, e.g.:

1. Nitrogenous bacterial metabolites absorbed from the colon, normally detoxicated in the liver, reach the brain cells via the blood

2. Metabolites, normally present in trace amounts, e.g., ammonia, may reach toxic concentrations and change the permeability of the cerebral blood vessels and the effectiveness of the blood–brain barrier
3. Hypoxia and electrolyte imbalance

Blood coagulation defects

The liver fails to synthesise substances needed for blood clotting, i.e., prothrombin, fibrinogen and Factors II, V, VII, IX and X. Platelet production is impaired but the cause is unknown.

Oliguria and renal failure

Portal hypertension may cause the development of oesophageal varices. If these rupture, bleeding may lead to a fall in blood pressure sufficient to reduce the renal blood flow, causing progressive oliguria and renal failure.

Oedema and ascites

These may be caused by the combination two factors:

1. Portal hypotension raises the capillary hydrostatic pressure in the organs drained by its tributaries (see Fig. 5:39)
2. Diminished production of serum albumin and clotting factors reduces the osmotic pressure

Together these changes cause the movement of excess fluid into the interstitial spaces where it causes *oedema*. Eventually free fluid accumulates in the peritoneal cavity and the resultant *ascites* may be severe.

Anaemia

This is usually due to the combined effect of a number of factors:

1. Upset in the metabolism of folic acid and vitamin B_{12}
2. Chronic blood loss from oesophageal varices, causing iron-deficiency anaemia
3. Increased breakdown of red blood cells in the congested spleen, causing haemolytic anaemia

Jaundice

The following factors may cause jaundice as liver failure develops:

1. Inability of the hepatocytes to conjugate and excrete bilirubin
2. Obstruction to the movement of bile through the bile channels by fibrous tissue that has distorted the structural framework of liver lobules.

TUMOURS OF THE LIVER

Benign tumours of the liver are very rare.

MALIGNANT TUMOURS

In many cases cancer of the liver is associated with cirrhosis but the relationship between them is not clear. It may be that both cirrhosis and cancer and caused by the same agents or that the carcinogenic action of other agents is promoted by cirrhotic changes. Malignancy develops in a number of cases of acute hepatitis caused by Type B virus. The most common sites of metastases are the abdominal lymph nodes, the peritoneum and the lungs.

Secondary malignant tumours in the liver are common, especially from primary tumours in the gastrointestinal tract, the lungs and the breast. The metastases tend to grow rapidly and are often the cause of death.

BILIARY TRACT

BILE DUCTS (Fig. 9:33)

The *right and left hepatic ducts* join to form the *common hepatic duct* just outside the portal fissure. The hepatic duct passes downwards for about 3 cm where it is joined at an acute angle by the *cystic duct* from the gall bladder. The cystic and hepatic ducts together form the *common bile duct* which passes downwards behind the head of the pancreas to be joined by the main pancreatic duct at the *ampulla of the bile duct*. The opening of the combined ducts

into the duodenum is controlled by the *sphincter of Oddi*. The common bile duct is about 7.5 cm long and has a diameter of about 6 mm.

Structure

The walls of the bile ducts have the same layers of tissue as those described in the *general plan* of the alimentary canal (see p. 139). In the cystic duct the mucous membrane lining is arranged in irregularly situated circular folds which have the effect of a *spiral valve*. Bile passes through the cystic duct twice — on its way into the gall bladder and again when it is expelled from the gall bladder to the common bile duct and thence to the duodenum.

GALL BLADDER

The gall bladder is a pear-shaped sac attached to the posterior surface of the liver by connective tissue. It has a *fundus* or expanded end, a *body* or main part and a *neck* which is continuous with the cystic duct.

Structure

The gall bladder has same layers of tissue as those described in the *general plan* of the alimentary canal, with some modifications.

Peritoneum covers only the inferior surface. The gall bladder is in contact with the posterior surface of the right lobe of the liver and is held in place by the visceral peritoneum of the liver (Fig. 9:30).

Muscle layer. There is an additional layer of oblique muscle fibres.

Mucous membrane displays small rugae when the gall bladder is empty that disappear when it is distended with bile.

Blood supply

The *cystic artery*, a branch of the hepatic artery, supplies blood to the gall bladder. Blood is drained away by the *cystic vein* which joins the portal vein.

Nerve supply

Nerve impulses are conveyed by sympathetic and parasympathetic nerve fibres. It has the same autonomic plexuses as those described in the *general plan* (p. 141).

Bile

Bile is secreted by the liver. In the gall bladder, mucus is added and water absorbed. It consists of:

Water
Mucus
Bile pigment: bilirubin
Bile salts: sodium taurocholate
                sodium glycocholate
Cholesterol

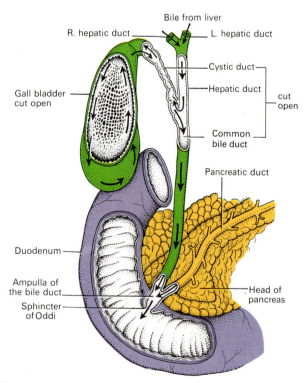

R. hepatic duct
Bile from liver
L. hepatic duct
Cystic duct
Hepatic duct
cut open
Gall bladder cut open
Common bile duct
Pancreatic duct
Duodenum
Ampulla of the bile duct
Sphincter of Oddi
Head of pancreas

Figure 9:33 Diagram of the flow of bile from the liver to the duodenum.

Functions of the gall bladder
1. It acts as a reservoir for bile
2. The lining membrane adds mucus to the bile
3. It absorbs water which concentrates the bile
4. By the contraction of the muscular walls bile is expelled from the gall bladder and passed via the bile ducts into the duodenum. The sphincter of Oddi must relax before bile can pass into the duodenum. Fat, the acidity of chyme in the duodenum and the hormone *cholecystokinin-pancreozymin* (CKK-PZ), secreted by the duodenum, stimulate the gall bladder to contract.

DISEASE OF THE GALL BLADDER AND BILE DUCTS

GALLSTONES

Gallstones consist of deposits of the constituents of bile, most commonly cholesterol. Many small or one large stone may form. The causes are not clear but predisposing factors include:
1. Changes in the composition of bile that affect the solubility of its constituents
2. High levels of blood and dietary cholesterol
3. Cholecystitis
4. Diabetes when associated with high blood cholesterol levels
5. Haemolytic disease

COMPLICATIONS

Pain
If a gallstone becomes impacted in the cystic duct or the common bile duct there is very strong spasmodic contraction of the wall of the duct to try to move stone onwards. This causes severe pain due to ischaemia of the duct wall over the stone during contraction (*biliary colic*).

Inflammation
Gallstones cause irritation and inflammation of the walls of the gall bladder and the cystic and common bile ducts. There may be superimposed microbial infection.

Impaction
Blockage of the cystic duct by a gallstone leads to distension of the gall bladder and cholecystitis. This does not cause jaundice because liver bile can pass directly into the duodenum. Obstruction of the common bile duct leads to retention of bile, jaundice and cholangitis (infection of the bile ducts).

ACUTE CHOLECYSTITIS

This is usually a complication of gallstones or an exacerbation of chronic cholecystitis, especially if there has been partial or intermittent obstruction of the cystic duct. Inflammation develops followed by secondary microbial infection spread from a focus of infection elsewhere in the body, e.g., they may be blood-borne or pass directly from the adjacent colon. Those most commonly involved are *Escherichia coli* and *Streptococcus faecalis*. In severe cases there may be fibrinous exudate into the gall bladder and suppuration. The activity of the gall bladder is disrupted and the infection may spread to the bile ducts and the liver.

CHRONIC CHOLECYSTITIS

The onset is usually insidious, sometimes following repeated acute attacks. Gallstones are usually present and there may be accompanying biliary colic. Pain is due to the spasmodic contraction of muscle, causing ischaemia when the gall bladder is packed with gallstones. There is usually secondary infection with suppuration. Ulceration of the tissues between the gall bladder and the duodenum or colon may occur with fistula formation and later, fibrous adhesions.

TUMOURS OF THE BILIARY TRACT

Benign tumours are rare.

Malignant tumours are relatively rare but when they do occur the most common sites are:
1. The neck of the gall bladder
2. The junction of the cystic and bile ducts
3. In the ampulla of the bile duct

Local spread to the liver, the pancreas and other adjacent organs is common. Lymph and blood spread leads to widespread metastases. Early sites include the liver, lungs, abdominal lymph nodes and the peritoneum.

JAUNDICE

This is not a disease in itself. It is a sign of abnormal bilirubin metabolism and excretion. Serum bilirubin may rise to 34 μmol/1 before the yellow colouration of jaundice is evident in the skin and conjunctiva (normal 3 to 13 μmol/1).

Jaundice develops when there is an abnormality at some stage in the metabolic sequence caused by one or more factors, e.g.:
1. Excess haemolysis of red blood cells with the production of more bilirubin than the liver can deal with

2. Abnormal liver function that may cause:
   a. incomplete uptake of unconjugated bilirubin by hepatocytes
   b. ineffective conjugation of bilirubin
   c. interference with bilirubin secretion

TYPES OF JAUNDICE

Haemolytic jaundice (acholuric)

This is due to increased haemolysis of red blood cells in the spleen. The amount of bilirubin is increased and if hypoxia develops the efficiency of hepatocyte activity is reduced.

Neonatal haemolytic jaundice occurs in many babies, especially in prematurity where the normal high haemolysis is coupled with shortage of conjugating enzymes in the hepatocytes.

Unconjugated bilirubin which is fat soluble has a toxic effect on brain cells. However, it is unable to cross the blood–brain barrier until the plasma level rises above 340 μmol/l, but when it does, it may cause neurological damage, fits and mental handicap.

Obstructive jaundice

Obstruction of the biliary tract is caused by, e.g.:
1. Gallstones
2. Tumour of the head of the pancreas
3. Fibrosis of the bile ducts, following inflammation or injury by cholangitis or the passage of gallstones

Toxic jaundice

This is the result of damage to the liver by, e.g.:
1. Viral infection
2. Toxic substances, such as drugs
3. Amoebiasis (amoebic dysentery)
4. Cirrhosis of the liver

The damaged cells may be unable to:
1. Remove unconjugated bilirubin from the blood
2. Conjugate the bilirubin
3. Secrete bilirubin into the bile canaluculi

SMALL INTESTINE (Figs 9:34 and 9:35)

The small intestine is continuous with the stomach at the *pyloric sphincter* and leads into the large intestine at the *ileocaecal valve*. It is a little over 5 metres long and lies in the abdominal cavity surrounded by the large intestine. In the small intestine the chemical digestion of food is completed and most of the absorption of nutrient materials takes place.

The small intestine is described in three parts which are continous with each other.

The *duodenum* is about 25 cm long and curves around the head of the pancreas. At its midpoint there is an opening, common to the pancreatic duct and the common bile duct, guarded by the *sphincter of Oddi*.

The *jejunum* is the middle part of the small intestine and is about 2 metres long.

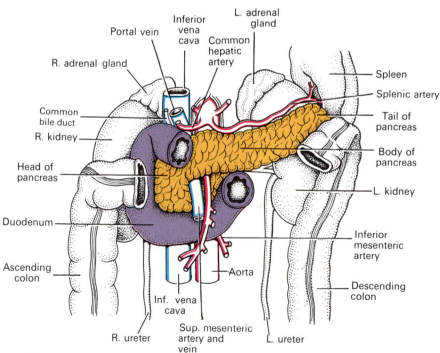

Figure 9:34   The duodenum and its associated structures.

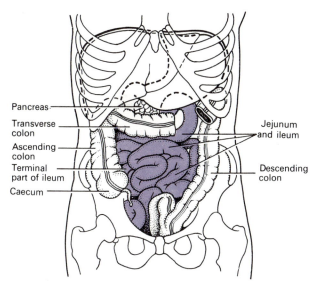

Figure 9:35 The jejunum and ileum and their associated structures.

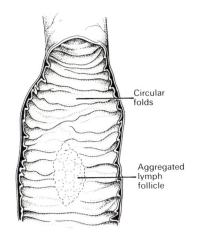

Figure 9:36 Small intestine cut open

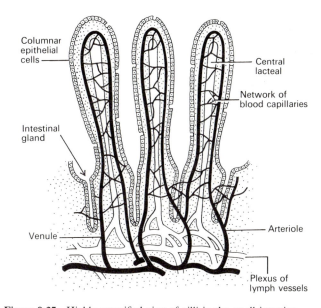

Figure 9:37 Highly magnified view of villi in the small intestine.

The *ileum* or terminal part, is about 3 metres long and ends at the *ileocaecal valve* which controls the flow of material from the ileum to the large intestine and prevents regurgitation.

STRUCTURE

The walls of the small intestine are composed of the four layers of tissue described in the *general plan* (see p. 139). There are some modifications of the peritoneum and mucous membrane lining.

Peritoneum
A double layer of peritoneum called the *mesentery* attaches the jejunum and ileum to the posterior abdominal wall (Fig. 9:25A). The attachment is quite short in comparison with the length of the small intestine, therefore it is fan-shaped. The large blood vessels and nerves lie on the posterior abdominal wall and the branches to the small intestine pass between the two layers of the mesentery.

Mucous membrane
The surface area of the small intestine mucosa is greatly increased by circular folds and villi.

The *circular folds*, unlike the rugae of the stomach, are not smoothed out when the small intestine is distended (Fig. 9:36).

The *villi* are tiny finger-like projections into the intestinal lumen, about 0.5 mm to 1 mm long (Fig. 9:37). Their walls consist of columnar epithelial cells with tiny microvilli (1 μm long) on their free border. These epithelial cells enclose a network of blood and lymph capillaries. The lymph capillaries are called *lacteals* because absorbed fat gives the lymph a milky appearance. Absorption and some

final stages of digestion of nutrient materials take place in the cells of the villi before entering the blood and lymph capillaries.

Intestinal glands. These are simple tubular glands situated below the surface between the villi. The cells of the glands migrate upwards to form the walls of the villi replacing those at the tips as they are rubbed off by the intestinal contents. During migration the cells form digestive enzymes that lodge in the microvilli and, together with intestinal juice, complete the chemical digestion of carbohydrates, protein and fats.

Lymph nodes. There are numerous lymph nodes in the mucous membrane at irregular intervals throughout the length of the small intestine. The smaller ones are known as *solitary lymphatic follicles*, and about 20 or 30 larger nodes situated towards the distal end of the ileum are called *aggregated lymphatic follicles* (Peyer's patches).

BLOOD SUPPLY (Figs 5:37 and 5:39, pp. 82–83)

The *superior mesenteric artery* supplies the whole of the small intestine, and venous drainage is by the superior mesenteric vein which joins other veins to form the portal vein.

NERVE SUPPLY

Sympathetic and parasympathetic.

INTESTINAL JUICE (SUCCUS ENTERICUS)

This is secreted by the glands of the small intestine and consists of:
  Water
  Mucus
  Enzymes: enteropeptidase (enterokinase), peptidases, lipase, sucrase, maltase, lactase

FUNCTION

The functions of the small intestine are:

1. Onward movement of its contents which is produced by peristaltic, segmental and pendular movements.

2. Secretion of intestinal juice.

3. Completion of digestion of carbohydrates, protein and fats in the cells of the villi.

4. Protection against infection by microbes that have survived the antimicrobial action of the hydrochloric acid in the stomach, by the solitary lymph follicles and aggregated lymph follicles.

5. Secretion of the two hormones cholecystokinin-pancreozymin and secretin.

6. Absorption of nutrient materials.

DIGESTION IN THE SMALL INTESTINE

When acid chyme passes into the small intestine it is mixed first with the *pancreatic juice* and *bile* then with *intestinal juice.* Although intestinal juice is secreted throughout the length of the small intestine its action is very limited in the duodenum. It is the juice which *completes* the digestion of carbohydrates to monosaccharides, proteins to amino acids and fats to fatty acids and glycerol.

PANCREATIC JUICE

Pancreatic juice enters the duodenum at the ampulla of the bile duct and consists of:
  Water
  Mineral salts
  Enzymes: amylase, lipase and peptidases, including trypsinogen and chymotrypsinogen.

Pancreatic juice is strongly alkaline (pH 8). When acid stomach contents enter the duodenum they are mixed with pancreatic juice and bile and the pH is raised to between 6 and 8. This is the pH at which the pancreatic enzymes act most effectively.

Functions

Trypsinogen and *chymotrypsinogen* are inactive enzymes until they come in contact with *enteropeptidase* (enterokinase), an enzyme in intestinal juice, which converts them into *trypsin* and *chymotrypsin.* These enzymes convert some polypeptides to amino acids and some to smaller molecule polypeptides (dipeptides and tripeptides). It is important that they are produced as inactive precursors of protein-splitting enzymes otherwise they would digest the pancreas.

Pancreatic amylase converts *all digestible* polysaccharides (starches) not affected by salivary amylase to disaccharides (sugars).

Lipase converts fats to fatty acids and glycerol. To aid the action of lipase, *bile salts* emulsify fats, i.e., reduce the size of the globules.

Secretions

The secretion of pancreatic juice is stimulated by the two hormones *secretin* and *cholecystokinin-pancreozymin* (CCK-PZ) produced by cells in the walls of the small intestine. The presence in the duodenum of acid material from the stomach stimulates the production of these hormones.

BILE

Bile, secreted by the liver, is unable to enter the duodenum when the sphincter of Oddi is closed, therefore it passes from the *hepatic duct* along the *cystic duct* to the gall bladder where it is stored. When a meal has been taken, the gall bladder contracts, the sphincter of Oddi relaxes and bile passes through the cystic duct and the common bile duct into the duodenum together with pancreatic juice. The hormone *cholecystokinin-pancreozymin* (CCK-PZ), produced by the walls of the duodenum, stimulated this activity. A more marked activity is noted if chyme entering the duodenum contains a high proportion of fat.

There is a slight difference between the composition of bile produced by the liver and that entering the duodenum. In the gall bladder water is absorbed and mucus is added by the goblet cells in the mucous membrane lining, therefore the bile is concentrated and becomes more viscid. The constituents of bile from the gall bladder are:
  Water
  Mineral salts
  Mucus
  Bile salts   — sodium taurocholate
                  sodium glycocholate
  Bile pigment — bilirubin
  Cholesterol

Functions
1. The bile salts, *sodium taurocholate* and *sodium glyco-cholate*, emulsify fats in the small intestine.
2. Bile pigment, *bilirubin*, is the waste product of the breakdown of erythrocytes. Bilirubin is altered by microbes in the intestine. The resultant *urobilin* is reabsorbed and excreted in urine, and *stercobilin* is exereted in faeces.
3. The presence of bile in the small intestine is necessary for the absorption of vitamin K and digested fats.
4. It colours and deodorises the faeces.
5. It has an aperient effect.

INTESTINAL SECRETIONS

Intestinal juice is secreted by the glands of the small intestine. It consists of:
Water
Mucus
Enzymes: enteropeptidase (enterokinase)

Traces of other enzymes found in succus entericus are believed to the released following the breakdown of cells brushed off the villi.

Most of the digestive enzymes in the small intestine are contained in the microvilli of the cells of the walls of the villi. The digestion of carbohydrate, protein and fat is completed by direct contact between these nutrients and the microvilli and within the cells of the walls of the villi.

The enzymes involved in completing the digestion of food in the cells of the villi are:
Peptidases
Lipase
Sucrase, maltase and lactase

Functions
Alkaline intestinal juice assists in raising the pH of the intestinal contents to between 6.5 and 7.5.
Enteropeptidase activates pancreatic peptidases which convert some polypeptides to amino acids and some to smaller molecule peptides. The final stage of breakdown to amino acids of all peptides occurs inside the cells of the villi.
Lipase completes the digestion of fats to *fatty acids* and *glycerol* partly in the intestine and partly in the cells of the villi.
Sucrase, maltase and *lactase* complete the digestion of carbohydrates by converting dissacharides to *monosaccharides*, inside the cells of the villi.

Secretion
Mechanical stimulation of the intestinal glands by chyme is believed to be the main stimulus to the secretion of intestinal juice, although the hormone secretin may be involved.

ABSORPTION OF NUTRITIONAL MATERIALS
(Fig. 9:38)

The processes involved in the absorption of nutrients include:
1. Carbohydrates as monosaccharides, proteins as amino acids and fats as fatty acids and glycerol may be slowly absorbed by diffusion but more rapidly by active transport
2. Carbohydrates as disaccharides and proteins as dipeptides and tripeptides are actively transported into the microvilli where digestion is completed to monosaccharides and amino acids before transfer to the capillaries in the villi.

Monosaccharides and amino acids pass in to the capillarides in the villi and fatty acids and glycerol into the lacteals.

Some proteins are absorbed unchanged. The evidence for this lies in the fact that oral vaccines, consisting of protein molecules, can only stimulate the formation of antibodies when they are absorbed unchanged. The extent of protein absorption is believed to be limited.

Other nutritional materials such as vitamins, mineral salts and water are absorbed from the small intestine into the blood capillaries.

The surface area through which absorption takes place in the small intestine is greatly increased by the *circular folds* of mucous membrane and by the very large number of *villi* present. It has been calculated that the surface area of the small intestine is about five times that of the whole body.

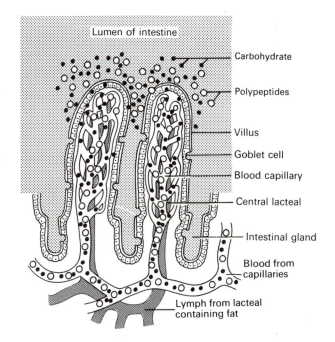

Figure 9:38 Diagram of the absorption of nutrient materials.

Summary of chemical digestion of food

Chemical digestion of protein		Chemical digestion of carbohydrates		Chemical digestion of fats	
PROTEIN FOOD (large molecule polypeptides)		POLYSACCHARIDES (Starches)		FAT	
	PEPSIN (Gastric juice)		SALIVARY AMYLASE (Ptyalin)		BILE SALTS (Bile)
SMALLER POLYPEPTIDES				EMULSIFIED FATS	
	TRYPSIN AND CHYMOTRYPSIN (Pancreatic juice)	PANCREATIC AMYLASE			
		DISACCHARIDES (Sugars)			LIPASE (Pancreatic juice and in microvilli)
SMALL PEPTIDES					
	PEPTIDASES (In microvilli)		MALTASE, LACTASE, SUCRASE (In microvilli)	FATTY ACIDS AND GLYCEROL	
AMINO ACIDS		MONOSACCHARIDES (Glucose)			

LARGE INTESTINE (Colon), RECTUM AND ANAL CANAL

The large intestine is about 1.5 metres long, beginning at the *caecum* in the right iliac fossa and terminating at the *rectum* and *anal canal* deep in the pelvis. Its lumen is larger than that of the small intestine. It forms an arch round the coiled-up small intestine (Fig. 9:39).

For descriptive purposes the colon is divided into the caecum, ascending colon, transverse colon, descending colon, sigmoid or pelvic colon, rectum and anal canal.

The *caecum* is the first part of the colon. It is a dilated portion which has a blind end inferiorly and is continuous with the *ascending colon* superiorly. Just below the junction of the two the *ileocaecal valve* opens from the ileum. The *vermiform appendix* is a fine tube, closed at one end, which leads from the caecum. It is usually about 13 cm long and

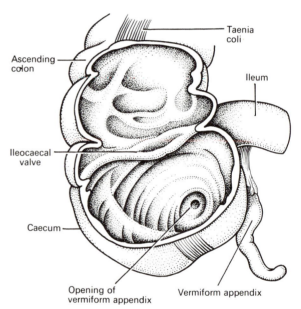

Figure 9:40   Interior of the caecum.

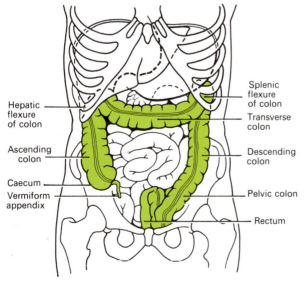

Figure 9:39   Diagram showing the parts of the large intestine (colon) and their positions.

has the same structure as the walls of the colon but contains more lymphoid tissue (Fig. 9:40).

The *ascending colon* passes upwards from the caecum to the level of the liver where it bends acutely to the left at the *right colic flexure* (hepatic flexure) to become the *transverse colon*.

The *transverse colon* is a loop of colon which extends across the abdominal cavity in front of the duodenum and the stomach to the area of the spleen where it forms the *left colic flexure* (splenic flexure) by bending acutely downwards to become the *descending colon*.

The *descending colon* passes down the left side of the abdominal cavity then curves towards the midline. After it enters the true pelvis it is known as the *pelvic colon*.

The *pelvic colon* describes an S-shaped curve in the pelvis then continues downwards to become the *rectum*.

The rectum is a slightly dilated part of the colon which is about 13 cm long. It leads from the pelvic colon and terminates in the *anal canal*.

The anal canal is a short canal about 3.8 cm long in the adult and leads from the rectum to the exterior. There are two sphincter muscles which control the anus; the *internal sphincter* consisting of smooth muscle fibres is under the control of the *autonomic nervous system* and the *external sphincter*, formed by striated muscle, is under *voluntary nerve* control.

STRUCTURE

The four layers of tissue described in the *general plan* are present in the colon, the rectum and the anal canal. The arrangement of the *longitudinal muscle fibres* is modified in the colon. They do not form a smooth continuous layer of tissue but are collected into three bands, called *taeniae coli*, situated at regular intervals round the colon. They stop at the junction of the pelvic colon and the rectum. As these bands of muscle tissue are slightly shorter than the total length of the colon they give a sacculated or puckered appearance to the organ (Fig. 9:41).

The longitudinal muscle fibres completely surround the rectum and the anal canal. The anal sphincters are formed by thickening of the circular muscle layer.

In the submucous layer there is more lymphoid tissue than in any other part of the alimentary tract.

In the mucous membrane lining of the colon and the upper part of the rectum there are large numbers of goblet cells forming simple tubular glands which secrete mucus.

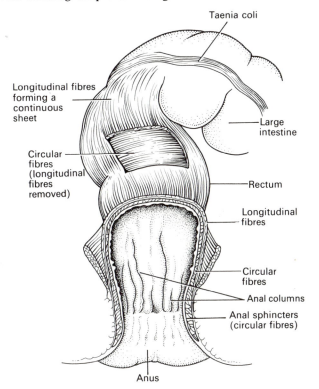

Figure 9:41   The arrangement of muscle fibres in the colon, rectum and anus. Sections have been removed to show the layers.

They are not present beyond the junction between the rectum and the anus. The lining membrane of the *anus* consists of stratified squamous epithelium which is continuous with the mucous membrane lining of the rectum above and merges with the skin beyond the external anal sphincter. In the upper part of the anal canal the mucous membrane is arranged in 6 to 10 vertical folds, the *anal columns*. Each column contains a terminal branch of the superior rectal artery and vein.

BLOOD SUPPLY

Arterial supply is mainly by the superior and inferior mesenteric arteries (Fig.9:6).

The *superior mesenteric artery* supplies the caecum, ascending and most of the transverse colon.

The *inferior mesenteric artery* supplies the remainder of the colon and the rectum.

The distal part of the rectum and the anus are supplied by branches from the *internal iliac arteries*.

Venous drainage is mainly by the *superior* and *inferior mesenteric veins* which drain blood from the parts supplied by arteries of the same names. These veins join the splenic and gastric veins to form the portal vein (Fig. 9:8). Veins draining the distal part of the rectum and the anus join the *internal iliac veins*.

FUNCTIONS OF THE LARGE INTESTINE, RECTUM AND ANAL CANAL

Absorption
The contents of the ileum which pass through the ileo-caecal valve into the caecum are fluid, even though some water has been absorbed in the small intestine. In the large intestine absorption of water continues until the familiar semisolid consistency of faeces is achieved. Mineral salts and some drugs are also absorbed into the blood capillaries from the large intestine.

Microbial activity
There are large numbers of microbes in the colon. They include *Escherichia coli*, *Enterobacter aerogenes*, *Streptococcus faecalis*, *Clostridium welchii*. These microbes are commensals in man when confined to the intestine. They synthesise vitamin K and folic acid.

Large numbers of microbes are present in the faeces.

Defaecation
The large intestine does not exhibit peristaltic movement as it is seen in other parts of the digestive tract. Only at fairly long intervals does a wave of strong peristalsis sweep along the transverse colon forcing its contents into the descending and pelvic colons. This is known as *mass movement* and it is often precipitated by the entry of food into the stomach. This combination of stimulus and response is called the *gastro-colic reflex*. Usually the rectum

is empty, but when a mass movement forces the contents of the pelvic colon into the rectum the nerve endings in its walls are stimulated by stretch. In the infant defaecation occurs by reflex action which is not under voluntary control. However, after the nervous system has fully developed, nerve impulses are conveyed to consciousness when the stretch receptors in the rectum are stimulated and the brain can inhibit the reflex until such times as it is convenient to defaecate. The external anal sphincter is under conscious control. Thus defaecation involves involuntary contraction of the muscle of the rectum and relaxation of the internal anal sphincter and voluntary relaxation of the external anal sphincter. Contraction of the abdominal muscles and lowering of the diaphragm increases the intra-abdominal pressure and so assists the process of defaecation. When defaecation is voluntarily postponed the feeling of fullness and need to defaecate tends to fade. Repeated suppression of the reflex may lead to constipation.

Constituents of faeces. The faeces consist of a semisolid brown mass. The brown colour is due to the presence of stercobilin.

Even though absorption of water takes place in the large intestine it still makes up about 60 to 70% of the weight of the faeces. The remainder consists of undigestible cellular material (roughage), dead and live microbes, epithelial cells from the walls of the tract, some fatty acids, and mucus secreted by the lining mucosa of the large intestine. Mucus helps to lubricate the faeces and an adequate amount of roughage in the diet ensures that the contents of the colon are sufficiently bulky to stimulate defaecation.

DISEASES OF THE INTESTINES

Diseases of the small and large intestines will be described together because they have certain characteristics in common and some conditions affect both.

APPENDICITIS

The lumen of the appendix is very small and there is little scope for swelling when it becomes inflamed. The initial cause of inflammation is not always clear. Microbial infection is commonly superimposed on obstruction by, e.g., faecal matter, kinking, a foreign body. Inflammatory exudate, with fibrin and phagocytes, passes into the lumen, causing swelling and ulceration of the mucous membrane lining. In mild cases the inflammation subsides and healing takes place. In more severe cases microbial growth progresses, leading to suppuration, abscess formation and further congestion. The rising pressure inside occludes first the veins then the arteries and ischaemia develops, followed by gangrene and rupture.

COMPLICATIONS OF APPENDICITIS

Peritonitis. This occurs as a complication of appendicitis when:
1. Microbes spread through the wall of the appendix and infect the peritoneum
2. An appendix abscess ruptures and pus enters the peritoneal cavity
3. The appendix becomes gangrenous and ruptures, discharging its contents into the peritoneal cavity

Abscess formation. The most common abscesses are (Fig. 9:42)
1. Subphrenic abscess, between the liver and diaphragm, from which infection may spread upwards to the pleura, pericardium and mediastinal structures
2. Pelvic abscess from which infection may spread to adjacent structures

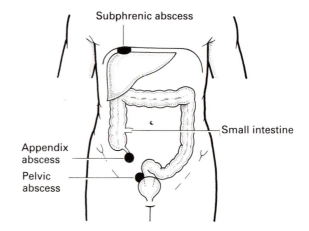

Figure 9:42   Abscess formation: complication of appendicitis.

Fibrous adhesions. When healing takes place fibrous tissue forms and later shrinkage may cause:
1. Stricture or obstruction of the bowel
2. Limitation of the movement of a loop of bowel which may twist around the adhesion, causing a type of bowel obstruction called a *volvulus*

ENTERIC FEVERS

TYPHOID FEVER

This type of enteritis is caused by the microbe *Salmonella typhi*, ingested in food and water. Man is its only host so the source of contamination is an individual who is either suffering from the disease or is a carrier.

After ingestion of microbes there is an incubation period of about 14 days before signs of the disease appear. During this period the microbes invade lymphoid tissue in the walls of the small and large intestine, especially the aggregated lymph follicles (Peyer's patches) and solitary lymph nodes. The microbes then enter the blood vessels and spread to the liver, spleen and gall bladder. In the bacteraemic period acute inflammation develops with necrosis of intestinal lymphoid tissue and ulceration of overlying mucosa. Other effects of *Salmonella typhi* or their endotoxins include:

1. Typhoid cholecystitis in which the microbes multiply and are excreted in bile, reinfecting the intestine
2. Red spots on the skin, especially of the chest and abdomen
3. Enlargement of the spleen
4. Myocardial damage and endocarditis
5. Liver and kidney damage
6. Reduced resistance to other infections, especially of the respiratory tract, e.g., laryngitis, bronchitis, pneumonia

Uncomplicated recovery takes place in about 5 weeks with healing of intestinal ulcers and very little fibrosis.

COMPLICATIONS

1. The ulcers may penetrate a blood vessel, causing haemorrhage, or erode the intestinal wall, leading to acute peritonitis
2. The individual may become a carrier. When this happens the typhoid fever becomes a chronic, asymptomatic infection of the biliary and urinary tracts. Microbes continue to be excreted indefinitely in urine and faeces. Contamination of food and water by carriers is the usual source of infection.

PARATYPHOID FEVER

This disease is caused by *Salmonella paratyphi A* or *B* spread in the same way as typhoid fever, i.e., in food and drink contaminated by infected urine or faeces. The infection, causing inflammation of the intestinal mucosa, is usually confined to the ileum. Other parts of the body are not usually affected but occasionally chronic infection of the urinary and biliary tracts occurs and the individual becomes an asymptomatic carrier, excreting the microbes in urine and faeces.

OTHER SALMONELLA INFECTIONS

Salmonella typhimurium and *S. enteritidis* are the most common infecting microbes in this group. In addition to man their hosts are domestic animals and birds. The microbes may be present in meat, poultry, eggs and milk, causing infection if cooking does not achieve sterilisation. Mice and rats also carry the organisms and may contaminate food before or after cooking.

The infection is usually of short duration but may be accompanied by acute diarrhoea, causing dehydration and electrolyte imbalance. Chronic infection of the biliary and urinary tracts may develop and the individual becomes a carrier, excreting the organisms in urine and faeces.

STAPHYLOCOCCAL FOOD POISONING

This is not an infection in the true sense. Acute gastroenteritis is caused by exotoxins produced by the *Staphylococcus aureus* before the ingestion of the contaminated food. The organisms are usually killed by cooking but the toxins may remain unchanged because they can withstand higher temperatures.

There is usually short-term acute inflammation with violent vomiting and diarrhoea, causing dehydration and electrolyte imbalance. In most cases recovery is uneventful after the toxin has been removed.

CLOSTRIDIUM PERFRINGENS (*Cl. welchii*) FOOD POISONING

These microbes, although normally present in the intestines of man and animals, cause food poisoning when ingested in large numbers. Meat may be contaminated at any stage between slaughter and the consumer. Outbreaks of food poisoning are associated with large-scale cooking, e.g., in institutions. The spores survive the initial cooking and if the food is cooled slowly they change to the vegetative phase and multiply between the temperatures of 50°C and 20°C and also during re-heating. After being eaten, microbes that remain vegetative die and release endotoxins that cause gastroenteritis.

CAMPYLOBACTER

These Gram-negative bacilli are a common cause of gastroenteritis accompanied by fever, acute pain and sometimes bleeding. They affect mainly young adults and children under 5 years. The microbes are present in the intestines of birds and animals and are spread in undercooked poultry and meat. They may also be spread in water and milk. Pets, e.g., cats and dogs, may be a source of infection.

CHOLERA

The disease is caused by *Vibrio cholerae* and is spread by contaminated water, food, hands and fomites. The only known host is man. A very powerful exotoxin is produced which stimulates the intestinal glands to secrete large quantities of water, bicarbonates and chlorides, leading to persistent diarrhoea, severe dehydration and electrolyte imbalance. The microbes occasionally spread to the gall bladder where they multiply. They are then excreted in bile and faeces. This carrier state usually lasts for a maximum of about 4 years.

DYSENTERY

BACILLARY DYSENTERY

This infection of the colon is caused by microbes of the *Shigella* group. The severity of the condition depends on the organisms involved. In Britain it is usually a relatively mild condition caused by *Shigella sonnei*. Outbreaks may reach epidemic proportions, especially in institutions. Children and elderly debilitated adults are particularly susceptible. The only host is man and the organisms are spread by faecal contamination of food, drink, hands and fomites. There is inflammation, ulceration and oedema of the intestinal mucosa with excess mucus secretion. In severe infections, the acute diarrhoea, containing blood and excess mucus, causes dehydration, electrolyte imbalance and anaemia. When healing occurs the mucous membrane is fully restored. Occasionally a chronic infection develops and the individual becomes a carrier, excreting the microbes in faeces.

AMOEBIC DYSENTERY

This disease is caused by *Entamoeba histolytica*. The only known host is man and it is spread by faecal contamination of food, water, hands and fomites. Before ingestion the amoeba are inside cysts. When these reach the colon the amoeba are released and invade the cells of the mucosa, causing inflammation and ulceration. Further development of the disease may result in destruction of the mucosa over a large area and sometimes perforation occurs. Diarrhoea containing mucus and blood is persistent and debilitating.

The disease may progress in a number of ways:
1. Healing may produce fibrous adhesions, causing partial or complete obstruction
2. The amoeba may spread to the liver, causing amoebic hepatitis and abscesses
3. Chronic dysentery may develop with intermittent diarrhoea and amoeba in the faeces

CROHN'S DISEASE (Regional enteritis)

This chronic inflammatory condition of the alimentary tract usually occurs in young adults. The terminal ileum is most commonly affected but the disease may be more widespread. There is chronic patchy inflammation with oedema of the full thickness of the intestinal wall, causing partial obstruction of the lumen. There are periods of remission of varying duration. The cause of Crohn's disease is not entirely clear but it may be that immunological abnormality renders the individual susceptible to infection, especially by viruses. Complications include:
1. Secondary infections, occurring when inflamed areas ulcerate
2. Fibrous adhesions and subsequent intestinal obstruction caused by the healing process

ULCERATIVE COLITIS

This is a chronic inflammatory disease of the mucosa of the colon and rectum which may ulcerate and become infected. There are periods of remission lasting weeks, months or years. The cause is not known but there is an association with arthritis, iritis, some skin lesions and haemolytic anaemia. In longstanding cases cancer sometimes develops.

FULMINATING ULCERATIVE COLITIS

This is an extremely serious, toxic form of the disease. There is a sudden onset of acute diarrhoea, with severe blood loss, leading to dehydration, electrolyte imbalance, hypovolaemic shock and possibly death.

DIVERTICULAR DISEASE

Diverticula are small pouches of mucosa that protrude into the peritoneal cavity through the circular muscle fibres of the colon between the taeniae coli (Fig. 9:43). The walls consists of mucous membrane with a covering of visceral peritoneum. They occur at the weakest points of the intestinal wall, i.e., where the blood vessels enter. Faeces tend to impact in the diverticula and the walls become inflamed and oedematous, as secondary infection develops. This reduces the blood supply, causing ischaemic pain. The causes of diverticulosis are not known but it is associated with low-residue diet and abnormally active peristalsis.

TUMOURS OF THE SMALL AND LARGE INTESTINES

Benign and malignant tumours of the small intestine are rare, compared with their occurrence in the stomach and colon.

BENIGN TUMOURS

Benign neoplasms may form a broad-based mass or develop a pedicle. Occasionally those with pedicles twist upon themselves, causing ischaemia, necrosis and possibly gangrene. Malignant changes may occur.

MALIGNANT TUMOURS

Small intestine malignant tumours tend not to obstruct the lumen and may remain unnoticed until symptoms caused by metastases appear. The most common sites of metastases are local lymph nodes, the liver, lungs and brain.

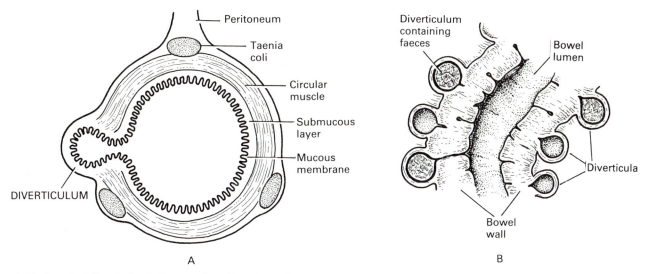

Figure 9:43 Intestinal diverticula. A. Cross-section of bowel showing one diverticulum. B. Longitudinal section of bowel showing numerous diverticula.

The colon is the most common site of malignancy in the alimentary tract in Western countries. The tumour may be a:

1. Soft friable mass, projecting into the lumen of the colon with a tendency to ulceration, infection and haemorrhage
2. Hard fibrous mass encircling the colon, causing reduced elasticity and peristalsis and narrowing of the lumen
3. Gelatinous mucoid mass that thickens the wall and tends to ulcerate and become infected

The causes of cancer of the colon are not known. Predisposing diseases include ulcerative colitis and some benign tumours.

Local spread of intestinal tumours occurs early but may not be evident until there is severe ulceration and haemorrhage or obstruction.

Lymph-spread metastases occur in mesenteric lymph nodes, the peritoneum and other abdominal and pelvic organs. Pressure caused by enlarged lymph nodes may cause obstruction or damage other structures.

Blood-spread metastases are most common in the liver, brain and bones.

CARCINOID TUMOURS (Argentaffinomas)

These tumours are considered, on clinical evidence, to be benign but they spread into the tissues around their original site. They grow very slowly and rarely metastasise. The parent cells are hormone-secreting cells widely dispersed throughout the body, not situated in endocrine glands. They are called APUD cells, an acronym for some of their chemical characteristics. The cells react with silver compounds, hence the name argentaffinomas. Common sites in the intestines for these *apudomas* are the appendix, ileum, stomach, colon and rectum. The tumours are frequently multiple and may spread locally, causing obstruction.

Carcinoid syndrome

This is the name given to the effects of the variety of substances secreted by apudomas in the intestine and elsewhere. The secretions include serotonin (5-hydroxytryptamine), histamine and bradykinin.

HERNIAS

A hernia is a protrusion of bowel though a weak point in the musculature of the anterior abdominal wall or an existing opening (Fig. 9:44). It occurs when there are intermittent increases in intra-abdominal pressure, most commonly in men who lift heavy loads at work. The underlying causes of the abdominal wall weakness are not known. Possible outcomes include:

1. Spontaneous reduction, i.e., the loop of bowel slips back to its correct place when the intra-abdominal pressure returns to normal
2. Manual reduction, i.e., by applying slight pressure over the abdominal swelling
3. Strangulation, when the venous drainage from the herniated loop of bowel is impaired, causing congestion, ischaemia and gangrene. In addition there is intestinal obstruction.

Sites of hernias

Inguinal hernia. The weak point is the inguinal canal which contains the spermatic cord in the male and the

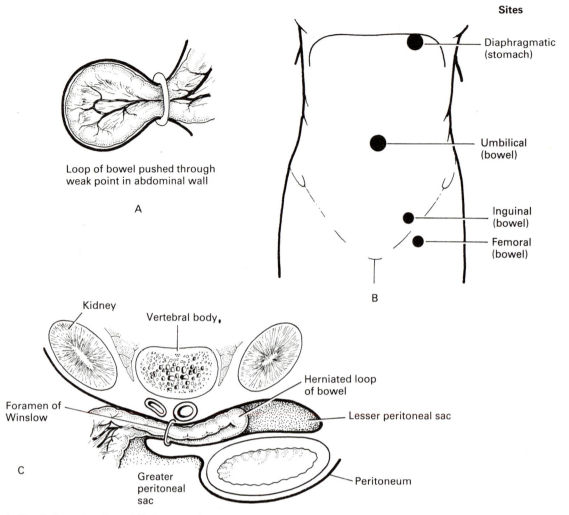

Figure 9:44 A. Hernia formation. B. and C. Common sites.

round ligament in the female. It occurs more commonly in males than in females.

Femoral hernia. The weak point is the femoral canal through which the femoral artery, vein and lymph vessels pass from the pelvis to the thigh.

Umbilical hernia. The weak point is the umbilicus where the umbilical blood vessels from the placenta enter the fetus.

Incisional hernia. This is caused by repeated stretching of the fibrous tissue formed during the repair of a surgical wound.

Diaphragmatic or hiatus hernia. This is the protrusion of a part of the fundus of the stomach through the oesophageal opening in the diaphragm. The main complication is irritation caused by reflux of acid gastric juice, especially when the individual lies flat or bends down. The long-term effects may be oesophagitis, fibrosis and narrowing of the oesophagus, causing dysphagia. Strangulation does not occur.

Sliding hiatus hernia. An unusually short oesophagus that ends above the diaphragm pulls a part of the stomach upwards into the thorax. The abnormality may be congen-ital or be caused by shrinkage of fibrous tissue formed during healing of a previous oesophageal injury. The sliding movement of the stomach in the oesophageal opening is due to normal shortening of the oesophagus by muscular contraction during swallowing.

Rolling hiatus hernia. An abnormally large opening in the diaphragm allows a pouch of stomach to 'roll' upwards into the thorax beside the oesophagus. This is associated with obesity and increased intra-abdominal pressure.

Peritoneal hernia. A loop of bowel may herniate through the foramen of Winslow, the opening in the lesser omentum that separates the greater and lesser peritoneal sacs.

VOLVULUS

This occurs when a loop of bowel twists through 180 degrees, cutting off its blood supply, causing gangrene and obstruction. It occurs in parts of the intestine that are attached to the posterior abdominal wall by a long double fold of visceral peritoneum. The most common site in adults is the pelvic colon and in children the small intestine. The causes are unknown but predisposing factors include:

1. An unusually long mesocolon or mesentery
2. Heavy loading of the pelvic colon with faeces
3. A slight twist of a loop of bowel, causing gas and fluid to accumulate and promote further twisting
4. Adhesions formed following surgery or peritonitis

INTUSSUSCEPTION

In this condition a length of intestine is invaginated into itself (Fig. 9:45). It occurs most commonly in children when a piece of terminal ileum is pushed through the ileocaecal valve. In a child, infection, usually by viruses, causes swelling of the lymphoid tissue in the intestinal wall. The overlying mucosa bulges into the lumen, creating a partial obstruction and a rise in pressure inside the intestine proximal to the swelling. Strong peristaltic waves develop in an attempt to overcome the partial obstruction. These push the swollen piece of bowel into the lumen of the section immediately distal to it, creating the intussusception. The pressure on the veins in the invaginated portion is increased, causing congestion, further swelling, ischaemia and possibly gangrene.

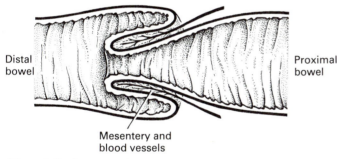

Distal bowel

Proximal bowel

Mesentery and blood vessels

Figure 9:45 Intussusception.

Complete intestinal obstruction may occur. In adults tumours that bulge into the lumen together with strong peristalsis may be the cause e.g., polypi.

INTESTINAL OBSTRUCTION

This is not a disease in itself. The following is a summary of the main causes of obstruction with some examples.

MECHANICAL CAUSES OF OBSTRUCTION

1. Constriction of the intestine by, e.g., strangulated hernia, intussusception, volvulus, peritoneal adhesions. Partial obstruction may suddenly become complete.

2. Stenosis and thickening of the intestinal wall, e.g., in diverticulosis, Crohn's disease and malignant tumours. There is usually a gradual progression from partial to complete obstruction.

3. Obstruction by, e.g., a large gallstone or a tumour growing into the lumen.

4. Pressure on the intestine from outside, e.g., a large tumour in any pelvic or abdominal organ, such as a uterine fibroid. This type is most likely to occur inside the confined space of the bony pelvis.

NEUROLOGICAL CAUSES OF OBSTRUCTION

Partial or complete loss of peristaltic activity produces the effects of obstruction. *Paralytic ileus* is the most common form but the paralysis may be more widespread. The causes is either excessive sympathetic stimulation or lack of paraysmpathetic stimulation. The mechanisms are not clear but there are well-recognized predisposing conditions including:

1. General peritonitis, especially when large amounts of exotoxin are produced
2. Following surgery when there has been a considerable amount of handling of the intestines
3. Severe intestinal infection, especially if there is acute toxaemia

Secretion of water and electrolytes continues although intestinal mobility is lost and absorption impaired. This causes distension and electrolyte imbalance. There is also growth and multiplication of microbes.

VASCULAR CAUSES OF OBSTRUCTION

When the blood supply to a segment of bowel is cut off, ischaemia is followed by infarction, gangrene and obstruction. The causes may be:

1. Atheromatous changes in the blood vessel walls, with thrombosis
2. Embolism
3. Mechanical obstruction of the bowel, e.g., strangulated hernia

MALABSORPTION

Impaired absorption of nutrient materials and water from the intestines is not a disease in itself. It is the result of diseases causing one or more of the following changes:

1. Atrophy of the villi of the mucosa of the small intestine
2. Incomplete digestion of food
3. Interference with the transport of absorbed nutrients from the small intestine to the blood

PRIMARY MALABSORPTION

DISEASE OF THE INTESTINAL MUCOUS MEMBRANE

Atrophy of the villi is the main cause, varying in severity from minor abnormality to almost complete loss of function. The most common underlying diseases are coeliac disease and tropical sprue.

Coeliac disease (idiopathic steatorrhoea)

This disease is believed to be due to a genetically determined abnormal immunological reaction to the protein *gluten*, present in wheat. When it is removed from the diet, recovery is complete. There is marked villous atrophy and malabsorption characterised by the passage of loose, pale-coloured, fatty stools.

There may be abnormal immune reaction to other antigens.

Tropical sprue

In this disease there is partial villous atrophy with malabsorption, chronic diarrhoea, macrocytic anaemia and severe wasting. The cause is unknown but it may be that bacterial growth in the small intestine is a factor. The disease is endemic in subtropical and tropical countries except Africa south of the Sahara. After leaving the endemic area most people suffering from sprue recover, but others may not develop symptoms until months or even years later.

SECONDARY MALABSORPTION

This is associated with incomplete digestion of food or impaired transport of absorbed nutrients.

DEFECTIVE DIGESTION

This occurs in a variety of conditions:
1. Disease of the liver and pancreas
2. Following a resection of small intestine
3. Following surgery when bacterial growth occurs in a blind end of intestine

IMPAIRED TRANSPORT OF NUTRIENTS

This occurs when there is:
1. Lymphatic obstruction by, e.g., lymph node tumours, removal of nodes at surgery, tubercular disease of lymph nodes
2. Impairment of mesenteric blood flow by, e.g., arterial or venous thrombosis, pressure caused by a tumour
3. Obstruction of blood flow through the liver, e.g., in cirrhosis of liver

METABOLISM

When nutritional materials are oxidised in the cells of the body, energy is released, some in the form of heat. Energy may be used immediately to do work, e.g., to synthesise new muscle cells from amino acids, or it may be stored in chemical form as adenosine triphosphate (ATP). Heat is used to maintain the body temperature at the optimum level for chemical activity (36.8° C or 98.4° F). Excess heat is disposed of through the skin and the body excreta.

The energy produced in the body may be measured and expressed in units of work (*joules*) or units of heat (*Calories*).

A *Calorie* (capital C) is the amount of heat required to raise the temperature of 1 litre of water through 1 degree Celsius.

1 Calorie = 4184 joules (J) = 4.184 kilojoules (kJ)

The nutritional value of carbohydrates, protein and fats eaten in the diet may be expressed in *kilojoules per gram* or *Calories per gram*.

1 gram of carbohydrate provides 17 kilojoules (4 Calories)
1 gram of protein provides 17 kilojoules (4 Calories)
1 gram of fat provides 38 kilojoules (9 Calories).

The minimum energy requirement is described as the *basal metabolic rate* (BMR). This is the rate at which the sum of the metabolic processes takes place when the individual is at rest and has eaten no food for a period of at least 12 hours, i.e., after the last meal has been digested and absorbed. The individual is then described as being at *rest* and in the *postabsorptive state*. In this state the energy output is that needed to maintain the essential body processes, e.g., breathing, heart contraction, body temperature, urine secretion.

An indirect measure of the BMR is made by measuring either the amount of oxygen taken into the body or the amount of carbon dioxide excreted in a given number of minutes. These are reliable measures because when energy is released, oxygen is utilised and carbon dioxide is produced as a waste product. The surface area of the body in square metres (m^2) is also taken into account when estimating the BMR, calculated from measurements of the height and weight of the individual because this influences heat loss (Fig. 9:46).

There is a wide range of normal basal metabolic rate but generally it is higher in men than in women, so they require more energy foods.

Men: BMR is about 170 kJ (40 Calories) per m^2 per hour
Women: BMR is about 155 kJ (37 Calories) per m^2 per hour

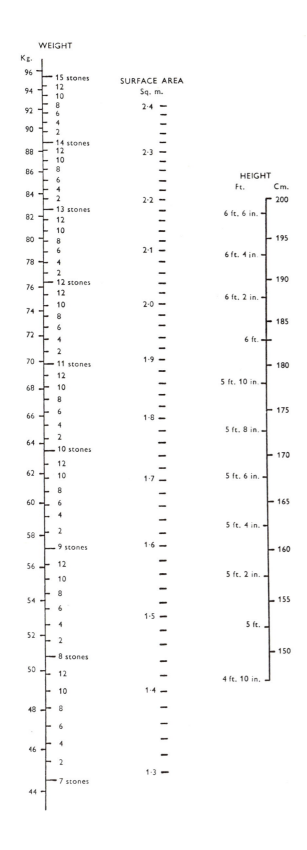

Figure 9:46 Conversion table from height and weight to square metres of surface area of the body. (*From* Green J H 1976 An introduction to human physiology, 4th SI edn. Oxford University Press, London. Reproduced with permission.)

Example

A woman weighing 57 kg (9 stones), 161 cm (5 ft 3 inches) tall has a surface area of 1.6 m². Her basal metabolic requirements, if she remains at rest and in the post-absorptive state throughout a 24-hour period, are:

$$155 \times 1.6 \times 24 = 5952 \text{ kJ } (5.952 \text{ MJ} \star)$$
$$\text{or}$$
$$37 \times 1.6 \times 24 = 1420 \text{ Calories}$$

In normal life most of the basal energy requirement is used during the 8 hours of sleep because even during sleep metabolism is above the basal level. This means that the normal diet must provide considerably more kilojoules to supply the individual's energy needs for 24 hours. The amount of extra energy needed depends on the amount of physical work being done (mental work requires little if any additional energy).

Most people would need at least an additional 4200 kJ (1000 Calories) but if they are involved in strenuous exercise or hard physical work their needs would be greater.

Using the above example a busy woman's energy output would be:

$$5952 + 4200 = 10\ 152 \text{ kJ } (10.152 \text{ MJ})$$
$$\text{or}$$
$$1420 + 1000 = 2420 \text{ Calories}$$

Most foods contain a mixture of different amounts of carbohydrate, protein, fat, minerals, vitamins, roughage and water. Carbohydrates, proteins and fats are the sources of energy and they are obtained from the variety of foods, usually in the following proportions:

Carbohydrates	55%
Proteins	15%
Fats	30%

The following table shows the application of these proportions to a 10 152 kJ (2421 Calorie) diet:

Foods	% of whole diet	kJ	Calories	Grams of food
Carbohydrates	55	5584	1332	330
Proteins	15	1522	363	90
Fats	30	3046	726	80
Total	100	10 152	2421	500

METABOLISM OF CARBOHYDRATE

When digested, carbohydrate, mainly glucose, is absorbed into the blood capillaries of the villi of the small intestine. It is transported by the portal circulation to the liver, where it is dealt with in several ways (see Fig. 9:47).

★ 1 MJ (megajoule) = 1000 kJ

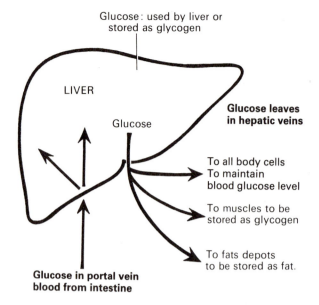

Glucose: used by liver or stored as glycogen

LIVER

Glucose

Glucose leaves in hepatic veins

To all body cells
To maintain
blood glucose level

To muscles to be
stored as glycogen

To fats depots
to be stored as fat.

Glucose in portal vein blood from intestine

Figure 9:47 Summary of the sources, distribution and utilisation of glucose.

1. Glucose may be used to provide the energy necessary for the considerable metabolic activity which takes place in the liver.

2. Some of the glucose may remain in the circulating blood to maintain the normal blood glucose of about 3.5 to 5.5 millimoles per litre (mmol/l) [60 to 100 mg%].

3. Some of the glucose may be converted to the insoluble polysaccharide, *glycogen*, in the liver and in the muscles. *Insulin* is the hormone necessary for this change to take place. The formation of glycogen inside cells is a means of storing carbohydrate without upsetting the osmotic equilibrium (see p. 59). Before it can be used it must be broken down again into its constituent monosaccharides. Liver glycogen constitutes a store of glucose used for liver activity and to maintain the blood glucose level. Muscle glycogen provides the glucose requirement of muscle activity. *Adrenaline*, *thyroxine* and *glucagon* are the main hormones associated with the conversion of glycogen to glucose.

4. Carbohydrate in excess of that required to maintain the blood glucose level and glycogen level in the tissues is converted to fat and stored in the fat depots.

All the cells of the body require energy to carry out their metabolic processes including: multiplication of cells for replacement of worn out cells; contraction of muscle fibres; synthesis of secretions produced by the cells of gland. The oxidation of carbohydrate and fat provides most of the energy required by the body.

OXIDATION OF CARBOHYDRATE

Complete oxidation of glucose requires an adequate supply of oxygen. This is the process by which energy is released during prolonged physical activity, e.g., the man who runs

1500 metres in 4 minutes depends upon *aerobic oxidation*. The energy release takes place slowly and is balanced by oxygen intake. Complete oxidation of carbohydrate in the body results in the production of energy, carbon dioxide and water.

Some energy can be provided by glucose in the absence of oxygen. This *anaerobic process* does not release all the energy from the glucose molecule, and the process can be maintained for only a limited period of time. This is the energy used in a sudden spurt of activity over a very short period of time, e.g., the man who runs 100 metres in 10 seconds could not take in enough oxygen in that time to provide energy by complete oxidation of glucose, so he has to depend on the anaerobic process. One of the end products of this process is lactic acid, and if it accumulates in excess in the muscle it causes the pain associated with unaccustomed exercise

Fate of the end products of carbohydrate metabolism

1. *Lactic acid*. Some of the lactic acid produced by anaerobic catabolism of glucose may be oxidised in the tissue to carbon dioxide and water but first it must be changed to pyruvic acid. If complete oxidation does not take place lactic acid passes to the liver in the circulating blood where it is converted to glucose and may then take any of the pathways open to glucose (see Fig. 9:47).

2. *Carbon dioxide* is excreted from the body as a gas by the lungs.

3. *Water*. The water of metabolism is added to the considerable amount of water already present in the body.

METABOLISM OF PROTEIN

Protein foods taken as part of the diet consist of a number of amino acids (see p. 130). About *20 amino acids* have been named and about 8 of these are described as *essential* because they cannot be synthesised in the body. The remainder are described as *non-essential* amino acids because they can be synthesised by many tissues. The enzymes involved in this process are called *transaminases*. Digestion breaks down the protein of the diet to its constituent amino acids in preparation for transfer into the blood capillaries of the villi in the wall of the small intestine. In the portal circulation amino acids are transported to the liver then into the general circulation, thus making them available to all the cells and tissues of the body. Different cells choose from those available the particular amino acids required for building or repairing their specific type of tissue.

Amino acids not required for building and repairing body tissues are broken down in the liver:

1. The *nitrogenous part* is converted to *urea* by the process of *deamintion* and excreted in the urine.

2. The remaining part is used to provide energy, or stored as fat, if in excess of immediate requirements

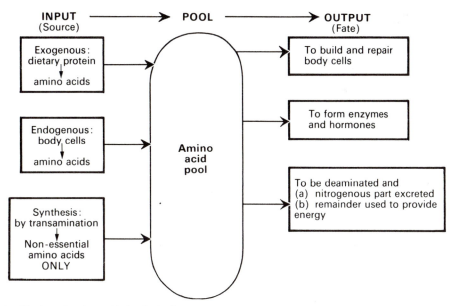

Figure 9:48 Sources and utilisation of amino acids in the body.

AMINO ACID POOL

A pool of amino acids is maintained within the body. This is the source from which the different cells of the body draw the amino acids they need to synthesise their own materials, e.g., new cells, secretion such as enzymes and hormones, blood proteins.

Sources of amino acids (Fig. 9:48)

1. *Exogenous*. These are derived from the protein eaten in the diet.

2. *Endogenous*. These are obtained from the breakdown of body protein. In an adult about 80 to 100 g of protein are broken down and replaced each day.

Loss of amino acids

1. *Deamination*. Amino acids not needed by the body are deaminated. The nitrogenous part is excreted as urea by the kidneys, and the remainder is used to provide energy and heat.

2. *Excretion*. The faeces contain a considerable amount of protein consisting of desquamated cells from the lining of the alimentary tract.

Endogenous and exogenous amino acids are mixed in the 'pool' and the body is said to be in *nitrogen balance* when the rate of removal from the pool is equal to the additions to it. Unlike carbohydrates, the body has no capacity for the storage of amino acids except for this relatively small pool. Figure 9.49 depicts what happens to amino acids in the body.

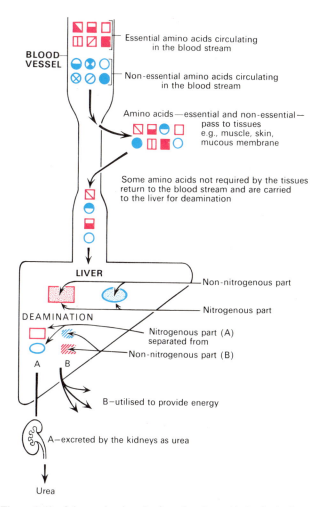

Figure 9:49 Scheme showing the fate of amino acids in the body.

METABOLISM OF FAT (Fig. 9:50)

Fats which have been digested and absorbed into the *lacteals* are transported via the receptaculum chyli and the thoracic duct to the blood stream and so, by a circuitous route, to the liver. Fatty acids and glycerol circulating in the blood are used by organs and glands to provide energy and in the synthesis of some of their secretions. In the liver some fatty acids and glycerol are used to provide energy and heat, and some reorganised and recombined to form a variety of fatty compounds and human fat which is stored in the fat depots of the body. These include subcutaneous fat, fat in the omentum and that supporting some organs such as the kidneys. Before this fat can be used for metabolic purposes it has to be transported back to the liver, and undergo further change. It is, then passed into the circulation in a form that can be oxidised. The end products of fat metabolism are energy, heat, carbon dioxide and water.

Ketone bodies are the *ketoacids* produced during the process of oxidation of fats and are always present in the blood in very small amounts. They are excreted in the urine and in the expired air as *acetone*. When there has been an insufficient intake of carbohydrate foods in the diet, fat is used in excessive quantities to provide energy and heat, thus a state of *ketosis* arises due to the increase in the amount of ketone acids in the blood.

Fat is synthesised from carbohydrates and proteins which are taken into the body in excess of its needs.

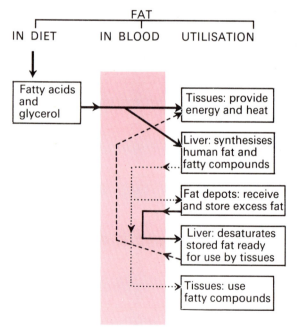

Figure 9:50 Sources, distribution and utilisation of fats in the body.

RELATIONSHIPS BETWEEN CARBOHYDRATES, FATTY ACIDS, GLYCEROL AND DEAMINATED AMINO ACIDS AS ENERGY-RELEASING SUBSTANCES

The degradation of carbohydrates, fatty acids, glycerol and the residue after amino acids are deaminated occurs inside the cells, releasing *energy* and forming the waste products *carbon dioxide* and *water*. The catabolism of these molecules occurs in a series of steps, a little energy being released at each stage. Up to a certain point each nutrient passes through a series of separate and distinct stages but thereafter, they all follow a *common pathway of degradation*. This final common pathway is called the *citric acid cycle* or *Krebs cycle*.

Figure 9:51 provides a diagrammatic representation of the processes involved and only a few of the many steps are shown.

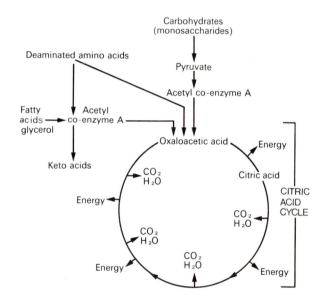

Figure 9:51 Diagram showing the relationship between carbohydrate, deaminated amino acids and fats as energy-releasing substances.

Carbohydrates go through a series of stages to *pyuvate* and *acetyl coenzyme A*. It is in this form that it joins *oxaloacetic acid* in the citric acid cycle.

Fatty acids pass through a series of oxidative stages to *acetyl coenzyme A* and, under normal circumstances, progress to oxaloacetic acid and the citric acid cycle. If, however, an excessive amount of acetyl coenzyme A is produced some of it develops into *ketoacids*.

Deaminated amino acids are of two types: those which go through a series of stages to *oxaloacetic acid* and so to the citric acid cycle and those which follow a different series of changes to become *acetyl coenzyme A* and thereafter take the pathway either to oxaloacetic acid or to ketoacids.

The formation of abnormal amounts of ketoacids occurs

in starvation and in *diabetes mellitus* when excessive amounts of fat and amino acids are used to provide energy, that is, when acetyl coenzyme A is produced more rapidly than it can be used in the citric acid cycle. In both these examples there is an insufficiency of carbohydrate inside the cells. In diabetes this is due to a shortage in the supply of the hormone *insulin* which facilitates the transportation of carbohydrate from the extracellular fluid across the cell membrane and its subsequent metabolism. Excess ketoacids are excreted in the urine and in expired air as acetone.

SUMMARY OF DIGESTION, ABSORPTION AND UTILISATION OF CARBOHYDRATES, PROTEINS AND FATS

CARBOHYDRATES

Digestion

Organ	Digestive juice	Enzyme and action
Mouth	Saliva	Amylase converts cooked starches to *disaccharide*
Stomach	Gastric juice	Hydrochloric acid stops the action of salivary amylase
Small intestine	Pancreatic juice	Amylase converts all starches to *disaccharides* (sugars)
Small intestine	In microvilli	Sucrase / Maltase / Lactase } Convert all sugars to *monosaccharides*, mainly *glucose*

Absorption

Glucose is absorbed into the capillaries of the villi and transported in the portal circulation to the liver.

Utilisation

Monosaccharides provide energy needed by all cells. To achieve this a fairly constant blood glucose level is maintained. Oxygen is needed to obtain all the energy they contain and the waste products are carbon dioxide and water. In addition:

1. Some of the excess is converted to glycogen in the presence of insulin and stored in the liver and in the muscle.

2. Any remaining glucose is converted into fat and stored in the fat depots.

PROTEIN

Digestion

Organ	Digestive juice	Enzyme and action
Mouth	Saliva	No action
Stomach	Gastric juice	Hydrochloric acid converts pepsinogen to *pepsin* Pepsin converts all proteins to smaller molecule polypeptides
Small intestine	Pancreatic juice	Enteropeptidase of intestinal juice converts trypsinogen and chymotrypsinogen to *trypsin* and *chymotrypsin* which convert all polypeptides to di- and tripeptides
Small intestine	In microvilli	Peptidases complete the conversion of peptides to *amino acids*

Absorption

Amino acids are absorbed into the capillaries of the villi and transported in the portal circulation to the liver.

Utilisation

1. In the liver to form albumin, globulin, prothrombin and fibrinogen.

2. In various combinations by cells of the body for cell replacement, cell repair, the production of secretions, e.g., hormones, enzymes, antibodies

3. To maintain the amino acid pool.

4. Amino acids not required are deaminated in the liver. The nitrogenous part is converted into urea and excreted in the urine. The remainder is used to provide energy and heat or deposited as fat in the fat depots.

FATS

Digestion

Organ	Digestive juice	Enzyme and action
Mouth	Saliva	No action
Stomach	Gastric juice	No action
Small intestine	Bile	Bile salts emulsify fats
Small intestine	Pancreatic juice	Lipase converts fats to *fatty acids* and *glycerol*
Small intestine	In microvilli	Lipase completes the digestion of fats to *fatty acids* and *glycerol*

Absorption

Fatty acids and glycerol are absorbed into the lacteals of the villi and are transported via the receptaculum chyli and the thoracic duct to the left subclavian vein. In this way they are transported by the circulating blood to the liver where fatty acids and glycerol are reorganised and recombined.

Utilisation

1. Utilised in the presence of oxygen to provide energy and heat, the waste products carbon dioxide and water being produced.

2. Stored in the fat depots.

3. When depot fat is required for oxidation it must first be desaturated by the liver.

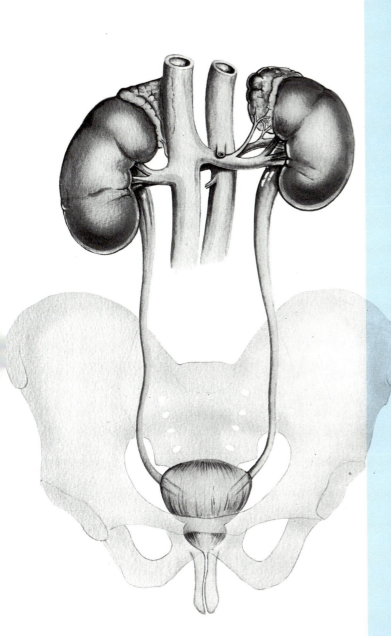

10

The Urinary System

10. The Urinary System

The urinary system is one of the excretory systems of the body. It consists of the following structures:

2 *kidneys* which secrete urine

2 *ureters* which convey the urine from the kidneys to the urinary bladder

1 *urinary bladder* where urine collects and is temporarily stored

1 *urethra* through which the urine is discharged from the urinary bladder to the exterior

Figure 10:1 shows an overview of the urinary system.

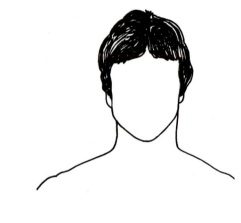

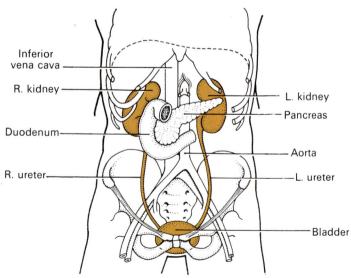

Figure 10:1 The parts of the urinary system and some associated structures.

Inferior vena cava
R. kidney
Duodenum
R. ureter
L. kidney
Pancreas
Aorta
L. ureter
Bladder

KIDNEYS

The kidneys lie on the posterior abdominal wall, one on each side of the vertebral column, behind the peritoneum and below the diaphragm. They extend from the level of the 12th thoracic vertebra to the 3rd lumbar vertebra. The right kidney is usually slightly lower than the left, probably because of the considerable space occupied by the liver.

Kidneys are bean-shaped organs, about 11 cm long, 6 cm wide and 3 cm thick. They are embedded in, and held in position by, a mass of fat. A sheath of fibroelastic *renal fascia* encloses the kidney and the renal fat.

ORGANS ASSOCIATED WITH THE KIDNEYS
(Figs 10:1, 10:2 and 10:3)

As the kidneys lie on either side of the vertebral column each is associated with a different group of structures.

Right kidney

Superiorly — the *right adrenal gland.*

Anteriorly — the *right lobe of the liver,* the *duodenum* and the *right colic flexure.*

Posteriorly — the *diaphragm,* and *muscles of the posterior abdominal wall*

Left kidney

Superiorly — the *left adrenal gland*

Anteriorly — the *spleen, stomach, pancreas, jejunum* and *left colic flexure*

Posteriorly — the *diaphragm* and *muscles of the posterior abdominal wall*

GROSS STRUCTURE OF THE KIDNEY

There are three areas of tissue which can be distinguished when a longitudinal section of the kidney is viewed with the naked eye (Fig. 10:4):

1. *A fibrous capsule,* surrounding the kidney
2. *The cortex* is the reddish-brown layer of tissue immediately under the capsule and between the pyramids

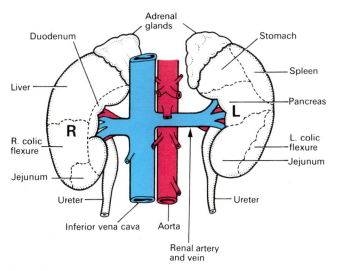

Figure 10:2 Anterior view of the kidneys showing the areas of contact with associated structures.

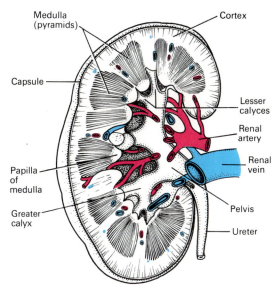

Figure 10:4 A longitudinal section of the right kidney.

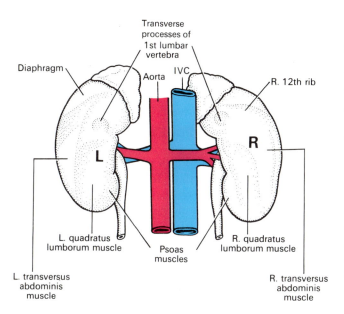

Figure 10:3 Posterior view of the kidneys showing the areas of contact with associated structures.

3. *The medulla* is the innermost layer, consisting of pale conical-shaped striations, the *renal pyramids*

The hilus is the concave medial border of the kidney where the renal blood and lymph vessels and nerves enter.

The renal pelvis is the funnel-shaped structure which acts as a receptacle for the urine formed by the kidney. It has a number of branches called *calyces* at its upper end, each of which surrounds the apex of a renal pyramid. Urine formed in the kidney passes through a *papilla* at the apex of a pyramid into a lesser calyx, then into a greater calyx before passing through the pelvis into the ureter.

MICROSCOPIC STRUCTURE OF THE KIDNEYS

The kidney substance is composed of about 1 million functional units, the *nephrons*, and a smaller number of *collecting tubules*. The uriniferous tubules are supported by a small amount of connective tissue, containing blood vessels, nerves and lymph vessels.

The nephron

The nephron consists of a tubule closed at one end, the other end opening into a collecting tubule. The closed or blind end is indented to form the cup-shaped *glomerular capsule* (Bowman's capsule) which almost completely encloses a network of arterial capillaries, the *glomerulus*. Continuing from the glomerular capsule the remainder of the nephron is described in three parts: the *proximal convoluted tubule*, the *loop of Henle* and the *distal convoluted tubule*, leading into a *collecting tubule* (Fig. 10:5).

After entering the kidney at the hilus the renal artery divides into smaller arteries and arterioles. In the cortex an arteriole, the *afferent arteriole*, enters each glomerular capsule then subdivides into a cluster of capillaries, forming the glomerulus. Between the capillary loops there are connective tissue phagocytic *mesangeal cells*. The blood vessel leading away from the glomerulus is the *efferent arteriole*; it breaks up into a second capillary network to supply oxygen and nutritional materials to the remainder of the nephron. Venous blood drained from this capillary bed eventually leaves the kidney in the renal vein which empties into the inferior vena cava (Fig. 10:6). The blood pressure in the glomerulus is higher than in other capillaries because the calibre of the afferent arteriole is greater than that of the efferent arteriole.

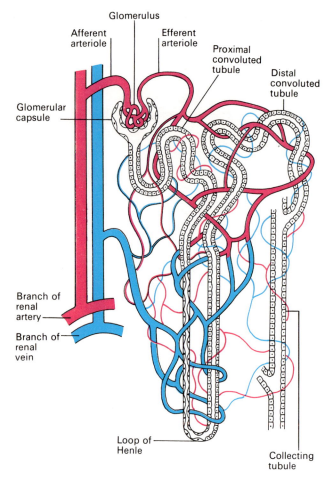

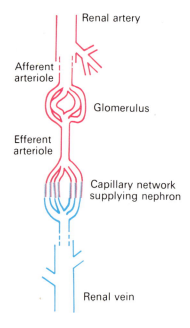

Figure 10:5   Diagram of a nephron including the arrangement of the blood vessels.

Figure 10:6   Diagram of the series of blood vessels in the kidney.

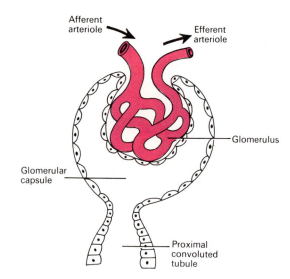

Figure 10:7   Diagram of the glomerulus and glomerular capsule.

The walls of the glomerulus and the glomerular capsule consist of a single layer of *flattened epithelial cells* (Fig. 10:7). The glomerular walls are more permeable than those of other capillaries. The remainder of the nephron and the collecting tubule are formed by a single layer of highly specialised cells.

The nerve supply consists of sympathetic and parasympathetic nerves.

FUNCTIONS OF THE KIDNEY

The kidneys form urine which passes through the ureters to the bladder for excretion. In doing so, vital functions in relation to the maintenance of fluid and electrolyte balance and the disposal of waste material from the body are carried out.

FORMATION OF URINE

There are three phases to urine formation:
  Simple filtration
  Selective reabsorption
  Secretion

Simple filtration (Fig. 10:8)
Filtration takes place through the *semipermeable* walls of the glomerulus and glomerular capsule. Water and a large number of small molecules pass through, some of which are reabsorbed later. Blood cells, plasma proteins and other large molecules are unable to filter through and remain in the capillaries.

The main factor assisting filtration is the difference between the blood pressure in the glomerulus and the pressure of the filtrate in the glomerular capsule. Because the calibre of the efferent arteriole is less than that of the afferent arteriole, a *capillary hydrostatic pressure* of about

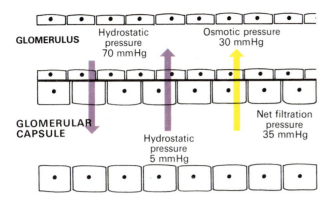

Figure 10:8 Diagram of filtration in the nephron.

70 mmHg builds up in the glomerulus. This pressure is opposed by the *osmotic pressure* of the blood, about 30 mmHg, and by *filtrate hydrostatic pressure* of about 5 mmHg in the glomerular capsule. The net *filtration pressure* is, therefore:

$$70 - (30 + 5) = 35 \text{ mmHg}$$

About 100 to 150 litres of dilute filtrate are formed each day by the two kidneys. Of these 1 to 1.5 litres are excreted as urine. The difference in volume and concentration is due to selective reabsorption and secretion in the tubules.

Blood constituents forming the glomerular filtrate	Blood constituents remaining in the glomerulus
water	leukocytes
mineral salts	erythrocytes
amino acids	platelets
ketoacids	blood proteins
glucose	
hormones	
urea	
uric acid	
toxins	
drugs	

Selective reabsorption (Fig. 10:9)

Selective reabsorption is the process by which the composition and volume of the glomerular filtrate are altered during its passage through the convoluted tubules, the loop of Henle and the collecting tubule. The general purpose of this process is to reabsorb those filtrate constituents needed by the body to maintain fluid and electrolyte balance and blood alkalinity.

Some constituents of glomerular filtrate do not normally appear in urine. These are called *high threshold substances*. They are completely reabsorbed unless they are present in blood in excessive quantities. The kidneys' maximum capacity for reabsorption of a substance is the *renal threshold*, e.g., normal blood glucose level is 3.3 to 5 mmol/l (60 to 90 mg/100 ml). If the level rises above the renal threshold of about 9 mmol/l (160 mg/100 ml) glucose

appears in the urine because the mechanism for active transfer out of the tubules is overloaded.

The threshold level of some substances varies according to the body's need for them *at the time*, i.e., in order to maintain homeostasis. In some cases reabsorption is regulated by hormones.

Parathormone from the parathyroid glands and *calcitonin* from the thyroid gland together regulate reabsorption of calcium and phosphate.

Antidiuretic hormone (ADH) from the posterior lobe of the pituitary gland affects the permeability of the distal convoluted tubules and collecting tubules, regulating water reabsorption.

Aldosterone, secreted by the cortex of the adrenal gland, influences the reabsorption of sodium and excretion of potassium.

Waste products, such as urea and uric acid, are absorbed only to a slight extent. These are called *low threshold substances*.

Secretion (Fig. 10:9)

Filtration occurs as the blood flows through the glomerulus. Non-threshold substances and foreign materials, e.g., drugs, may not be cleared from the blood by filtration because of the short time it remains in the glomerulus. Such substances are cleared by *secretion into the convoluted tubules* and passed from the body in the urine.

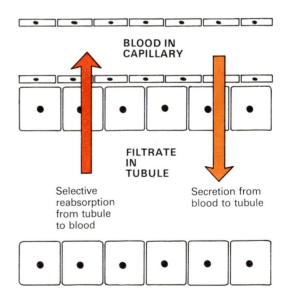

Figure 10:9 Diagram of selective reabsorption and secretion in the nephron.

COMPOSITION OF URINE

water	96%
urea	2%
uric acid	
creatinine	
ammonia	
sodium	
potassium	2%
chlorides	
phosphates	
sulphates	
oxalates	

Urine is amber in colour due to the presence of urobilin, a bile pigment altered in the intestine, reabsorbed then excreted by the kidneys. The specific gravity is between 1020 and 1030, and the reaction is acid. A healthy adult passes 1000 to 1500 ml per day. The amount of urine secreted and the specific gravity vary according to the fluid intake and the amount of solute excreted. During sleep and muscular exercise urine production is decreased.

WATER BALANCE AND URINE OUTPUT

Water is taken into the body through the alimentary tract and a small amount is formed by the metabolic processes. Water is excreted in saturated expired air, as a constituent of the faeces, through the skin as sweat and in the urine. The amount lost in expired air and in the faeces is fairly constant and the amount of sweat produced is associated with the maintenance of normal body temperature (see p. 205).

The balance between fluid intake and output is controlled by the kidneys. The minimum urinary output, consistent with the essential removal of waste material, is about 500 ml per day. The amount produced in excess of this is controlled mainly by the *antidiuretic hormone* (ADH) released into the blood by the *posterior lobe of the pituitary gland*. There is a close link between the posterior pituitary and the *hypothalamus* in the brain.

There are cells in the hypothalamus (*osmoreceptors*) sensitive to changes in the osmotic pressure of the blood. Nerve impulses from the osmoreceptors stimulate the posterior lobe of the pituitary gland to release ADH. When the osmotic pressure is raised, ADH output is increased and as a result, water reabsorption is increased, reducing the blood osmotic pressure and ADH output. This feedback mechanism maintains the blood concentration within normal limits (see Fig. 10:10).

Although this feedback mechanism is the most important means by which the water balance of the body is maintained, it is not the only one. If there is an excessive amount of any dissolved substance in the blood which must be excreted through the kidneys, extra water is excreted with it, e.g., in diabetes mellitus when glucose

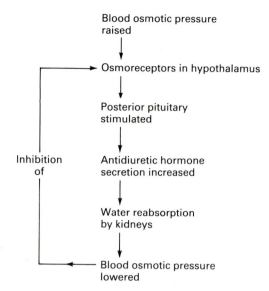

Figure 10:10   Feedback mechanism for the control of antidiuretic hormone (ADH) secretion.

is present in urine. This *polyuria* may lead to dehydration in spite of increased production of ADH but it is usually accompanied by acute thirst and increased water intake.

ELECTROLYTE BALANCE

Changes in the *concentration of electrolytes* in the body fluids may be due to changes in the amounts of water or of electrolytes. There are several methods of maintaining the balance between water and electrolyte concentration.

SODIUM AND POTASSIUM CONCENTRATION

Sodium is the most common cation (positively charged ion) in extracellular fluid and potassium is the most common intracellular cation.

Sodium is a constituent of almost all foods and it is often added to food during cooking. This means that the intake is usually in excess of the body's needs. There are two main routes of sodium excretion.

1. *In urine*. Sodium is a normal constituent of urine and the amount excreted is regulated by the hormone *aldosterone*, secreted by the cortex of the *adrenal gland* (suprarenal gland). When blood flow through the kidneys is low they secrete an enzyme, *renin*, which catalyses the conversion of inactive *angiotensinogen* to active *angiotensin* and this, in turn, stimulates the release of aldosterone. Water is reabsorbed with sodium and together they increase the blood volume, leading to reduced renin secretion. When *sodium reabsorption* is increased *potassium excretion* is increased, indirectly reducing intracellular potassium (see Fig. 10:11).

2. *In sweat*. The amount of sodium excreted in sweat is insignificant except when sweating is excessive. This

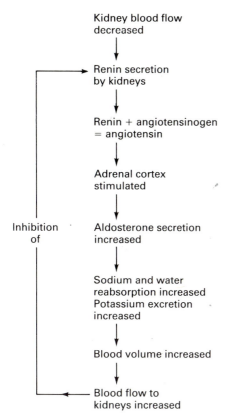

Figure 10:11   Summary of the relationship between renal blood flow and selective reabsorption by the nephron.

may occur when there is a high environmental temperature or during sustained physical exercise. Normally the renal mechanism described above maintains the cation concentration within physiological limits. When excessive sweating is sustained, e.g., living in a hot climate or working in a hot environment, aclimatisation occurs in about 7 to 10 days and the amount of electrolyte lost in sweat is reduced.

Sodium and potassium occur in high concentrations in digestive juices — sodium in gastric juice and potassium in pancreatic and intestinal juice. Normally these ions are reabsorbed by the colon but following acute and prolonged diarrhoea they may be excreted in large quantities with resultant electrolyte imbalance.

DISEASES OF THE KIDNEYS

ACUTE GLOMERULONEPHRITIS

The damage to the glomerulus is believed to be caused by antigen/antibody (immune) complexes being deposited in the walls of the glomerular capillaries, glomerular capsule and/or mesangial cells, causing local inflammatory reactions. About 7 days after the initial encounter with an antigen, antibodies appear in the blood and combine with the antigen. As more antibodies are produced, the number of immune complexes increases, until all the antigen is used up. Immune complexes are deposited in the walls of the glomeruli and usually cause an inflammatory reaction. Although these complexes circulate widely in the blood they cause more damage to the glomeruli than to other tissues because the blood flows relatively slowly over their large surface area, allowing more immune complexes to come into contact with the glomeruli (see Fig. 10:12).

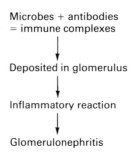

Figure 10:12   Stages of development of glomerulonephritis.

ACUTE DIFFUSE PROLIFERATIVE GLOMERULONEPHRITIS

This type usually follows pharyngitis caused by group A beta-haemolytic streptococci, and all the glomeruli are affected. Immune complexes deposited in the glomerular wall stimulate an inflammatory reaction and the walls become swollen and infiltrated with granulocytes. There is increased permeability with leakage of plasma proteins and red blood cells into the filtrate, leading to haematuria and proteinuria. If the glomeruli are obstructed by inflammatory swelling, renal blood flow and the production of filtrate are reduced, causing oliguria, fluid retention, oedema and raised blood urea. There are various outcomes, e.g.:

1. Complete recovery after the immune complexes have been either phagocytosed or have passed through the glomerular membrane into the filtrate and been excreted in urine
2. A persistent mild form of the disease of varying duration, followed either by complete recovery or chronic glomerulonephritis
3. Death due to:
   a. hypertension, leading to heart failure or acute renal failure early in the disease
   b. infection and hypertension after 1 to 2 years
   c. chronic glomerulonephritis several years later

MINIMAL-CHANGE GLOMERULONEPHRITIS

This occurs mainly in children between 1 and 4 years of age, often following a chest infection, but it is not associated with specific antigen/antibody complexes. There is usually spontaneous recovery but recurrences are fairly common. In children this is the most common cause of *the nephrotic syndrome* (see p. 193). In adults it may lead to chronic glomerulonephritis.

FOCAL GLOMERULONEPHRITIS

This is a relatively rare condition, found mainly in male children, in which proliferation of mesangial cells occurs, i.e., connective tissue cells that lie between the capillary loops in glomeruli. Only parts of some glomeruli are affected. In time, adhesions form between the glomerular loops and the Bowman's capsule, and the associated tubules atrophy. There is short-term haematuria and proteinuria but repeated recurrences may lead to renal failure. No specific antigen/antibody complexes have been identified.

DIFFUSE MEMBRANOUS GLOMERULONEPHRITIS

Immune complexes of unknown origin pass through the glomerular epithelium and are deposited in the basement membrane. No inflammatory reaction occurs but there is diffuse thickening of the walls of the glomeruli. This reduces the amount of filtrate and increases the membrane permeability, causing proteinuria, oliguria, marked progressive oedema and pleural and pericardial effusion. Slow recovery may occur but in the majority of cases there is increased development of glomerular sclerosis, leading to the nephrotic syndrome, chronic renal failure and death.

CHRONIC GLOMERULONEPHRITIS

This condition may develop with no previous history of kidney disease or:
1. Following acute post-streptococcal membranous and recurrent focal glomerulonephritis
2. In association with viral diseases elsewhere in the body

The nephrons gradually become sclerosed and there is progressive loss of function. The rate at which the disease progresses may vary considerably. Hypertension develops which may be either malignant and lead to early renal failure or benign, leading to cardiac failure, cerebral haemorrhage or renal failure.

DIABETIC KIDNEY

Renal failure is the cause of death in many diabetics, especially early-onset insulin-dependent diabetes. The underlying cause is not known but there may be an immune reaction to insulin. The effects include:
1. Progressive glomerulosclerosis followed by atrophy of the tubules

2. Pyelonephritis with papillary necrosis
3. Atheroma of the renal arteries and their branches, leading to renal ischaemia and hypertension
4. Development of the nephrotic syndrome

HYPERTENSION AND THE KIDNEYS

The types of hypertension that affect the kidneys are:
1. Benign (chronic) essential hypertension
2. Malignant (accelerated) essential hypertension
3. Persistent secondary hypertension

In all cases there is renal arteriosclerosis and arteriolosclerosis, causing ischaemia. The reduced blood flow stimulates the renin–angiotensin mechanism, raising the blood pressure still further.

Benign hypertension
This causes gradual and progressive sclerosis and fibrosis of the glomeruli, leading to renal failure or, more commonly, to malignant hypertension.

Malignant hypertension
This causes rapidly developing arteriolosclerosis which spreads to the glomeruli with subsequent destruction of nephrons, leading to:
1. Further rise in blood pressure
2. Reduction in renal blood flow and the amount of filtrate
3. Increased permeability of the glomeruli, with the passage of plasma proteins and red blood cells into the filtrate
4. Progressive oliguria and renal failure

Persistent secondary hypertension
This is caused by kidney diseases such as chronic glomerulonephritis and pyelonephritis and leads to renal ischaemia, further hypertension and renal failure.

ACUTE PYELONEPHRITIS

This is an acute microbial infection of the kidney pelvis and calyces, spreading to the kidney substance. The infection may travel up the urinary tract from the perineum or be blood-borne.

Ascending infection
This is the most common route of infection, usually by microbes that are commensals of the bowel, e.g., *Escherichia coli*, *Streptococcus faecalis*, *Pseudomonas pyocyanea*. Pyelonephritis is preceded by cystitis (inflammation of the bladder). Reflux of infected urine into the ureters when the bladder contracts during micturition allows infection to

spread upwards to the renal pelves and kidney substance. The bladder may be infected by:

1. Microbes from the perineum, especially in females because of the wide short urethra, its proximity to the anus and the moist perineal conditions
2. Blood-borne microbes, especially if there is stasis of urine in the bladder due to urethral obstruction
3. A catheter or other instrument

Blood-borne infection

The source of microbes may be:

1. Elsewhere in the body, e.g., boils, carbuncles, wound infections, abscesses
2. The lower urinary tract, especially following surgical procedures to relieve urethral strictures due to, e.g., fibrosis following previous infection or enlarged prostate gland in the male

When the infection spreads into the kidney substance there is suppuration and the destruction of nephrons. The outcome depends on the amount of healthy kidney remaining after the infection subsides. Necrotic tissue is eventually replaced by fibrous tissue but there may be some hypertrophy of healthy nephrons. This is a common cause of renal failure.

CHRONIC PYELONEPHRITIS

This usually follows repeated attacks of acute pyelonephritis with scar tissue formation. The progressive loss of functioning nephrons leads to renal failure and uraemia. Concurrent hypertension is common.

NEPHROTIC SYNDROME

This is not a disease in itself but is an important feature of several kidney diseases. The main characteristics are hypoalbuminaemia, generalised oedema and hyperlipidaemia. When glomeruli are damaged, the permeability of the glomerular membrane is increased and plasma proteins pass through in the filtrate. Albumin is the main protein lost because it is the most common and also has the smallest plasma protein molecule. When the amount lost per day exceeds 10 g there is a significant fall in the total plasma protein level. The consequent low plasma osmotic pressure leads to widespread oedema and reduced plasma volume. This reduces the renal blood flow and stimulates the renin–angiotensin–aldosterone mechanism, causing increased reabsorption of water and sodium from the renal tubules. The reabsorbed water further reduces the osmotic pressure, increasing the oedema. The key factor is the loss of albumin across the glomerular membrane and as long as this continues the vicious circle is perpetuated (see Fig. 10:13). Hyperlipidaemia also occurs but the cause is unknown.

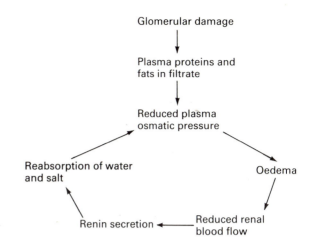

Figure 10:13   Stages of development of nephrotic syndrome.

The nephrotic syndrome occurs in a number of diseases. In children the most common cause is minimal-change glomerular nephritis, although it may follow acute diffuse and focal glomerulonephritis. In adults it is caused by:

1. Most forms of glomerulonephritis
2. Diabetes mellitus
3. Systemic lupus erythematosus
4. Rheumatoid arthritis

ACUTE RENAL FAILURE

There is severe reduction in glomerular filtrate, occurring as a complication of a variety of conditions not primarily associated with the kidneys. These include:

1. Severe and prolonged shock due to, e.g., haemorrhage, severe trauma, marked dehydration, acute intestinal obstruction, prolonged and complicated surgical procedures, extensive burns, incompatible blood transfusions
2. Toxic chemicals, e.g., carbon tetrachloride (used in dry cleaning), chromic acid, ethylene glycol (antifreeze), mercurial compounds, radioactive material, some drugs, trilene

The outstanding effect on the kidneys is *acute tubular necrosis* due to ischaemia or chemical injury. Oliguria (less than 400 ml of urine per day) and anuria (less than 100 ml of urine per day) may last for a few weeks, followed by diuresis. There is reduced glomerular filtrate and tubular selective reabsorption and secretion, leading to:

1. Generalised and pulmonary oedema
2. Retention of urea and other metabolic waste products
3. Electrolyte imbalance which may be exacerbated by the retention of potassium released from cells following severe injury and extensive tissue damage elsewhere in the body
4. Acidosis due to disrupted excretion of hydrogen ions

Diuresis occurs during the healing phase when the epithelial cells of the tubules have regenerated but are still incapable of selective reabsorption and secretion. Diuresis may lead to acute dehydration, complicating the existing high plasma urea, acidosis and electrolyte imbalance. If the patient survives the initial acute phase, a considerable degree of renal function is usually restored over a period of months.

END-STAGE KIDNEY

This is reached when irreversible damage to nephrons is so severe that the kidney cannot function effectively. The main causes are chronic glomerulonephritis, chronic pyelonephritis and hypertension. The effects are a reduced filtration rate, selective reabsorption and secretion, and fibrosis which interferes with blood flow. These changes have a number of effects on the body.

Uraemia develops after about 7 days of anuria because of the reduced glomerular filtration rate and impaired tubular secretion of urea.

Polyuria is caused by defective reabsorption of water in spite of the reduced glomerular filtrate rate (GFR) (see Table 10:1).

Table 10:1 End-stage kidney polyuria

	Normal kidney	End-stage kidney
GFR	125 ml/min or 180 l/day	10 ml/min or 14 l/day
Reabsorption of water	>99%	approx. 0.7%
Urine output	<1 ml/min or 1.5 l/day	approx 7 ml/min or 10 l/day

Fixed specific gravity. The specific gravity of the urine is similar to that of glomerular filtrate, i.e., about 1010 (normal = 1020 to 1030). It remains low and fixed because of defective tubular reabsorption of water.

Acidosis. Control of the pH of body fluids is defective mainly because the tubules fail to remove hydrogen ions by forming ammonia and hydrogen phosphates.

RENAL CALCULI

Stones form in the kidneys and bladder when urinary constituents normally in solution are precipitated. The solutes involved are oxalates, phosphates, urates and uric acid, and stones usually consist of more than one substance, deposited in layers. Most originate in collecting tubules or in renal papillae. They then pass into the pelvis of the kidney where they may increase in size. Some become too large to pass through the ureter and may obstruct the flow of urine. Others pass to the bladder and are either excreted or increase in size and obstruct the urethra. Occasionally stones originate in the bladder.

Predisposing factors include:

Dehydration. This leads to increased reabsorption of water from the tubules but does not change solute reabsorption, resulting in a reduced volume of highly concentrated filtrate in the collecting tubules.

pH of urine. When the normally acid filtrate becomes alkaline some substances may be precipitated, e.g., phosphates. This occurs when the kidney buffering system is defective, and in some infections.

Infection. Necrotic material and pus provide foci upon which solutes in the filtrate may be deposited and the products of infection may alter the pH of the urine.

Tumours. Pressure caused by tumours in the kidney or adjacent to the urinary tract may restrict the flow of urine. This may cause ischaemia and necrosis or predispose to infection. Necrotic debris and tumour fragments provide foci for the deposition of solutes in the urine.

SMALL CALCULI

These may pass through or become impacted in a ureter and damage the epithelium, leading to fibrosis and stricture. In ureteric obstruction, usually unilateral, there is excessive spasmodic peristaltic contraction of the ureter, causing acute intermittent ischaemic pain (*renal colic*) as the ureter contracts over the stone. Stones reaching the bladder may be passed in urine or increase in size and eventually block the urethra, leading to bilateral hydronephrosis, infection proximal to the blockage, pyelonephritis and severe kidney damage.

LARGE CALCULI (Staghorn calculus)

One large stone may form, filling the renal pelvis and the calyces. It causes stagnation of urine, predisposing to infection, and may cause irreversible kidney damage.

CONGENITAL ABNORMALITIES OF KIDNEYS

MISPLACED KIDNEY

One or both kidneys may develop in abnormally low positions. Misplaced kidneys function normally if the blood vessels are long enough to provide an adequate blood supply but a kidney in the pelvis may cause problems during pregnancy if the fetus compresses the renal blood vessels. There may also be difficulties during parturition.

POLYCYSTIC DISEASE

This disease, caused by genetic abnormality, occurs in infantile and adult forms. The infantile form is very rare and the child usually dies soon after birth.

Adult polycystic kidney disease usually becomes apparent at between 30 and 50 years of age. Both kidneys are affected. Dilatations (cysts) form at the junction of the distal convoluted and collecting tubules. The cysts slowly enlarge and pressure causes ischaemia and necrosis of nephrons, resulting in their destruction. The disease is progressive and secondary hypertension usually develops. Death may be due to chronic renal failure, uraemia, cardiac failure or cerebral haemorrhage. Other associated abnormalities include polycystic liver disease, cysts in the spleen, pancreas and lungs and berry aneurysms in cerebral arteries.

TUMOURS OF THE KIDNEY

Benign tumours of the kidney are relatively uncommon.

MALIGNANT TUMOURS

Clear-cell carcinoma (Grawitz tumour or hypernephroma)
This is a tumour of tubular epithelium. Local spread involves the renal vein and leads to early blood-spread of tumour fragments, most commonly to the lungs and bones. The causes of the malignant changes are not known.

Nephroblastoma (Wilms' disease)
This is one of the most common malignant tumours in children, occurring in the first 3 years. It is usually unilateral but becomes very large and invades the renal blood vessels, causing early blood-spread of the malignancy to the lungs and bones. It is believed that the cell abnormalities occur before birth but the cause is not known.

URETERS (Fig. 10:14)

The ureters are the tubes that convey urine from the kidneys to the urinary bladder. They are about 25 to 30 cm long with a diameter of about 3 mm.

The ureter is continuous with the funnel-shaped *renal pelvis*. It passes downwards through the abdominal cavity, behind the peritoneum in front of the psoas muscle into the pelvic cavity, and passes obliquely through the posterior wall of the bladder (Fig. 10:15). Because of this arrangement the ureters are compressed and the opening occluded when the pressure rises in the bladder. This prevents reflux of urine as the bladder fills and during micturition, when the bladder wall contracts.

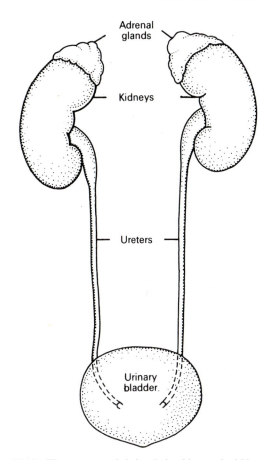

Figure 10:14 The ureters and their relationships to the kidneys and bladder.

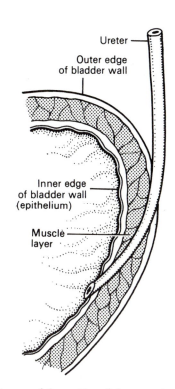

Figure 10:15 Diagram of the position of the ureter in relation to the bladder wall.

STRUCTURE

The ureters consist of three layers of tissue:
1. An outer covering of *fibrous tissue*, continuous with the fibrous capsule of the kidney
2. A middle *muscular layer* consisting of interlacing muscle fibres that form a syncytium spiralling round the ureter, some in clockwise and some in anticlockwise directions
3. An inner lining of *mucous membrane*

FUNCTION

The ureters propel the urine from the kidneys into the bladder by peristaltic contraction of the muscular wall. This is an intrinsic function not under nerve control. The waves of contraction originate in a pacemaker in the minor calyces. Peristaltic waves occur at about 10 second intervals, sending little spurts of urine into the bladder.

URINARY BLADDER

The urinary bladder is a reservoir for urine. It lies in the pelvic cavity and its size and position vary, depending on the amount of urine it contains. When distended the bladder rises into the abdominal cavity.

ORGANS ASSOCIATED WITH THE BLADDER

In the female (Fig. 10:16A)
Anteriorly — the symphysis pubis
Posteriorly — the uterus

Superiorly — the small intestine
Inferiorly — the urethra and the muscles forming the pelvic floor

In the male (Fig. 10:16B)
Anteriorly — the symphysis pubis
Posteriorly — the rectum and seminal vesicles

Superiorly — the small intestine
Inferiorly — the urethra and prostate gland

STRUCTURE (Fig. 10:17)

The bladder is roughly pear-shaped, but becomes more oval as it fills with urine. It has anterior, superior and posterior surfaces. The posterior surface is the *base*. The bladder opens into the urethra at its lowest point, *the neck*. It is composed of four layers of tissue:

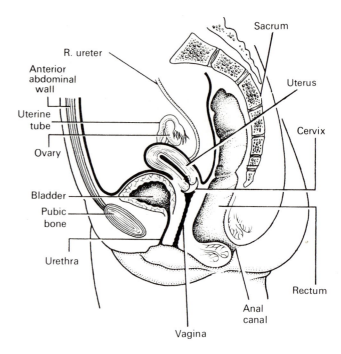

Figure 10:16A The pelvic organs associated with the bladder and the urethra in the female.

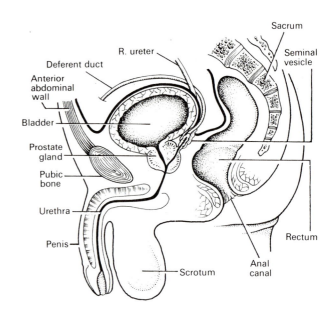

Figure 10:16B The pelvic organs associated with the bladder and the urethra in the male.

1. *Peritoneum* covers only the superior surface before it is reflected upwards to become the parietal peritoneum, lining the anterior abdominal wall. Posteriorly it is reflected on to the uterus in the female and the rectum in the male (see Fig. 9:3).

2. *The muscle layer* consists of a mass of interlacing smooth muscle fibres.

3. *The submucous layer* joins the inner lining and the muscular layer and is made up of areolar tissue containing blood vessels, lymph vessels and sympathetic and parasympathetic nerves.

4. *Mucous membrane* forms the inner lining. When the bladder is empty or contracted the inner lining is arranged in folds and these gradually disappear as the bladder fills.

The three orifices in the bladder wall form a triangle or *trigone*. The upper two orifices on the posterior wall are the openings of the ureters. The lower orifice is the point of origin of the urethra. Where the urethra commences there is a thickening of the smooth muscle layer which acts as a sphincter and controls the passage of urine from the bladder into the urethra. The *internal sphincter*, like the muscle layer, is under autonomic nerve control.

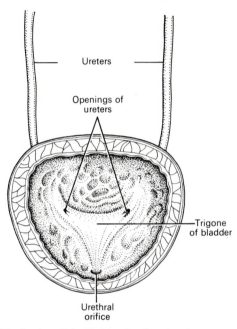

Figure 10:17 Section of the bladder showing the trigone.

URETHRA

The urethra is a canal extending from the neck of the bladder to the exterior. Its length differs in the male and in the female. The male urethra is associated with the urinary and the reproductive systems, and is described in Chapter 15.

The female urethra is approximately 4 cm long. It runs downwards and forwards behind the symphysis pubis and opens at the *external urethral orifice* just in front of the vagina. The external urethral orifice is guarded by the *external sphincter* which is under voluntary control. Except during the passage of urine, the walls of the urethra are in close apposition.

The urethra is composed of three layers of tissue:

1. A *muscular layer* which is continuous with that of the bladder. At its origin there is an *internal sphincter*, composed mainly of elastic tissue and smooth muscle fibres, under autonomic control. Near the external urethral orifice the smooth muscle is replaced by striated muscle which forms the *external sphincter*, under voluntary control.

2. A *thin spongy coat* containing large numbers of blood vessels.

3. A *lining of mucous membrane* continuous with that of the bladder in the upper part of the urethra. The lower part consists of stratified squamous epithelium, continuous externally with the skin of the vulva.

FUNCTIONS OF THE BLADDER AND MICTURITION

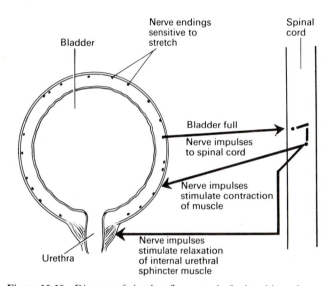

Figure 10:18 Diagram of simple reflex control of micturition when inhibition of the reflex action is not possible.

The urinary bladder acts as a reservoir for urine. When 200 to 300 ml of urine have accumulated, autonomic nerve fibres in the bladder wall sensitive to stretch are stimulated. In the infant this initiates a *spinal reflex action* (see p. 224) and micturition occurs (Fig. 10:18). When the nervous system is fully developed the micturition reflex is stimulated but sensory impulses pass upwards to the brain and there is a awareness of the desire to micturate. By conscious effort, reflex contraction of the bladder wall and relaxation of the internal sphincter can be inhibited for a limited period of time (Fig. 10:19).

Micturition occurs when the muscular wall of the bladder contracts, there is reflex relaxation of the internal sphincter and voluntary relaxation of the external sphincter. It can be assisted by increasing the pressure within the pelvic cavity, achieved by lowering the diaphragm and contracting the abdominal muscles. Overdistension of the bladder is extremely painful, and when

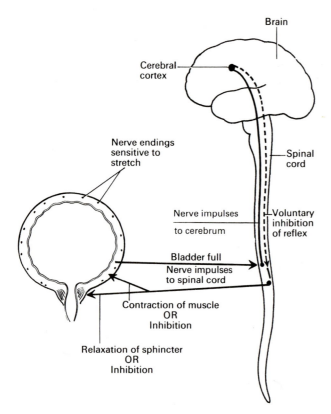

Figure 10:19   Diagram of the nerve control of micturition when inhibition of the reflex action is possible.

this stage is reached there is a tendency for involuntary relaxation of the external sphincter to occur and a small amount of urine to escape, provided there is no mechanical obstruction.

DISEASES OF THE RENAL PELVIS, URETERS, BLADDER AND URETHRA

These structures are considered together because their combined functions are to collect and store urine and discharge it from the body. Obstruction and infection are the main causes of dysfunction.

OBSTRUCTION TO THE FLOW OF URINE

HYDRONEPHROSIS

This is dilatation of the kidney pelvis and calyces caused by accumulation of urine, leading to destruction of the nephrons, fibrosis and atrophy of the kidney. One or both kidneys may be involved, depending on the cause, e.g., obstruction of a ureter or the urethra or disturbance of bladder nerve supply.

Complete sustained obstruction

In this condition hydronephrosis develops quickly, pressure in the nephrons rises and urine production stops. The most common causes are a large calculus or tumour. The outcome depends on whether one or both kidneys are involved (homeostasis can be maintained by one kidney).

Partial or intermittent obstruction

This may lead to progressive hydronephrosis caused by, e.g.:

1. A succession of renal calculi in a ureter, eventually moved onwards by peristalsis
2. A calculus that only partially blocks the ureter
3. Fibrous constriction of a ureter or the urethra, following epithelial inflammation caused by the passage of a stone or by infection
4. Pressure caused by:
   a. a tumour in the tract or in the abdominal or pelvic cavity
   b. an enlarged prostate gland in the male

Spinal lesions

The immediate effect of transverse spinal cord lesions that damage the nerve supply to the bladder is that micturition does not occur. When the bladder fills the rise in pressure causes overflow incontinence, back pressure into the ureters and hydronephrosis. Reflex micturition is usually re-established after a time, but loss of voluntary control may be irreversible.

COMPLICATIONS OF URINARY TRACT OBSTRUCTION

Infection. Urine stasis predisposes to infection and pyelonephritis. The microbes usually spread upwards in the walls of the urinary tract or are blood-borne.

Calculus formation. Infection and urine stasis predispose to calculus formation when:

1. The pH of urine changes from acid to alkaline, promoting the precipitation of some solutes, e.g., phosphates
2. Cell debris and pus provide foci upon which solutes in the urine may be deposited

INFECTIONS OF THE URINARY TRACT

Infection of any part of the tract may lead to pyelonephritis and severe kidney damage.

URETERITIS

This is usually due to the upward spread of infection in cystitis.

ACUTE CYSTITIS

This may be due to:
1. Spread of coliform microbes (*Escherichia coli, Streptococcus faecalis* and *Pseudomonas pyocyanea*) from the perineum, especially in women because of the short wide urethra, its proximity to the anus and the moist perineal conditions
2. Blood-borne infection
3. A mixed infection of coliform and other organisms may follow the passage of a catheter or other instrument

The effects are:
1. Inflammation, small haemorrhages and oedema of the mucosa
2. Hypersensitivity of sensory nerve endings in the bladder wall, stimulated by inflammation before the bladder has filled to its usual capacity, leading to frequency of micturition accompanied by pain

PREDISPOSING FACTORS

The most important predisposing factors are coliform microbes in the perineal region and stasis of urine in the bladder. In the female, hormones associated with pregnancy cause relaxation of perineal muscle and in the late stages pressure caused by the fetus may obstruct the flow of urine. In the male, prostatitis provides a focus of local infection or an enlarged prostate gland may cause progressive urethral obstruction.

CHRONIC CYSTITIS

This may follow repeated attacks of acute cystitis. It occurs most commonly in males when compression of the urethra by an enlarged prostate gland prevents the bladder from emptying completely. Calculus formation is common, especially if the normally acid urine becomes alkaline due to microbial action or kidney damage.

URETHRITIS

A common cause is *Neisseria gonorrhoeae* (gonococcus) spread by sexual intercourse directly to the urethra in the male and indirectly from the perineum in the female. Many cases of urethritis have no known cause, i.e., *nonspecific urethritis*. (See Ch. 15.)

TUMOURS OF THE BLADDER

It is not always clear whether bladder tumours are benign or malignant.

PAPILLOMAS

These tumours consist of a stalk with fine-branching fronds which tend to break off, causing bleeding. Papillomas commonly recur, even when they are benign.

SOLID TUMOURS

These are all malignant to some degree. At an early stage the more malignant rapidly invade the bladder wall and spread in lymph and blood to other parts of the body. If the surface ulcerates there may be haemorrhage and necrosis. The causes are not known but predisposing factors include cigarette smoking and exposure to chemicals used in some industries, e.g., manufacture of analine dyes, rubber industry, benzidene-based industries.

The Skin

11. The Skin

The skin completely covers the body and, is continuous with the membranes lining the body orifices.

It protects the underlying structures from injury and from invasion by microbes.

It contains sensory (*somatic*) nerve endings of pain, temperature and touch.

It is involved in the regulation of the body temperature.

STRUCTURE OF THE SKIN

There are two main layers to the skin:
  Epidermis
  Dermis or corium
Between the skin and underlying structures there is a layer of subcutaneous fat.

EPIDERMIS (Fig. 11:1)

The epidermis is the most superficial layer of the skin and is composed *stratified epithelium* which varies in thickness in different parts of the body. It is thickest on the palms of the hands and soles of the feet. There are no blood vessels or nerve endings in the epidermis, but its deeper layers are bathed in interstitial fluid which is drained away as lymph.

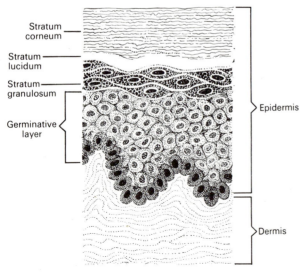

Figure 11:1   The skin showing the main layers of the epidermis.

There are several layers of cells in the epidermis which extend from the superficial *stratum corneum* (horny layer) to the deepest *germinative layer*. The cells on the surface are flat, thin, non-nucleated, dead cells in which the protoplasm has been replaced by *keratin*. These cells are constantly being rubbed off and replaced by cells which originated in the germinative layer and have undergone gradual change as they progressed towards the surface.

The maintenance of healthy epidermis depends upon three processes being synchronised:

1. Desquamation of the keratinised cells from the surface
2. Effective keratinisation of the cells approaching the surface
3. Continual cell division in the deeper layers with cells being pushed to the surface

Hairs, secretion from sebaceous glands and ducts of sweat glands pass through the epidermis to reach the surface.

The surface of the epidermis is ridged by projections of cells in the dermis called the *papillae*. The pattern of ridges is different in every individual and the impression made by them is called the 'fingerprint'. The downward projections of the germinative layer between the papillae are believed to aid nutrition of epidermal cells and stabilise the two layers, preventing damage due to shearing forces.

DERMIS (Fig. 11:2)

The dermis is tough and elastic. It is composed of *collagen fibres* interlaced with *yellow elastic fibres*. Underlying its deepest layer there is areolar tissue and varying amounts of fat. The structures in the dermis are:
  Blood vessels
  Lymph vessels
  Sensory (somatic) nerve endings
  Sweat glands and their ducts
  Hair roots, hair follicles and hairs
  Sebaceous glands
  The arectores pilorum — involuntary muscles attached to the hair follicles

1. *Blood vessels.* Arterioles form a fine network with capillary branches supplying sweat glands, sebaceous glands, hair follicles and the dermis. The epidermis has no

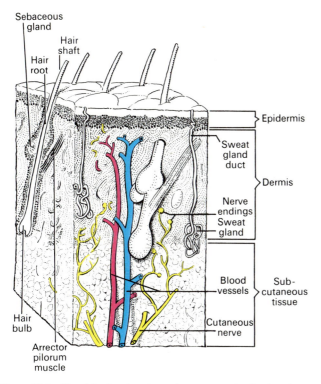

Figure 11:2   The skin showing the main structures in the dermis.

blood supply. It obtains nutrition and oxygen from interstitial fluid derived from blood vessels in the papillae of the dermis.

2. *Lymph vessels* form a network throughout the dermis and the deeper layers of the epidermis.

3. *Sensory nerve endings*. Nerve endings which are sensitive to *touch, change in temperature, pressure and pain* are widely distributed in the dermis. The skin is an important sensory organ through which the individual is aware of his/her environment. Nerve impulses that originate in the nerve endings in the dermis are conveyed to the spinal cord by sensory (*cutaneous*) nerves, then to the sensory area of the cerebrum where the sensations are perceived (see Ch. 12).

4. *Sweat glands* are found widely distributed throughout the skin and are most numerous in the palms of the hands, soles of the feet, axillae and groins. They are composed of epithelial cells. The bodies of the glands lie coiled in the subcutaneous tissue. Some ducts open on to the skin surface at tiny depressions, or pores, and others open into hair follicles. Glands opening into hair follicles do not become active until puberty. In the axilla they secrete an odourless milky fluid which, if decomposed by surface microbes, causes an unpleasant odour. The functions of this secretion are not known.

The most important function of sweat secreted by glands opening on to the skin surface is in the regulation of body temperature. Evaporation of sweat on the surface takes heat from the body and the amount produced is governed by the temperature-regulating centre in the

hypothalamus. Excessive sweating may lead to dehydration and serious depletion of body sodium chloride unless intake of water and salt is appropriately increased. After 7 to 10 days' exposure to high environmental temperatures the amount of salt lost is substantially reduced but water loss remains high.

5. *Hair follicles* consist of a down-growth of epidermal cells into the dermis or subcutaneous tissue. At the base of the follicle there is a cluster of cells called *the bulb*. The hair is formed by the multiplication of cells of the bulb and as they are pushed upwards, away from their source of nutrition, the cells die and are converted to keratin. The part of the hair above the skin is *the shaft* and the remainder, *the root* (Fig. 11:3).

The colour of the hair depends on the amount of melanin present. White hair is the result of the replacement of melanin by tiny air bubbles.

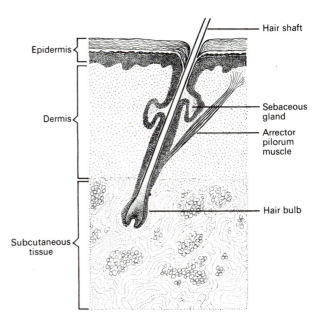

Figure 11:3   A hair in the skin.

6. *The sebaceous glands* (Fig. 11:3) consist of secretory epithelial cells derived from the same tissue as the hair follicles. They pour their secretion, *sebum*, into the hair follicles so they are present in the skin of all parts of the body except the palms of the hands and the soles of the feet. They are most numerous in the skin of the scalp, face, axillae and groins.

Sebum is an oily substance that keeps the hair soft and pliable and gives it a shiny appearance. On the skin it provides some water-proofing and acts as a bactericidal and fungicidal agent, preventing the successful invasion of microbes. It also prevents drying and cracking of skin, especially on exposure to heat and sunshine.

7. *The arrectores pilorum* (Fig. 11:3) are little bundles of involuntary muscle fibres attached to the hair follicles. Contraction makes the hair stand erect and raises the skin around the hair, causing 'goose flesh'. The muscles are

stimulated by sympathetic nerve fibres in response to fear and cold. Although each muscle is very small the contraction of a large number generates an appreciable amount of heat, especially when accompanied by shivering, i.e., involuntary contraction of skeletal muscles.

PIGMENTATION OF THE SKIN

When no pigment is present the skin looks pinkish white in colour due to the blood in the capillaries of the dermis. In most individuals this colour is modified by varying amounts and proportions of several pigments secreted by epidermal cells. The most important are melanin and carotene.

NAILS (Fig. 11:4)

The nails in human beings are equivalent to the claws, horns and hoofs of animals. They are derived from the same cells as epidermis and hair and consist of a hard, horny type of keratinised dead cell. They protect the tips of the fingers and toes.

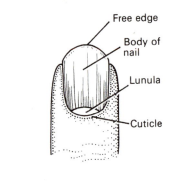

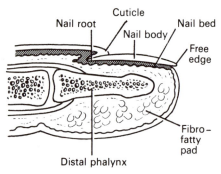

Figure 11:4 The nail and related structures.

The root of the nail is embedded in the skin, is covered by the *cuticle* and forms the hemispherical pale area called the *lunula*.

The body of the nail is the exposed part which has grown out from the germinative zone of the epidermis called the *nail bed*.

Finger nails grow more quickly than toe nails and growth is quicker when the environmental temperature is high.

FUNCTIONS OF THE SKIN

PROTECTION

The skin in one of the main protective organs of the body. It protects the deeper and more delicate organs and acts as the main barrier against the invasion of microbes and other harmful agents.

Due to the presence of the sensory nerve endings in the skin the body reacts by reflex action to unpleasant or painful stimuli, protecting it from further injury (see p. 224).

FORMATION OF VITAMIN D₃

There is a fatty substance, *7-dehydrocholesterol*, in the skin and ultraviolet light from the sun converts it to vitamin D. This circulates in the blood and is used, with calcium and phosphorus, in the formation and maintenance of bone. Any vitamin D in excess of immediate requirements is stored in the liver.

REGULATION OF BODY TEMPERATURE

Human beings are warm-blooded animals and the body temperature is maintained at an average of 36.8 °C (98.4 °F). In health, variations are usually limited to between 0.5 and 0.75 °C, although it may be found that the temperature in the evening is a little higher than in the morning. This is the optimum temperature for the many chemical processes in the body. If the temperature is raised the metabolic rate is increased and if it is lowered the rate of metabolism is reduced. To ensure this constant temperature a fine balance is maintained between heat produced in the body and heat lost to the environment.

HEAT PRODUCTION

Some of the energy released in the cells when carbohydrates, fats and deaminated amino acids are metabolised is in the form of heat. Because of this the most active organs, chemically and physically, produce the most heat. The principal organs involved are:

1. *The muscles.* Contraction of voluntary muscles produces a large amount of heat and the more strenuous the muscular exercise the greater the heat produced. Shivering involves muscle contraction and produces heat when there is the risk of the body temperature falling below normal.

2. *The liver* is very chemically active, and heat is produced as a by-product.

3. *The digestive organs* produce heat by the contraction of the muscle of the alimentary tract and by the chemical reactions involved in digestion.

HEAT LOSS

Most of the heat loss from the body occurs through the skin. Small, amounts are lost in expired air, urine and faeces.

Only the heat lost through the skin can be regulated to maintain a constant body temperature. There is no control over heat lost by the other routes.

Heat loss through the skin is affected by the difference between body and environmental temperatures, the amount of the body surface exposed to the air and the type of clothes worn. Air is a poor conductor of heat and when layers of air are trapped in clothing and between the skin and clothing they act as effective insulators against excessive heat loss. For this reason several layers of lightweight clothes provide more effective insulation against a low environmental temperatures than one heavy garment. A fine balance is maintained between heat production and heat loss. Control is achieved mainly by hypothalamic and skin thermoreceptors.

CONTROL OF BODY TEMPERATURE

Nervous control

The *temperature regulating centre* in the hypothalamus is responsive to the temperature of circulating blood. This centre affects body temperature through antonomic nerve stimulation of the sweat glands.

The *vasomotor centre* in the medulla oblongata controls the calibre of the small arteries and arterioles, and they control the amount of blood which circulates in the capillaries in the dermis. When the skin capillaries are dilated the extra blood near the surface increases heat loss by radiation, conduction and convection. Arteriolar constriction conserves heat.

Activity of the sweat glands

If the temperature of the body is increased by 0.25 to 0.5 °C the sweat glands are stimulated to secrete sweat, which is conveyed to the surface of the body by ducts. The body is cooled by loss of the heat used to *evaporate* the water in sweat. When sweat droplets can be seen on the skin the rate of production is exceeding the rate of evaporation. This is most likely to happen when the environmental air is humid and the temperature high.

Loss of heat from the body by *evaporation* also occurs by *insensible water loss*. In this case heat is being continuously lost by evaporation, even though the sweat glands are not active. Water diffuses upwards from the deeper layers of the skin to the surface of the body and evaporates into the air. The individual is unaware of this process.

Effects of vasodilation

The amount of heat lost from the skin depends to a great extent on the amount of blood in the vessels in the dermis. As heat production increases the arterioles become dilated and more blood pours into the capillary network in the skin. In addition to increasing the amount of sweat produced the *temperature of the skin is raised* and there is an increase in the amount of heat lost by radiation, conduction and convection.

In *radiation* the exposed parts of the body radiate heat away from the body.

In *conduction* the clothes in contact with the skin take up heat.

In *convection* the air passing over the exposed parts of the body is heated and rises, cool air replaces it and convection currents are set up. Heat is also lost from the clothes by convection.

If the external environmental temperature is low or if heat production is decreased, vasoconstriction occurs, stimulated by sympathetic nerves. This decreases the blood flow near the body surface, conserving heat.

Hypothermia

At a rectal temperature below 32 °C (89.6 °F), compensatory mechanisms to restore body temperature usually fail, e.g., shivering is replaced by muscle rigidity and cramps, vasoconstriction fails to occur and blood pressure and heart rate are raised. Death usually occurs when the temperature falls below 25 °C (77 °F).

WOUND HEALING

PRIMARY HEALING (Healing by first intention)

This method of healing follows minimal destruction of tissue when the damaged edges of wound are in close apposition. There are several stages in the repair process (see Fig. 11:5)

1. The cut surfaces become inflamed and blood clot and cell debris fill the gap between them (first few hours)
2. Phagocytes and fibroblasts migrate into the blood clot
   a. phagocytes begin to remove the clot and cell debris
   b. fibroblasts secrete collagen fibres which begin to bind the surfaces together
   c. epithelial cells of the dermis begin to spread across the gap through the clot and the clot above the new skin cells becomes the scab (stage 2 takes 3 to 5 days)
3. The clot between the cut surfaces is completely removed and the scab separates
4. Layers of epithelial cells grow upwards until the full thickness of skin is restored
5. Fibrous tissue continues to grow, binding the edges of the wound more strongly

6. The inflammation resolves but scar tissue remains vascular (stages 3, 4, 5 and 6 progress concurrently and take 2 to 4 weeks)

7. The wound gradually becomes stronger, the fibrous tissue is reduced by the action of fibrolytic enzymes, the scar becomes less vascular, appearing eventually as a pale line

The channels left when stitches are removed heal by the same process.

SECONDARY HEALING (Healing by second intention)

This method of healing follows destruction of a substantial amount of tissue or when the edges of a wound cannot be brought into apposition, e.g., following chronic inflammation or in decubitous ulcers. The time taken for healing depends on the effective removal of the cause and on the size of the wound. There are several recognised stages in the repair process, e.g., of decubitous ulcers (Fig. 11:6):

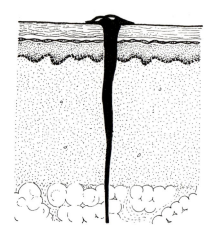

A. Cut surfaces separated by blood clot and cell debris

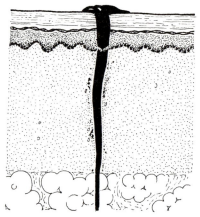

B. Cut surfaces become inflamed
Phagocytosis of clot begins
Epidermal cells spread through clot
Fibrous tissue begins to form

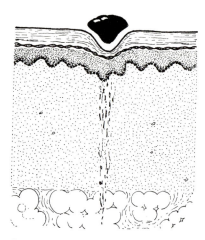

C. Epidermal cells bridge gap
Scab separates
Fibrous tissue binds surfaces together

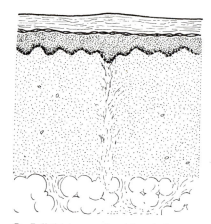

D. Full thickness of epidermis is restored
Wound becomes stronger
Fibrous tissue gradually absorbed

Figure 11:5   Some stages in primary healing of skin.

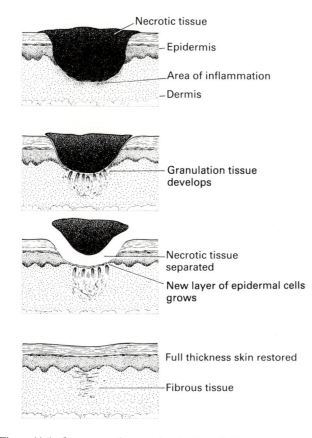

Necrotic tissue

Epidermis

Area of inflammation

Dermis

Granulation tissue develops

Necrotic tissue separated

New layer of epidermal cells grows

Full thickness skin restored

Fibrous tissue

Figure 11:6 Some stages in secondary healing of skin.

1. Acute inflammation develops on the surface of the healthy tissue and separation of necrotic tissue (*slough*) begins, due mainly to the action of phagocytes in the inflammatory exudate
2. Granulation tissue, consisting of new budding capillaries and fibroblasts, begins to develop at the base of the cavity and grows towards the surface, probably stimulated by macrophages in the inflamed area
3. Phagocytes in the plentiful blood supply tend to prevent infection of the wound after separation of the slough
4. For unknown reasons some fibroblasts at the edges of the wound develop a limited ability to contract, reducing the size of the wound and the healing time
5. When granulation tissue reaches the level of the dermis, epithelial cells at the edges proliferate and grow towards the centre, forming a single layer of cells which gradually increase in number until, several months later, the full thickness of skin is restored
6. In time the scar tissue shrinks and becomes less vascular

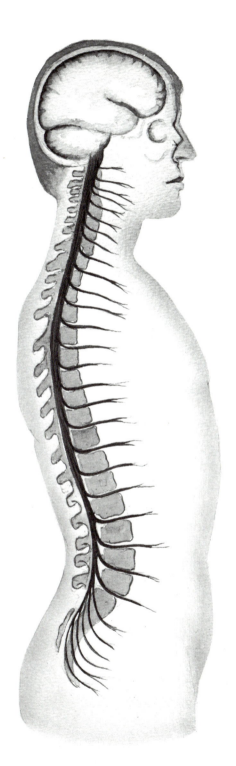

The Nervous System

12. The Nervous System

In a systematic study of anatomy and physiology the systems of the body are described separately, but it must be appreciated that they are dependent upon each other. In previous chapters the nerve supply to the organs has been mentioned but not described in detail.

For descriptive purposes the parts of the nervous system are grouped as follows:

1. The central nervous system consisting of the brain and the spinal cord
2. The peripheral nervous system consisting of
   31 pairs of spinal nerves
   12 pairs of cranial nerves
   the autonomic part of the nervous system

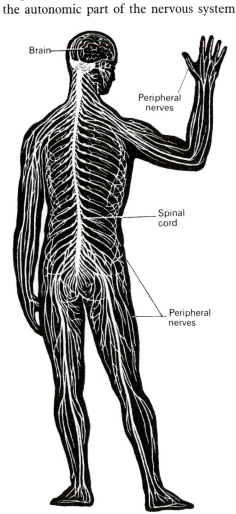

Figure 12:1 General view of the brain, spinal cord and spinal nerves.

NEURONES

The nervous system consists of a vast number of units called *neurones* (Fig. 12:2), supported by a special type of connective tissue, *neuroglia*. Each neurone consists of a *nerve cell* and its processes, *axons* and *dendrites*. Neurones are commonly referred to simply as nerves.

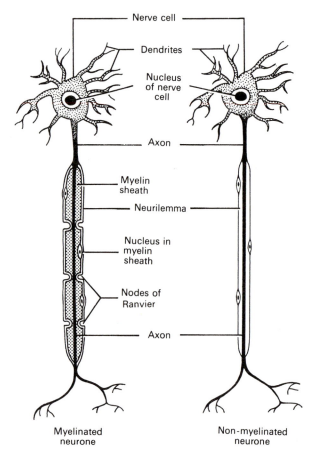

Figure 12:2 A neurone.

The physiological 'units' of the nervous system are *nerve impulses* which are akin to tiny electrical charges. However, unlike ordinary electrical wires, the neurones are actively involved in conducting nerve impulses. In effect the strength of the impulse is maintained throughout the length of the neurone.

The cells of some neurones initiate nerve impulses while others act as 'relay stations' where impulses are passed on and sometimes redirected.

NERVE CELLS

The nerve cells vary considerably in size and shape but they are all too small to be seen by the naked eye. They form the grey matter of the nervous system and are found at the periphery of the brain, in the centre of the spinal cord, in groups called *ganglia* outside the brain and spinal cord and as single cells in the walls of organs.

AXONS AND DENDRITES

Axons and dendrites are the processes of the nerve cells and form the *white matter* of the nervous system. They are found deep in the brain, at the periphery of the spinal cord and they are referred to as *nerves* or *nerve fibres* outside the brain and spinal cord.

AXONS

Each nerve cell has only one axon, carrying nerve impulses away from the cell. They are usually longer than the dendrites, sometimes about 100 cm long.

Structure of an axon
1. *Axolemma* is the membrane of the axon and the internal cytoplasma in *axoplasm*.
2. *Myelin* is a sheath of fatty material which surrounds most axons and gives them a white appearance, thus the term *white matter* used to describe collections of myelinated axons in the spinal cord and brain. Postganglionic autonomic fibres, some small fibres in the central nervous system and some fine peripheral sensory fibres are non-myelinated. The myelin sheath:

Acts as an insulator
Protects the axon from injury
Speeds the flow of nerve impulses through the axon

The myelin sheath is absent at intervals along the length of the axon and near its branching end. The breaks are the *nodes of Ranvier*. They aid the rapid transmission of nerve impulses along myelinated fibres.

3. *The neurilemma* is a very fine, delicate membrane that encases the axons of all peripheral nerves. It consists of a series of *Schwann cells* which surround the axon and myelin sheath. A group of non-myelinated fibres is enclosed within one series of Schwann cells.

DENDRITES

The dendrites are the processes or nerve fibres which carry impulses towards nerve cells. They have the same structure as axons but they are usually shorter and branching. Each neurone has many dendrites.

TYPES OF NERVES

SENSORY OR AFFERENT NERVES

These are the nerves that transmit impulses from the periphery of the body to the spinal cord. The impulses may then pass to the brain or to connector neurones of reflex arcs (see p. 224). The *somatic* or *common senses*, originating in the skin, are pain, touch, heat and cold. *Proprioceptor senses* originate in muscles and joints and contribute to the maintenance of balance and posture. *Special senses* are sight, hearing, smell and taste. *Autonomic afferent nerves* originate in internal organs and tissues and are associated with reflex regulation of activity and visceral pain.

MOTOR OR EFFERENT NERVES

Motor nerves originate in the brain, spinal cord and autonomic ganglia. They are involved in:
1. Voluntary and reflex skeletal muscle contraction
2. Involuntary (autonomic) smooth muscle contraction and glandular secretion

MIXED NERVES

In the spinal cord, sensory and motor nerves and arranged in separate groups, or *tracts*. Outside the spinal cord, when sensory and motor nerves are enclosed within the same sheath of connective tissue they are called *mixed nerves*.

SYNAPSE AND CHEMICAL TRANSMITTERS

There is always more than one neurone involved in the transmission of a nerve impulse from its origin to its destination, whether it is sensory or motor. There is no anatomical continuity between these neurones and the point at which the nerve impulse passes from one to another is the *synapse* (Fig. 12:3 A and B). At its free end the axon of one neurone breaks up into minute branches which terminate in small swellings called *presynaptic knobs* which are in close proximity to the dendrites and the cell body of the next neurone. The space between them is the *synaptic cleft*. At the ends of presynaptic knobs there are spherical *synaptic vesicles*, containing *chemical transmitters* which carry nerve impulses across synaptic clefts. Chemical transmitters are secreted by nerve cells, actively transported along axons and stored in synaptic vesicles. After release their action is short-lived as immediately they have stimulated the next neurone they are neutralised by enzymes. A knowledge of the action of different chemical transmitters has become important because of the drugs now available to neutralise them or prolong their effect.

The chemical transmitters and their modes of action in the brain and spinal cord are not yet fully understood. It is believed however that *noradrenalin, 5-hydroxytryptamine*

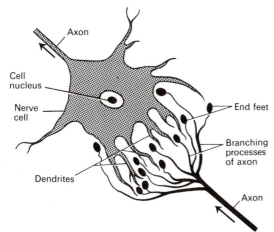

Figure 12:3A Diagram of a synapse.

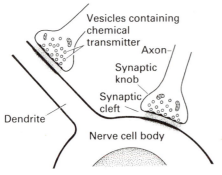

Figure 12:3B Magnified synapse.

(*serotonin*), *gamma aminobutyric acid* (*GABA*), *dopamine* and *acetylcholine* act as chemical transmitters.

Figure 12:4 summarises the chemical transmitters known to function outside the brain and spinal cord.

TERMINATION OF NERVES

The *sensory nerves*, e.g., in the skin, lose their myelin sheath and neurilemma and divide into fine branching filaments, the *sensory nerve endings* (Fig. 12:5). These are stimulated in the skin by touch, pain, heat and cold. The impulse is then transmitted to the brain where the sensation is perceived.

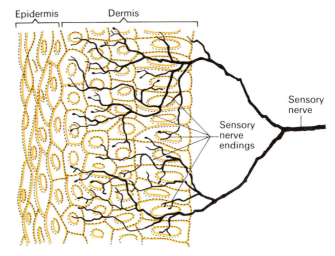

Figure 12:5 Sensory nerve endings in the skin.

The *motor nerves*, conveying impulses to skeletal muscle to produce contraction, divide into fine filaments terminating in minute pads called *motor end-plates* (Fig. 12:6). At the point where the nerve reaches the muscle the myelin sheath and neurilemma are absent and the fine filament passes to a sensitive area on the surface of the muscle fibre. Each muscle fibre is stimulated through a

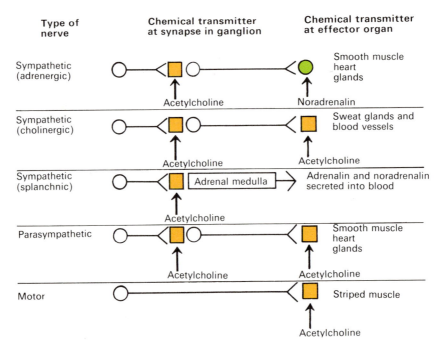

Figure 12:4 Chemical transmitters of nerve impulses at synapses outside the brain and spinal cord.

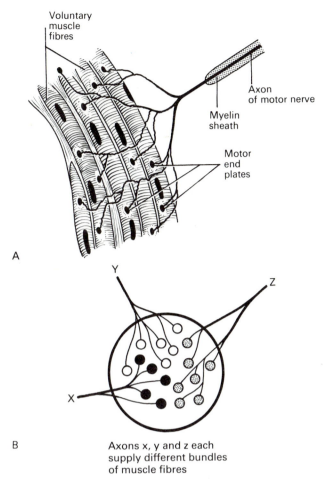

A

B

Axons x, y and z each
supply different bundles
of muscle fibres

Figure 12:6 A motor unit. A. Longitudinal section. B. Cross section.

single motor end-plate, and one motor nerve has many motor end-plates. The motor end-plate and the sensitive area of muscle fibre through which it is stimulated is analogous to the synapse between neurones. The nerve impulse is passed across the gap between the motor end-plate and the muscle fibre by the neurotransmitter, *acetylcholine*. The group of muscle fibres and the motor end-plates of the nerve fibre that supplies them constitute a *motor unit*. Nerve impulses cause serial contraction of motor units in a muscle and each unit contracts to its full capacity. The *strength* of the contraction depends on the *number* of motor units in action at a particular time.

The endings of *autonomic nerves* supplying smooth muscle and glands branch near their effector structure and secrete a transmitter substance which stimulates or depresses the activity of the structure.

THE PROPERTIES OF NERVE TISSUE

Nerve tissue has the characteristics of *irritability* and *conductivity*.

Irritability is the ability to initiate nerve impulses in response to stimuli from:

1. Outside the body, e.g., touch, light waves
2. Inside the body, e.g., a change in the concentration of carbon dioxide in the blood alters respiration; a thought may result in voluntary movement

In the body this stimulation may be described as partly electrical and partly chemical — electrical in that motor nerve cells and sensory nerve endings initiate nerve impulses, and chemical in the transmission of impulses across synapses.

Conductivity means the ability to transmit an impulse from:

1. One part of the brain to another
2. The brain to striated muscle, resulting in voluntary muscle contraction
3. Muscles and joints to the brain, contributing to the maintenance of balance
4. The brain to organs of the body, resulting in the contraction of smooth muscle or secretion by glands
5. Organs of the body to the brain in association with the regulation of body functions
6. The outside world to the brain through sensory nerve endings in the skin which are stimulated by temperature, touch, pain
7. The outside world to the brain through the special sense organs, i.e., eyes, ears, nose, tongue

CENTRAL NERVOUS SYSTEM

The central nervous system consists of the brain and the spinal cord.

NEUROGLIA

The neurones of the central nervous system are supported by three types of non-excitable *glial cells* that make up a quarter to a half of the volume of brain tissue. Unlike nerve cells these continue to replicate throughout life. They are *astrocytes*, *oligodendrocytes* and *microglia*.

ASTROCYTES

These cells form the main supporting tissue of the central nervous system. They are star-shaped with fine branching processes and they lie in a mucopolysaccharide ground substance. At the free ends of some of the processes there are small swellings called *foot processes*. Astrocytes are found in large numbers adjacent to blood vessels with their foot processes forming a sleeve round them. This means that the blood is separated from the neurones by the capillary wall and a layer of astrocyte foot processes which together constitute the *blood–brain barrier* (see Fig. 12:7). Their functions are analogous to those of fibroblasts elsewhere in the body.

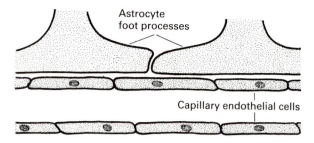

Figure 12:7 Blood brain barrier.

OLIGODENDROCYTES

These cells are smaller than astrocytes and are found:
1. In clusters round the neurone cells walls in grey matter
2. Adjacent to, and along the length of, myelinated nerve fibres

The oligodendrocytes form and maintain myelin, having the same functions as Schwann cells in peripheral nerves.

MICROGLIA

These cells are derived from monocytes that migrate from the blood into the nervous system before birth. They are found mainly in the area of blood vessels. They enlarge and become phagocytic in areas of inflammation and cell destruction.

MEMBRANES COVERING THE BRAIN AND SPINAL CORD

The brain and spinal cord are completely surrounded by three membranes, the *meninges*, lying between the skull and the brain and between the vertebrae and the spinal cord (Fig. 12:8). Named from without inwards they are:

Dura mater
Arachnoid mater
Pia mater

The dura and arachnoid maters are separated by a potential space, the *subdural space*, and the arachnoid and pia maters by the *subarachnoid space*, containing *cerebrospinal fluid*.

DURA MATER

The cerebral dura mater consists of two layers of dense fibrous tissue. The outer layer takes the place of the periosteum on the inner surface of the skull bones and the inner layer provides a protective covering for the brain. There is only a potential space between the two layers except where the inner layer sweeps inwards between the cerebral hemispheres to form the *falx cerebi*; between the cerebellar hemispheres to form the *falx cerebelli*; and between the cerebrum and cerebellum to form the *tentorium cerebelli*.

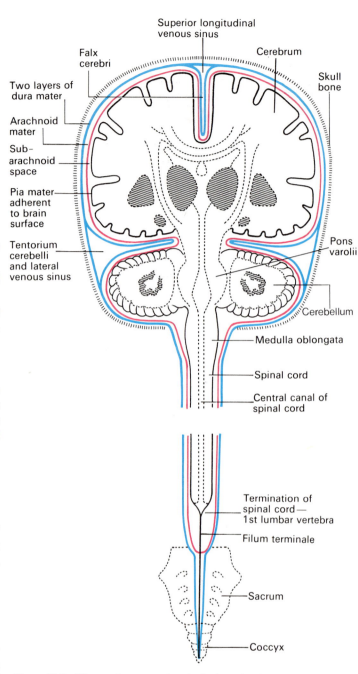

Figure 12:8 The meninges covering the brain and spinal cord.

Venous blood from the brain drains into venous sinuses between the layers of dura mater. The *superior sagittal sinus* is formed by the falx cerebri, and the tentorium cerebelli forms the *straight* and *transverse sinuses*.

The *spinal dura mater* corresponds to the inner layer of cerebral dura mater and forms a loose sleeve around the spinal cord. Between it and the periosteum and ligaments lining the vertebral canal, there is the *epidural* or *extradural space*, containing blood vessels and areolar tissue.

The spinal dura mater begins at the foramen magnum, where the spinal cord leaves the skull, and extends to the level of the second sacral vertebra. Thereafter it invests the *filum terminale* and fuses with the periosteum of the coccyx.

ARACHNOID MATER

This delicate serous membrane lies between the dura and pia maters. It is separated from the dura mater by the *subdural space*, and from the pia mater by the *subarachnoid space*, containing *cerebrospinal fluid*. The arachnoid mater passes over the convolutions of the brain and accompanies the inner layer of dura mater in the formation of the falx cerebri, tentorium cerebelli and falx cerebelli. It continues downwards to envelop the spinal cord and ends by merging with the dura mater at the level of the 2nd sacral vertebra.

PIA MATER

This is a fine connective tissue containing many minute blood vessels. It closely invests the brain, completely covering the convolutions and diping into each fissure. It continues downwards to invest the spinal cord. Beyond the end of the cord it continues as the *filum terminale*, pierces the arachnoid tube and goes on, with the dura mater, to fuse with the periosteum of the coccyx.

VENTRICLES OF THE BRAIN AND THE CEREBROSPINAL FLUID

Within the brain there are four irregular-shaped cavities, or *ventricles*, containing *cerebrospinal fluid* (CSF) (Fig.12:9). They are:

Right and left lateral ventricles
Third ventricle
Fourth ventricle

The lateral ventricles
These cavities lie within the cerebral hemispheres, one on each side of the median plane just below the *corpus callosum*. They are separated from each other by a thin membrane, the *septum lucidum*, and are lined with ciliated epithelium. They communicate with the third ventricle by *interventricular foramina*.

The third ventricle
The third ventricle is a cavity situated below the lateral ventricles between the two parts of the *thalamus*. It communicates with the fourth ventricle by a canal, the *cerebral aqueduct* or aqueduct of the midbrain.

The fourth ventricle
The fourth ventricle is a lozenge-shaped cavity situated below and behind the third ventricle, between the *cerebellum* and *pons varolii*. It is continuous below with the *central canal* of the spinal cord.

The fourth ventricle communicates with the subarachnoid space by foramina in its roof. Cerebrospinal fluid enters the subarachnoid space through these openings and through the open distal end of the central canal of the spinal cord.

CEREBROSPINAL FLUID (CSF) (Fig. 12:10)

Cerebrospinal fluid is secreted into each ventricle of the brain by *choroid plexuses*, consisting of areas where the lining membrane of the ventricle walls is thin and has a profusion of blood capillaries.

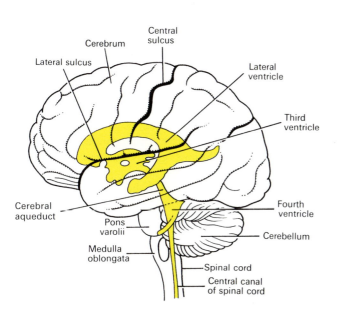

Figure 12:9 The positions of the ventricles of the brain superimposed on its surface. Viewed from the left side.

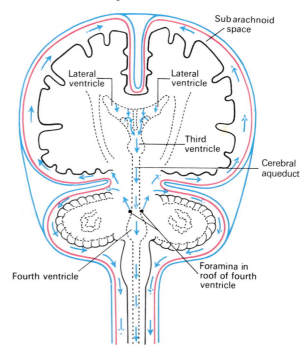

Figure 12:10 Arrows showing the flow of cerebrospinal fluid.

From the roof of the 4th ventricle CSF flows through foramina into the subarachnoid space and completely surrounds the brain and spinal cord. CSF is reabsorbed into blood capillaries in the arachnoid mater and returned to the circulating blood. It is a clear, slightly alkaline fluid with a specific gravity of 1005, consisting of:

Water
Mineral salts
Glucose
Plasma proteins: small amounts of albumin and globulin
Creatinine ⎱
Urea ⎰ small amounts

The normal hydrostatic pressure of CSF is about 10 mmHg when the individual is lying on his side and about 30 mmHg when sitting up.

Functions of the cerebrospinal fluid
1. It supports and protects the brain and spinal cord
2. It maintains a uniform pressure around these delicate structures
3. It acts as a cushion and shock absorber between the brain and the cranial bones
4. It keeps the brain and spinal cord moist and there may be interchange of substances between CSF and nerve cells

BRAIN

The brain constitutes about one-fiftieth of the body weight and lies within the cranial cavity. The parts are (Fig. 12:11):

Cerebrum of forebrain
Midbrain ⎫
Pons varolli ⎬ the brain stem
Medulla oblongata ⎭
Cerebellum or hindbrain

CEREBRUM

This is the largest part of the brain and it occupies the anterior and middle cranial fossae (see Fig. 16:4, p. 306). It is divided by a deep cleft, the *longitudinal cerebral fissure*, into *right* and *left cerebral hemispheres*, each containing one of the lateral ventricles. Deep within the brain the hemispheres are connected by a mass of white matter (nerve fibres) called the *corpus callosum*. The falx cerebri separates the two hemispheres and penetrates to the depth of the corpus callosum. The superficial (peripheral) part of the cerebrum is composed of nerve cells or grey matter, forming the *cerebral cortex* and the deeper layers consist of nerve fibres or white matter.

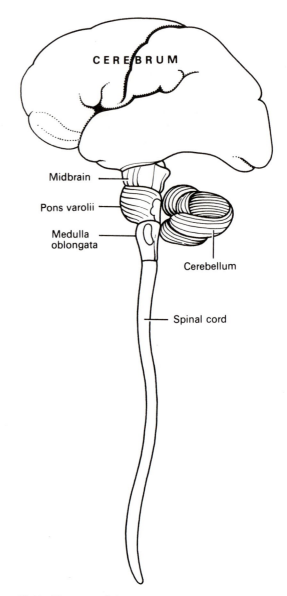

Figure 12:11   The parts of the central nervous system.

The cerebral cortex shows many infoldings or furrows of varying depth. The exposed areas of the folds are the *gyri* or *convolutions* and these are separated by *sulci* or *fissures*. These convolutions greatly increase the surface area of the cerebrum.

For descriptive purposes each hemisphere of the cerebrum is divided into *lobes* which take the names of the bones of the cranium under which they lie:

Frontal
Parietal
Temporal
Occipital

The boundaries of the lobes are marked by deep sulci (fissures). These are the *central*, *lateral* and *parieto-occipital sulci* (see Fig. 12:12).

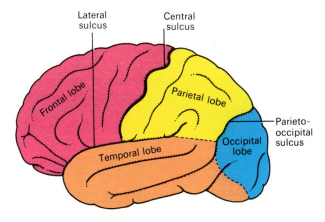

Figure 12:12   The lobes and sulci of the cerebrum.

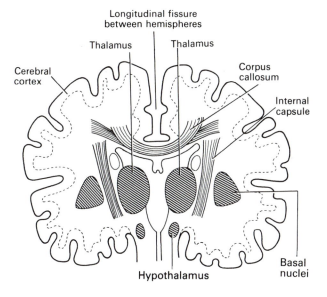

Figure 12:13   A section of the cerebrum showing some connecting nerve fibres.

basal nuclei and the thalamus. All nerve impulses passing to and from the cerebral cortex are carried by fibres that form the internal capsule.

Functions of the cerebrum

There are three main varieties of activity associated with the cerebral cortex:

1. Mental activities involved in memory, intelligence, sense of responsibility, thinking, reasoning, moral sense and learning are attributed to the *higher centres*
2. Sensory perception, including the perception of pain, temperature, touch, sight, hearing, taste and smell
3. Initiation and control of voluntary muscle contraction

Functional areas of the cerebrum (Fig. 12:14)

The main areas of the cerebrum associated with sensory perception and voluntary motor activity are known but it is unlikely that any area is associated exclusively with only one function. Except where specially mentioned, the different areas are active in both hemispheres.

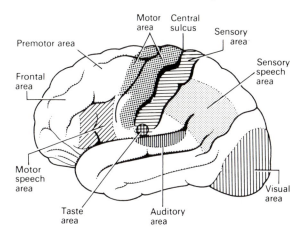

Figure 12:14   The cerebrum showing the functional areas.

Interior of the cerebrum (Fig. 12:13)

The cerebral cortex is composed of nerve cells on the surface. Within the cerebrum the lobes are connected by masses of nerve fibres, or tracts, which make up the white matter of the brain. The afferent and efferent fibres linking the different parts of the brain and spinal cord are:

1. *Arcuate (association) fibres* which connect different parts of the cerebral cortex by extending from one gyrus to another, some of which are adjacent and some distant
2. *Commissural fibres* which connect the two cerebral hemispheres (corpus callosum)
3. *Projection fibres* which connect the cerebral cortex with grey matter of lower parts of the brain and with the spinal cord, e.g., the internal capsule

The *internal capsule* is an important area consisting of projection fibres. It lies deep within the brain between the

The precentral (motor) area lies in the frontal lobe immediately anterior to the *central sulcus*. The nerve cells are pyramid-shaped (Betz cells) and they initiate the contraction of voluntary muscles. A nerve fibre from a Betz cell passes downwards through the internal capsule to the medulla oblongata where it crosses to the opposite side then descends in the spinal cord. At the appropriate level in the spinal cord the nerve impulse crosses a synapse to stimulate a second neurone which terminates at the motor end-plate of a muscle fibre. This means that the motor area of the *right hemisphere* of the cerebrum controls voluntary muscle movement on the left side of the body and vice versa. The neurone with its cell in the cerebrum is the *upper motor neurone* and the other, with its cell in the spinal cord, is the *lower motor neurone* (Fig. 12:15). Damage to either of these neurones may result in paralysis.

In the motor area of the cerebrum the body is represented upside down, i.e., the cells nearest the vertex

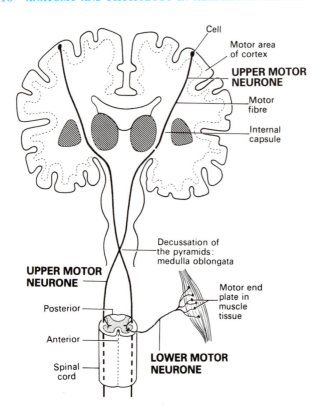

Figure 12:15 The motor nerve pathways: upper and lower motor neurones.

control the feet and those in the lowest part control the head, neck, face and fingers (Fig. 12:16A). The sizes of the areas of cortex representing different parts of the body are proportional to the *complexity of movement* of the part, not to its size. Figure 12:16A shows that, in comparison with the trunk, the hand, tongue and lips are represented by large cortical areas.

The premotor area lies in the frontal lobe immediately anterior to the motor area. The cells are thought to exert a controlling influence over the motor area, ensuring an orderly series of movements. For example, in tying a shoe lace or writing, many muscles contract but the movements must be co-ordinated and carried out in a particular sequence. Such a pattern of movement, when established, is described as *manual dexterity*.

In the lower part of this area just above the lateral sulcus there is a group of nerve cells known as the *motor speech (Broca's) area* which controls the movements necessary for speech. It is dominant in the *left hemisphere* in *right-handed people* and vice versa.

The frontal area or *pole* extends anteriorly from the premotor area to include the remainder of the frontal lobe. It is a large area and is more highly developed in humans than in other animals. It is thought that communications between this and the other regions in the cerebrum are responsible for the behaviour, character and emotional state of the individual. No particular behaviour, character or intellectual trait has, so far, been attributed to the activity of any one group of cells.

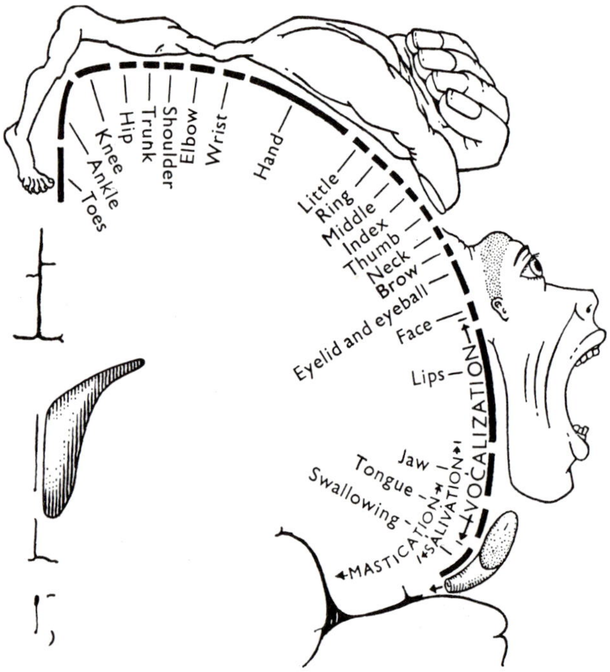

Figure 12:16A *The motor homonculus* showing how the body is represented in the motor area of the cerebrum.

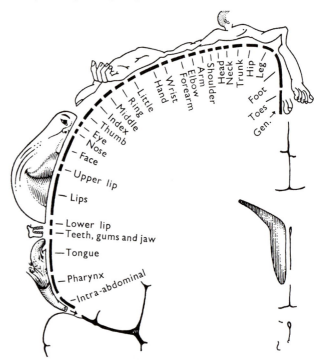

Figure 12:16B *The sensory homonculus* showing how the body is represented in the sensory area of the cerebrum. (Both A and B from Penfield W, Rasmussen T 1950 The cerebral cortex of man. Macmillan, New York. Reproduced with permission.)

The postcentral (sensory) area is the area behind the central sulcus. Here sensations of pain, temperature, pressure and touch, knowledge of muscular movement and the position of joints are perceived. The sensory area of the *right hemisphere* receives impulses from the *left side of the body* and vice versa. The size of the areas representing different parts of the body (Fig. 12:16B) is proportional to the *extent of sensory innervation*, e.g., the large area for the face is consistent with the extensive sensory nerve supply by the three branches of the trigeminal nerve (5th cranial nerve).

The parietal area lies behind the postcentral area and includes the greater part of the parietal lobe of the cerebrum. Its functions are believed to be associated with obtaining and retaining accurate knowledge of objects. It has been suggested that objects can be recognised by touch alone because of the knowledge from past experience retained in this area.

The sensory speech area is situated in the lower part of the parietal lobe and extends into the temporal lobe. It is here that the spoken word is perceived. There is a dominant area in the *left hemisphere* if the individual is *right-handed* and vice versa.

The auditory (hearing) area lies immediately below the lateral sulcus within the temporal lobe. The cells receive and interpret impulses transmitted from the inner ear by the vestibulocochlear (auditory) nerves.

The olfactory (smell) area lies deep within the temporal lobe where impulses received from the nose via the olfactory nerves are received and interpreted.

The taste area is thought to lie just above the lateral sulcus in the deep layers of the sensory area, and it is here that taste is perceived.

The visual area lies behind the parieto-occipital sulcus and includes the greater part of the occipital lobe. The optic nerves (nerves of the sense of sight) pass from the eye to this area which receives and interprets the impulses as visual impressions.

Deep within the cerebral hemispheres there are groups of nerve cells called *nuclei* or *ganglia* which act as relay stations where impulses are passed from one neurone to the next in a chain. Important masses of grey matter include:

Basal nuclei
Thalamus
Hypothalamus

Basal nuclei

This area of grey matter, lying deep within the cerebral hemispheres, is thought to influence skeletal muscle tone. If control is inadequate or absent movements are jerky, clumsy and unco-ordinated.

Thalamus

The thalamus consists of two masses of nerve cells and fibres situated within the cerebral hemispheres just below the corpus callosum, one on each side of the third ventricle. Sensory input from the skin, viscera and special sense organs are transmitted to the thalamus before redistribution to the cerebrum.

Hypothalamus

The hypothalamus is composed of a number of groups of nerve cells. It is situated below and in front of the thalamus, immediately above the pituitary gland. The hypothalamus is linked to the posterior lobe of the pituitary gland by nerve fibres and to the anterior lobe by a complex system of blood vessels. Through these connections the hypothalamus controls the output of hormones from both lobes of the gland (see p. 270).

Other functions with which the hypothalamus is concerned include control of the autonomic nervous system, e.g., control of hunger, thirst, body temperature, heart and blood vessels and defensive reactions, such as those associated with fear and rage.

MIDBRAIN

The midbrain is the area of the brain situated around the cerebral aqueduct between the cerebrum above and the *pons varolii* below. It consists of groups of nerve cells and nerve fibres which connect the cerebrum with lower parts of the brain and with the spinal cord. The nerve cells act as relay stations for the ascending and descending nerve fibres.

PONS VAROLII

The pons varolii is situated in front of the cerebellum, below the midbrain and above the medulla oblongata. It consists mainly of nerve fibres which form a bridge between the two hemispheres of the cerebellum, and of fibres passing between the higher levels of the brain and the spinal cord. There are groups of cells within the pons which act as relay stations and some of these are associated with the cranial nerves.

The anatomical structure of the pons varolii differs from that of the cerebrum in that the nerve cells lie deeply and the nerve fibres are on the surface.

MEDULLA OBLONGATA

The medulla oblongata extends from the pons varolii above and is continuous with the spinal cord below. It is about 2.5 cm long, is shaped like a pyramid with its base upwards and it lies just within the cranium above the foramen magnum. Its anterior and posterior surfaces are marked by central fissures. The outer aspect is composed of *white matter* which passes between the brain and the spinal cord and *grey matter lies* centrally. Some cells constitute relay stations for sensory nerves passing from the spinal cord to the cerebrum.

The vital centres, consisting of groups of cells associated with autonomic reflex activity, lie in its deeper structure. These are the:

Cardiac centre
Respiratory centre
Vasomotor centre
Reflex centres of vomiting, coughing, sneezing and swallowing

The medulla oblongata has several special features:

1. *Decussation of the pyramids*. In the medulla the majority of *motor nerves* descending from the motor area in the cerebrum to the spinal cord cross from one side to the other. This means that the left hemisphere of the cerebrum controls the right half of the body, and vice versa

2. *Sensory decussation*. Some of the *sensory nerves* ascending to the cerebrum from the spinal cord cross from one side to the other in the medulla. Others decussate at lower levels, i.e., in the spinal cord.

3. *The cardiac centre* controls the rate and force of cardiac contraction. Sympathetic and parasympathetic nerve fibres originating in the medulla pass to the heart. Sympathetic stimulation increases the rate and force of the heart beat and parasympathetic stimulation has the opposite effect.

4. *The respiratory centre* controls the rate and depth of respiration. From this centre, nerve impulses pass to the phrenic and intercostal nerves which stimulate contraction of the diaphragm and intercostal muscles, thus initiating inspiration. The respiratory centre is stimulated by excess carbon dioxide and, to a lesser extent, by deficiency of oxygen in the blood.

5. *The vasomotor centre* controls the diameter of the blood vessels, especially the small arteries and arterioles which have a large proportion of smooth muscle fibres in their walls. Vasomotor impulses reach the blood vessels through the autonomic nervous system. Stimulation may cause either constriction or dilatation of blood vessels depending on the site (see Figs 12:41 and 12:42).

The sources of stimulation of the vasomotor centre are the arterial baroreceptors, body temperature and emotions such as sexual excitement and anger. Pain usually causes vasoconstriction although severe pain may cause vasodilation, a fall in blood pressure and fainting.

6. *Reflex centres*. When irritating substances are present in the stomach or respiratory tract, nerve impulses pass to the medulla oblongata, stimulating the reflex centres which initiate the reflex actions of vomiting, coughing and sneezing.

CEREBELLUM

The cerebellum is situated behind the pons varolii and immediately below the posterior portion of the cerebrum occupying the posterior cranial fossa (Fig. 12:17). It is ovoid in shape and has two hemispheres, separated by a narrow median strip called the *vermis*. Grey matter forms the surface of the cerebellum, and the white matter lies deeply.

Functions

The cerebellum is concerned with voluntary muscular movement and balance. However, cerebellar activities are carried out below the level of consciousness, i.e., not under voluntary control. The cerebellum controls and co-ordinates the movements of various groups of muscles ensuring smooth, even, precise actions. It co-ordinates activities associated with the *maintenance of the balance and equilibrium* of the body. The sensory input for the functions is derived from the muscles and joints, the eyes and the ears. *Proprioceptor impulses* from the muscles and joints indicate their position in relation to the body as a whole and those impulses from the eyes and the semicircular canals in the ears provide information about the position of the head in space. Impulses from the cerebellum influence the contraction of skeletal muscle so that balance and posture are maintained.

Damage to the cerebellum results in clumsy uncoordinated muscular movement, staggering gait and inability to carry out smooth, steady, precise movements.

RETICULAR FORMATION

The reticular formation is a collection of neurones in the core of the brain stem, surrounded by neural pathways which pass nerve impulses between the brain and the spinal cord. It has a vast number of synaptic links with

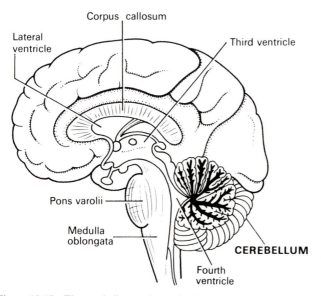

Figure 12:17 The cerebellum and associated structures.

other parts of the brain and is therefore constantly receiving 'information' being transmitted in ascending and descending tracts.

Functions

The reticular formation is involved in:

1. Co-ordination of skeletal muscle activity associated with voluntary motor movement and the maintenance of balance.

2. Co-ordination of activity controlled by the autonomic nervous system, e.g., cardiovascular, respiratory and gastro-intestinal activity.

3. Selective awareness that functions through the *reticular activating system* (RAS) which selectively blocks or passes sensory information to the cerebral cortex, e.g., the slight sound made by a sick child moving in bed may arouse his mother but the noise of regularly passing trains may be suppressed.

SPINAL CORD

The spinal cord is the elongated, almost cylindrical part of the central nervous system, which is suspended in the vertebral canal surrounded by the meninges and cerebrospinal fluid (see Fig. 12:18). It is continuous above with the medulla oblongata and extends from the *upper border of the atlas* to the lower border of the *1st lumbar vertebra* (Fig. 12:19). It is approximately 45 cm long in an adult caucasian male, and is about the thickness of the little finger. When a specimen of cerebrospinal fluid is required it is taken from a point beyond the end of the cord, i.e., below the level of the 2nd lumbar vertebra.

The spinal cord is the nervous tissue link between the brain and the rest of the body (Fig. 12:20). Nerves conveying impulses from the brain to the various organs

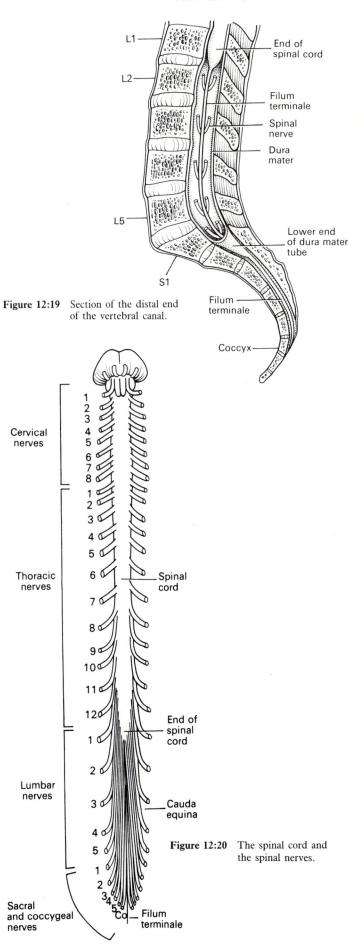

Figure 12:19   Section of the distal end of the vertebral canal.

Figure 12:20   The spinal cord and the spinal nerves.

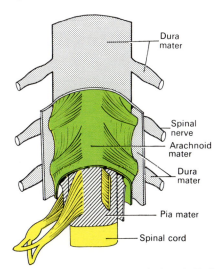

Figure 12:18   The meninges covering the spinal cord. Each cut away to show the various layers.

and tissues descend through the spinal cord. At the appropriate level they leave the cord and pass to the structure they supply. Similarly, sensory nerves from organs and tissues enter and pass upwards in the spinal cord to the brain.

Some activities of the spinal cord are independent of the brain, i.e., *spinal reflexes*. To facilitate these there are extensive neurone connections between sensory and motor neurones at the same or different levels in the cord.

STRUCTURE

The spinal cord is incompletely divided into two equal parts, anteriorly by a short, shallow *median fissure* and posteriorly by a deep narrow septum, the *posterior median septum*.

A cross-section of the spinal cord shows that it is composed of grey matter in the centre surrounded by white matter supported by neuroglia.

GREY MATTER

The arrangement of grey matter in the spinal cord resembles the shape of the letter H, having *two posterior, two anterior* and *two lateral columns*. The area of grey matter lying transversely is the *transverse commissure* and it is pierced by the central canal containing cerebrospinal fluid (see Fig. 12:21). The nerve cells may be:

1. *Sensory cells* which receive impulses from the periphery of the body
2. Cells of *lower motor neurones* which transmit impulses to the skeletal muscles

3. Cells of *connector neurones*, linking sensory and motor neurones, at the same or different levels, in the formation of spinal reflex arcs

At each point where nerve impulses are passed from one neurone to another there is a synaptic cleft and a chemical transmitter (see p. 212).

Posterior columns of grey matter

These are composed of nerve cells which are stimulated by *sensory impulses* from the periphery of the body. The nerve fibres of these cells contribute to the formation of the white matter of the cord and transmit the sensory impulses to the brain.

Anterior columns of grey matter

These are composed of the *cells of the lower motor neurones* which are stimulated by the axons of the upper motor neurones or by the *cells of connector neurones* linking the anterior and posterior columns.

The posterior root (spinal) ganglia are composed of nerve cells which lie just outside the spinal cord on the pathway of the sensory nerves. All sensory nerve fibres pass through these ganglia. The only function of the cells is to promote the onward movement of nerve impulses.

WHITE MATTER

The white matter of the spinal cord is arranged in three *columns* or *tracts*; anterior, posterior and lateral. These tracts are formed by *sensory nerve fibres* ascending to the brain, *motor nerve fibres* descending from the brain and fibres of *connector neurones*.

SENSORY NERVE TRACTS (Afferent or ascending) IN THE SPINAL CORD

There are two main sources of sensation transmitted to the brain via the spinal cord.

1. *The skin.* Sensory nerve endings in the skin, called *cutaneous receptors*, are stimulated by *pain, heat, cold* and *touch*. The nerve impulses are passed by three neurones to the sensory area in the *opposite hemisphere of the cerebrum* where the sensation and its location are perceived (Fig. 12:22). Crossing to the other side, or *decussation*, occurs either at the level of entry into the cord or in the medulla.

2. *The tendons, muscles and joints.* Sensory nerve endings in these structures, called *proprioceptors*, are stimulated by stretch. Together with impulses from the eyes and the ears they are associated with the maintenance of balance and posture and with perception of the position of the body in space. These nerve impulses have two destinations:

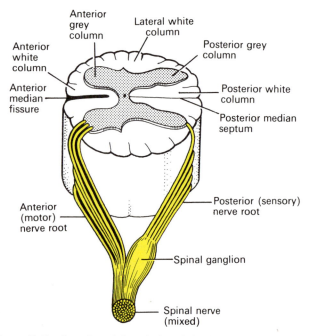

Figure 12:21   A section of the spinal cord showing nerve roots.

Anterior grey column
Lateral white column
Posterior grey column
Anterior white column
Anterior median fissure
Posterior white column
Posterior median septum
Posterior (sensory) nerve root
Anterior (motor) nerve root
Spinal ganglion
Spinal nerve (mixed)

Table 12:1 Sensory nerve impulses: origins, routes, destinations

Receptor		Route	Destination
Pain, touch, temperature	Neurone 1	to spinal cord by posterior root	
	Neurone 2	decussation on entering spinal cord then in anterolateral spinothalamic tract to thalamus	
		Neurone 3	to parietal lobe of cerebrum
Touch, proprioceptors	Neurone 1	to medulla in posterior spinothalamic tract	
	Neurone 2	decussation in medulla, transmission to thalamus	
		Neurone 3	to parietal lobe of cerebrum
Proprioceptors	Neurone 1	to spinal cord	
		Neurone 2	no decussation, to cerebellum in posterior spinocerebellar tract

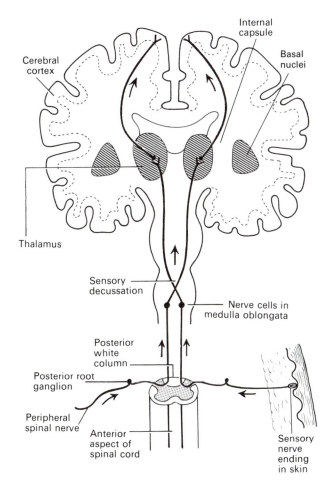

Figure 12:22 One of the sensory nerve pathways from the skin to the cerebrum.

Labels on figure: Internal capsule; Basal nuclei; Cerebral cortex; Thalamus; Sensory decussation; Nerve cells in medulla oblongata; Posterior white column; Posterior root ganglion; Peripheral spinal nerve; Anterior aspect of spinal cord; Sensory nerve ending in skin

a. By a three-neurone system the impulses reach the sensory area of the *opposite hemisphere of the cerebrum*.

b. By a two-neurone system the nerve impulses reach the *cerebellar hemisphere on the same side*.

Table 12:1 provides further information about the origins, routes of transmission and the destinations of sensory nerve impulses.

MOTOR NERVE TRACTS (Efferent or descending) IN THE SPINAL CORD

Neurones which transmit nerve impulses away from the brain are motor (efferent or descending) neurones. Motor neurone stimulation results in:

1. The contraction of voluntary (striated, skeletal) muscle
2. The contraction of smooth (involuntary) muscle and the secretion by glands controlled by nerves of the *autonomic part of the nervous system* (see p. 233)

Voluntary muscle movement

The contraction of the muscles which move the joints is, in the main, under the control of the will, which means that the stimulus to contract originates at the level of consciousness in the cerebrum. However, some nerve impulses which affect skeletal muscle contraction are initiated in the midbrain, brain stem and cerebellum. This activity occurs below the level of consciousness and is associated with co-ordination of muscle activity, e.g., when very fine movement is required and in the maintenance of posture and balance.

Efferent nerve impulses are transmitted from the brain to the body via bundles of nerve fibres or *tracts* in the spinal cord. The *motor pathways* from the brain to the muscles are made up of two *neurones*.

1. **The upper motor neurone** has its cell (Betz cell) in the *precentral sulcus area* of the cerebrum. The axons pass through the internal capsule, pons and medulla. In the spinal cord they form the *lateral corticospinal tracts* of white matter and the fibres terminate in close association with the cells of the *lower motor neurones* in the anterior columns of grey matter. The axons of most upper motor neurones decussate either in the medulla or in the spinal cord just before they terminate.

2. **The lower motor neurone** has its cell in the *anterior horn of grey matter* in the spinal cord. Its axon emerges from the spinal cord by the *anterior root*, joins with the incoming sensory fibres and forms the *mixed spinal nerve* which passes through the *intervertebral foramen*. Near its termination the axon branches into a variable number of tiny fibres which form *motor endplates*, each of which is in close association with a sensitive area on the wall of a muscle fibre. The motor end-plates of each nerve and the muscle fibres they supply form a *motor unit* (see Fig. 12:6). The chemical transmitter that conveys the nerve impulse across the gap to stimulate the muscle fibre is *acetylcholine*. Motor units contract as a whole and the strength of contraction depends on the number of motor units in action at a time.

The lower motor neurone has been described as the *final common pathway* for the transmission of nerve impulses to striated muscles. The cell of this neurone is influenced by a number of upper motor neurones originating from various sites in the brain and by some neurones which begin and end in the spinal cord. Some of these neurones stimulate the cells of the lower motor neurone while others have an inhibiting effect. The outcome of these influences is smooth, co-ordinated muscle movement, some of which is voluntary and some involuntary.

Involuntary muscle movement

1. *Upper motor neurones* which have their cells in the brain at a level *below* the cerebrum, i.e., in the midbrain,
brain stem, cerebellum or spinal cord, influence muscle activity in relation to the maintenance of posture and balance, the co-ordination of muscle movement and the control of muscle tone.

Table 12:2 shows details of the area of origin of these neurones and the tracts which their axons form before reaching the cell of the lower motor neurone in the spinal cord.

2. *Spinal reflexes*. These consist of three elements: sensory neurones, connector neurones in the spinal cord and lower motor neurones. In the simplest *reflex arc* there is only one of each (Fig. 12:23). A *reflex action* is an immediate motor response to a sensory stimulus. Many connector and motor neurones may be stimulated by afferent impulses from a small area of skin, e.g., the pain impulses initiated by touching a very hot surface with the finger are transmitted to the spinal cord by sensory nerves. These stimulate many connector and lower motor neurones in the cord which results in the contraction of many skeletal muscles of the hand, arm and shoulder, and the removal of the finger. Reflex action takes place very quickly, in fact, the motor response may have occurred simultaneously with the perception of the pain in the cerebrum. Reflexes of this type are invariably protective but they can on occasion be inhibited. For example, if it is a precious plate which is very hot when lifted every effort will be made to overcome the pain to prevent dropping the plate!

3. *Stretch reflexes*. Only two neurones are involved. The cell of the lower motor neurone is stimulated by the sensory neurone. There is no connector neurone involved. The *knee jerk* is one example, but this type of reflex can be demonstrated at any point where a stretched tendon crosses a joint. By tapping the tendon just below the knee when it is bent, the sensory nerve endings in the tendon and in the thigh muscles are stretched. This initiates a nerve impulse which passes into the spinal cord to the cell of the lower motor neurone in the anterior column of grey matter on the same side. As a result the thigh muscles suddenly contract and the foot kicks forward. This is used as a test of the integrity of the reflex arc. This type of reflex has a protective function — it prevents excessive joint movement that may damage tendons, ligaments and muscles.

4. *Autonomic reflexes*. See page 235.

Table 12:2 Upper motor neurones: origins and tract

Origin	Name of tract	Situation in spinal cord	functions
Midbrain and pons	Rubrospinal tract decussates in brain stem	Lateral column	Control of skilled muscle movement
Reticular formation	Reticulospinal tract does not decussate	Lateral column	Co-ordination of muscle movement. Maintenance of posture and balance
Midbrain and pons	Tectospinal tract decussates in midbrain	Anterior column	
Midbrain and pons	Vestibulospinal tract, some fibres decussate in the cord	Anterior column	

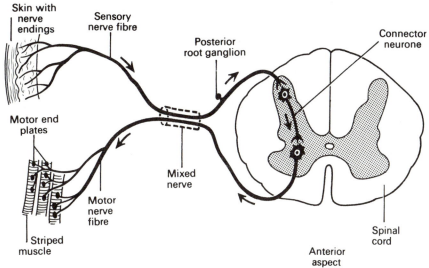

Figure 12:23 A single reflex arc.

PERIPHERAL NERVOUS SYSTEM

This part of the nervous system consists of:
   31 pairs of spinal nerves
   12 pairs of cranial nerves
   The autonomic part of the nervous system

Most of the nerves of the peripheral nervous system are composed of *sensory nerve fibres* conveying impulses from sensory end organs to the brain, and *motor nerve fibres* conveying impulses from the brain through the spinal cord to the effector organs, e.g., skeletal muscles, smooth muscle and glands.

Each nerve consists of numerous nerve fibres collected into bundles. Each bundle has several coverings of protective connective tissue (see Fig. 12:24):

1. *Endoneurium* is a delicate tissue, surrounding each individual fibre
2. *Perineurium* is a smooth tissue, surrounding each *bundle* of fibres

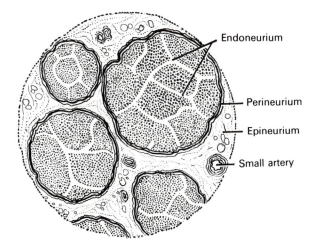

Figure 12:24   Transverse section of a peripheral nerve showing the protective coverings.

3. *Epineurium* is the tissue which surrounds and encloses a number of bundles of nerve fibres. Most large nerves are covered by epineurium.

SPINAL NERVES

There are *31 pairs of spinal nerves* that leave the vertebral canal by passing through the intervertebral foramina formed by adjacent vertebrae. They are named and grouped according to the vertebrae with which they are associated (see Fig. 12.20):

   8 cervical   5 sacral
   12 thoracic   1 coccygeal
   5 lumbar

Although there are only seven cervical vertebrae, there are eight nerves because the first pair leave the vertebral canal between the occipital bone and the atlas and the eighth pair leave below the last cervical vertebra. Thereafter the nerves are given the name and number of the vertebra immediately *above*.

The lumbar, sacral and coccygeal nerves leave the *spinal cord* near its termination at the level of the first lumbar vertebra, and extend downwards inside the vertebral canal in the subarachnoid space, forming a sheaf of nerves which resembles a horse's tail, the *cauda equina*. These nerves leave the vertebral canal at the appropriate lumbar, sacral or coccygeal level.

The spinal nerves arise from both sides of the spinal cord and emerge through the intervertebral foramina. Each nerve is formed by the union of a *motor and a sensory nerve root* and is, therefore, a *mixed nerve*. Each spinal nerve has a contribution from the sympathetic part of the autonomic nervous system in the form of a *preganglionic fibre*.

For details of the bones and muscles mentioned in the following section see Chapters 16 to 18. Bones and joints are supplied by adjacent nerves.

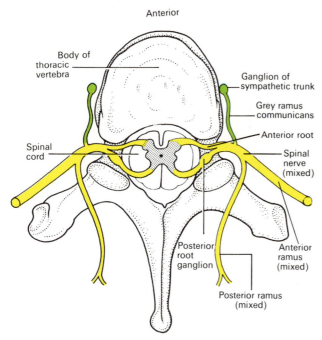

Figure 12:25 Diagram showing the relationship between sympathetic and mixed spinal nerves. Sympathetic part in green.

NERVE ROOTS (Fig. 12:26)

The anterior nerve root consists of *motor nerve fibres* which are the axons of the nerve cells is the anterior column of grey matter in the spinal cord and, in the thoracic and lumbar regions, *sympathetic nerve fibres* which are the axons of cells in the lateral columns of grey matter.

The posterior nerve root consists of *sensory nerve fibres*. Just outside the spinal cord there is a *spinal ganglion* (posterior root ganglion), consisting of a little cluster of nerve cells. Sensory nerve fibres pass through these ganglia before entering the spinal cord. The area of skin supplied by each nerve is called a *dermatome* (Figs 12:30 and 12:34).

For a very short distance after leaving the spinal cord the nerve roots have a covering of *dura* and *arachnoid maters*. These terminate before the two roots join to form the mixed spinal nerve. The nerve roots have no covering of pia mater.

Immediately after emerging from the intervertebral foramen each spinal nerve divides into a ramus communicans, a posterior ramus and an anterior ramus.

The rami communicans are part of preganglionic sympathetic neurones of the autonomic nervous system (see p. 234).

The posterior rami pass backwards and divide into medial and lateral branches to supply skin and muscles of relatively small areas of the posterior aspect of the head, neck and trunk.

The anterior rami supply the anterior and lateral aspects of the trunk and the upper and lower limbs.

In the cervical, lumbar and sacral regions the anterior rami unite near their origins to form large masses of

nerves, or *plexuses* where nerve fibres are regrouped and rearranged before proceeding to supply skin, bones, muscles and joints of a particular area.

In the thoracic region the anterior rami do not form plexuses.

There are five large plexuses of mixed nerves formed on each side of the vertebral column. They are the:

Cervical plexuses   Sacral plexuses
Brachial plexuses   Coccygeal plexuses
Lumbar plexuses

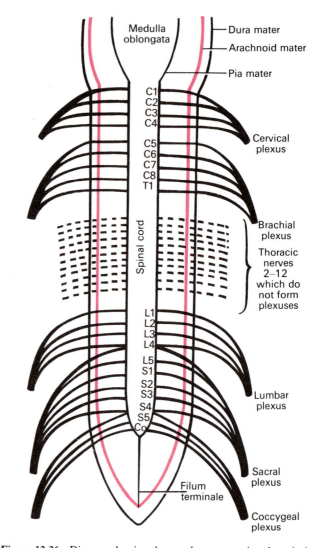

Figure 12:26 Diagram showing the membranes covering the spinal cord, spinal nerves and the plexuses they form.

Cervical plexus (Fig. 12:27)

This is formed by the anterior rami of the first four cervical nerves. It lies opposite the 1st, 2nd, 3rd and 4th cervical vertebrae under the protection of the sternocleido-mastoid muscle.

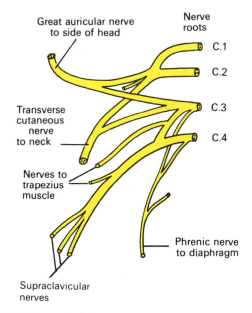

Figure 12:27   The cervical plexus.

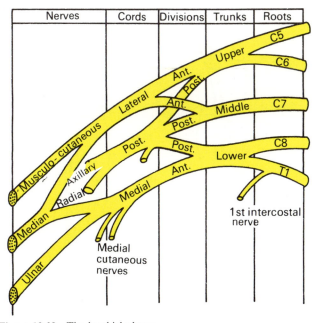

Figure 12:28   The brachial plexus.

The superficial branches supply the structures at the back and side of the head and the skin of the front of the neck to the level of the sternum.

The deep branches supply muscles of the neck, e.g., the sternocleidomastoid and the trapezius.

The phrenic nerve originates from cervical roots 3, 4 and 5 and passes downwards through the thoracic cavity in front of the root of the lung to supply the muscle of the diaphragm with impulses which stimulate contraction.

Brachial plexus

The anterior rami of the lower four cervical nerves and a large part of the first thoracic nerve from the brachial plexus. Figure 12:28 shows its formation and the nerves which emerge from it. The plexus is situated above and behind the subclavian vessels and in the axilla.

The branches of the brachial plexus supply the skin and muscles of the upper limbs and some of the chest muscles. Five large nerves and a number of smaller ones emerge from this plexus, each with a contribution from more than one nerve root, containing sensory, motor and autonomic fibres.

Axillary (circumflex) nerve: C5, 6
Radial nerve: C5, 6, 7, 8, T1
Musculocutaneous nerve: C5, 6, 7
Median nerve: C5, 6, 7, 8, T1
Ulnar nerve: C7, 8, T1
Medial cutaneous nerve: C8, T1

The axillary (circumflex) nerve winds round the humerus at the level of the surgical neck. It then breaks up into minute branches to supply the deltoid muscle, shoulder joint and overlying skin.

The radial nerve is the largest branch of the brachial plexus. It supplies the triceps muscle behind the humerus, crosses in front of the elbow joint then winds round to the back of the forearm to supply extensors of the wrist and finger joints. It continues into the back of the hand to supply the skin of the thumb, the first two fingers and the lateral half of the third finger.

The musculocutaneous nerve passes downwards to the lateral aspect of the forearm. It supplies the muscles of the upper arm and the skin of the forearm.

The median nerve passes down the midline of the arm in close association with the brachial artery. It passes in front of the elbow joint then down to supply the muscles of the front of the forearm. It continues into the hand where it supplies small muscles and the skin of the front of the thumb, the first two fingers and the lateral half of the third finger. It gives off no branches above the elbow.

The ulnar nerve descends through the upper arm lying medial to the brachial artery. It passes behind the medial epicondyle of the humerus to supply the muscles on the ulnar aspect of the forearm. It continues downwards to supply the muscles in the palm of the hand and the skin of the whole of the little finger and the medial half of the third finger. It gives off no branches above the elbow.

The main nerves of the arm are presented in Figure 12:29. The distribution and origins of the cutaneous sensory nerves of the arm are shown in Figure 12:30, i.e., the dermatomes.

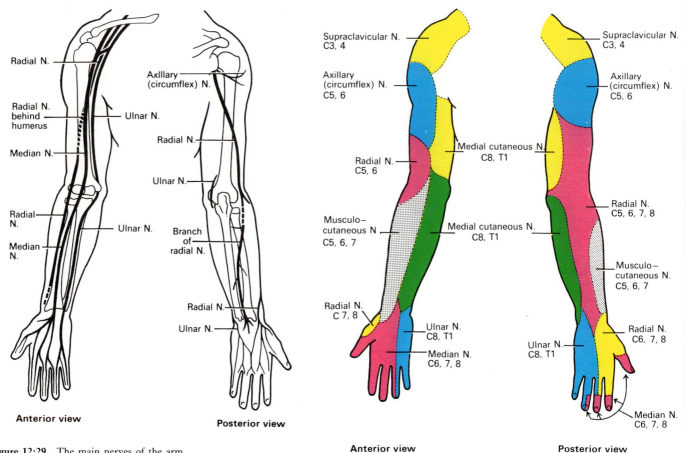

Figure 12:29 The main nerves of the arm.

Anterior view   Posterior view

Figure 12:30 The distribution and origins of the cutaneous nerves of the arm.

Lumbar plexus (Figs 12:31, 12:33 and 12:34)

The lumbar plexus is formed by the anterior rami of the first three and part of the fourth lumbar nerves. The plexus is situated in front of the transverse processes of the lumbar vertebrae and behind the psoas muscle. The main branches, and the nerve roots which contribute to them are:

    Iliohypogastric nerve: L1
    Ilioinguinal nerve: L1
    Genitofemoral: L1, 2
    Lateral cutaneous nerve of thigh: L2, 3
    Femoral nerve: L2, 3, 4
    Obturator nerve: L2, 3, 4
    Lumbosacral trunk: L4, (5)

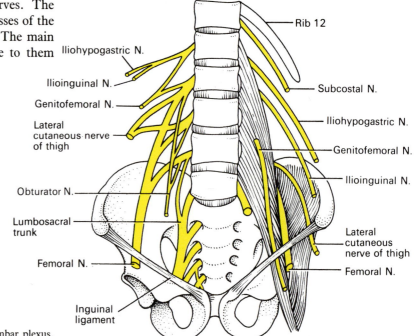

Figure 12:31 The lumbar plexus.

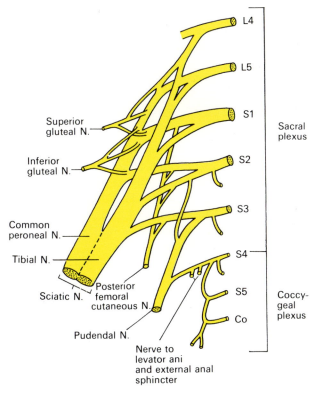

Figure 12:32   The sacral and coccygeal plexuses.

The iliohypogastric, ilioinguinal and *genitofemoral nerves* supply muscles and the skin in the area of the lower abdomen, upper and medial aspects of thigh and the inguinal region.

The lateral cutaneous nerve of the thigh supplies the skin of the lateral aspect of the thigh including part of the anterior and posterior surfaces.

The femoral nerve is one of the larger branches. It passes behind the inguinal ligament to enter the thigh in close association with the femoral artery. It divides into cutaneous and muscular branches to supply the skin and the muscles of the front of the thigh. It has one branch, the *saphenous nerve*, which supplies the medial aspect of the leg.

The obturator nerve supplies the adductor muscles of the thigh and skin of the medial aspect of the thigh. It ends just above the level of the knee joint.

The lumbosacral trunk descends into the pelvis and makes a contribution to the sacral plexus.

Sacral plexus (Figs 12:32, 12:33 and 12:34)
The sacral plexus is formed by the anterior rami of the first, second and third sacral nerves and by the *lumbosacral*

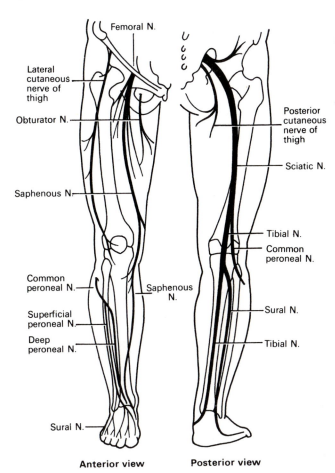

Figure 12:33   The main nerves of the leg.

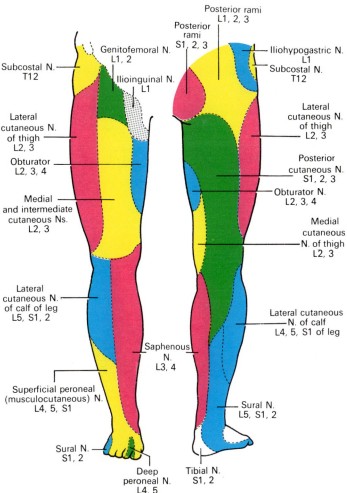

Figure 12:34   The distribution and origins of the cutaneous nerves of leg.

trunk, formed by the fifth and part of the fourth lumbar nerves. It lies in the posterior wall of the pelvic cavity.

The sacral plexus divides into a number of branches, supplying the muscles and skin of the pelvic floor, muscles around the hip joint and the pelvic organs. In addition to these it provides the *sciatic nerve* which contains fibres from L4, 5, S1, 2, 3.

The sciatic nerve is the largest nerve in the body. It is about 2 cm wide at its origin. It passes through the greater sciatic foramen into the buttock then descends through the posterior aspect of the thigh supplying the hamstring muscles. At the level of the middle of the femur it divides to form the *tibial* and the *common peroneal nerves*.

The tibial nerve descends through the popliteal fossa to the posterior aspect of the leg where it supplies muscles and skin. It passes under the medial maleolus to supply muscles and skin of the sole of the foot and toes. One of the main branches is the *sural nerve* which supplies the tissues in the area of the heel, the lateral aspect of the ankle and a part of the dorsum of the foot.

The common peroneal nerve descends obliquely along the lateral aspect of the popliteal fossa, winds round the neck of the fibula into the front of the leg where it divides into the *deep peroneal* (anterior tibial) and the *superficial peroneal* (musculocutaneous) nerves. These nerves supply the skin and muscles of the anterior aspect of the leg and the dorsum of the foot and toes.

The main nerves of the leg are shown in Figure 12:33. The distribution and origins of the cutaneous nerves of the leg are shown in Figure 12:34.

Coccygeal plexus (Fig. 12:32)
The coccygeal plexus is a very small plexus formed by part of the fourth sacral nerve and by the fifth sacral and the coccygeal nerves. The nerves from this plexus supply the skin in the area of the coccyx and the muscles of the pelvic floor.

Thoracic nerves
The thoracic nerves *do not* intermingle to form plexuses. There are 12 pairs and the first 11 are the *intercostal nerves*. They pass between the ribs supplying them, the intercostal muscles and overlying skin. The 12th pair are the subcostal nerves. The 7th to the 12th thoracic nerves also supply the muscles and the skin of the posterior and anterior abdominal walls.

CRANIAL NERVES (Fig. 12:35)

There are 12 pairs of cranial nerves originating from nuclei in the brain. Some sensory, some motor and some mixed. They have names and numbers.

    I. Olfactory: sensory
   II. Optic: sensory
 III. Oculomotor: motor
 IV. Trochlear: motor
  V. Trigeminal: mixed
 VI. Abducent: motor
VII. Facial: mixed

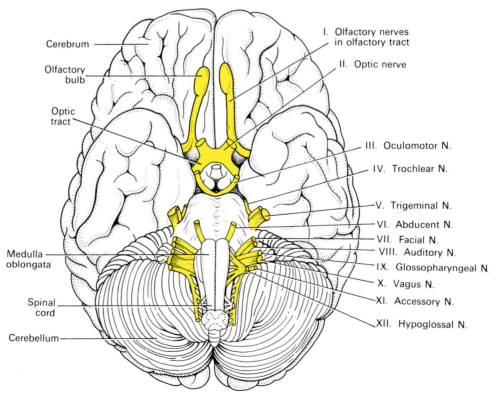

Figure 12:35 The inferior surface of the brain showing the cranial nerves.

VIII. Vestibulocochlear (auditory): sensory
IX. Glossopharyngeal: mixed
X. Vagus: mixed
XI. Accessory: motor
XII. Hypoglossal: motor

I. Olfactory nerves (sensory)

These are the nerves of the *sense of smell*. Their nerve endings and fibres arise in the upper part of the mucous membrane of the nose and pass upwards through the cribriform plate of the ethmoid bone (Fig. 12:36). These nerves pass to the *olfactory bulb*, a group of nerve cells of the second neurones. The nerves then proceed backwards as the olfactory tract, to the area for the perception of smell in the temporal lobe of the cerebrum.

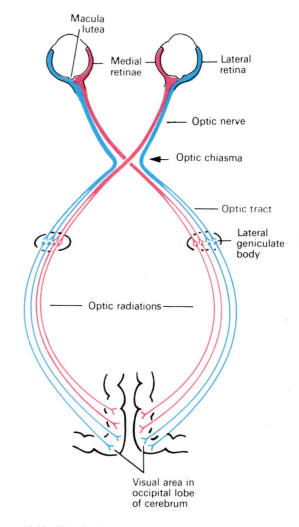

Figure 12:37   The visual pathway.

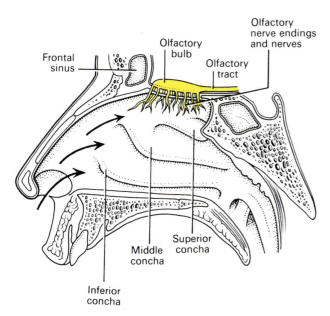

Figure 12:36   The olfactory nerve.

II. Optic nerves (sensory) (Fig. 12:37)

These are the nerves of the *sense of sight*. The fibres originate in the retinae of the eyes and they combine to form the optic nerves. They are directed backwards and medially through the posterior part of the orbital cavity. They then pass through the *optic foramina* of the sphenoid bone into the cranial cavity and join at the *optic chiasma* just above the pituitary gland. The nerves proceed backwards as the *optic tracts* to the *lateral geniculate body*. Impulses pass from these to the centre for sight in the occipital lobes of the cerebrum and to the cerebellum. In the occipital lobe sight is perceived, and in the cerebellum the impulses from the eyes contribute to the maintenance of balance (see p. 220).

The central retinal artery and vein enter the eye enveloped by the fibres of the optic nerve.

III. Oculomotor nerves (motor)

These nerves arise from nerve cells near the cerebral aqueduct. They supply:
1. Four extraocular muscles, i.e., the *superior, medial* and *inferior recti* and the *inferior oblique muscle*
2. Intraocular muscles;
   a. *ciliary muscles* which alter the shape of the lens, changing its refractive power
   b. *circular muscles of the iris* which constrict the pupil
3. The *levator palpebrae* muscle which raises the upper eyelid

IV. Trochlear nerves (motor)

These nerves arise from nerve cells near the cerebral aqueduct. They supply the *superior oblique muscles* of the eyes.

V. Trigeminal nerves (mixed)

These nerves contain motor and sensory fibres and are among the largest of the cranial nerves. They are the chief sensory nerves for the face and head, receiving impulses of *pain*, *temperature* and *touch*. The motor fibres stimulate the *muscles of mastication*.

There are three main branches of the trigeminal nerves. The dermatomes supplied by the sensory fibres on the right side are shown in Figure 12:38.

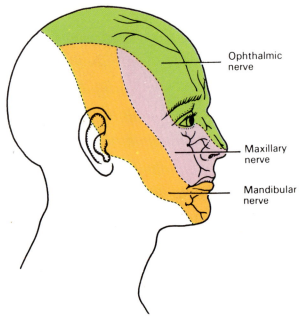

Ophthalmic nerve

Maxillary nerve

Mandibular nerve

Figure 12:38   The cutaneous distribution of the main branches of the right trigeminal nerve.

1. *The ophthalmic nerves* are sensory only and supply the *lacrimal glands*, *conjunctiva of the eyes*, *forehead*, *eyelids*, *anterior aspect of the scalp* and *mucous membrane of the nose*.

2. *The maxillary nerves* are sensory only and supply the *cheeks*, *upper gums*, *upper teeth* and *lower eyelids*.

3. *The mandibular nerves* contain both sensory and motor fibres. These are the largest of the three divisions and they supply the *teeth* and *gums of the lower jaw*, *pinna of the ears*, *lower lip* and *tongue*. The motor fibres supply the *muscles of mastication*.

VI. Abducent nerves (motor)

These nerves arise from a group of nerve cells lying under the floor of the fourth ventricle. They supply the *lateral rectus muscles* of the eyeballs.

VII. Facial nerves (mixed)

These nerves are composed of both motor and sensory nerve fibres, arising from nerve cells in the lower part of the pons varolii. The motor fibres supply the *muscles of facial expression*. The sensory fibres convey impulses from the *taste buds* in the anterior two-thirds of the tongue to the taste perception area in the cerebral cortex.

VIII. Vestibulocochlear (auditory) nerves (sensory)

These nerves are composed of two distinct sets of fibres, vestibular nerves and cochlear nerves.

The vestibular nerves arise from the semicircular canals of the inner ear and convey impulses to the cerebellum. They are associated with the *maintenance of posture and balance*.

The cochlear nerves originate in the organ of corti in the inner ear and convey impulses to the hearing areas in the cerebral cortex where *sound is perceived*.

IX. Glossopharyngeal nerves (mixed)

These nerves arise from nuclei in medulla oblongata. The motor fibres stimulate the *muscles of the pharynx* and the *secretory cells of the parotid* glands.

The sensory fibres convey impulses to the cerebral cortex from the posterior third of the tongue, the tonsils and pharynx and from *taste buds* in the tongue and pharynx.

X. Vagus nerves (mixed) (Fig. 12:39)

These nerves have a more extensive distribution than any other cranial nerves. They arise from nerve cells in the

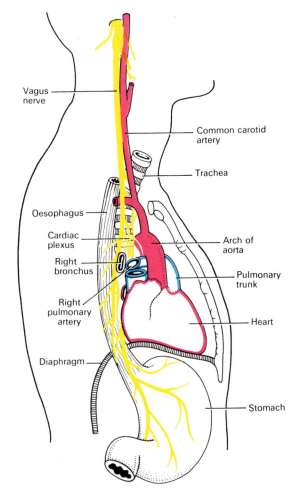

Vagus nerve

Common carotid artery

Trachea

Oesophagus

Cardiac plexus

Right bronchus

Right pulmonary artery

Diaphragm

Arch of aorta

Pulmonary trunk

Heart

Stomach

Figure 12:39   The vagus nerve in the thorax viewed from the side.

medulla oblongata and other nuclei, and pass down through the neck into the thorax and the abdomen.

The motor fibres supply the *smooth muscles* and *secretory glands* of the pharynx, larynx, trachea, heart, oesophagus, stomach, intestine, pancreas, gall bladder, bile ducts, spleen, kidneys, ureter and blood vessels in the thoracic and abdominal cavities.

The sensory fibres convey impulses from the *lining membranes* of the same structures to the brain.

XI. Accessory nerves (motor)

These nerves arise from nerve cells in the medulla oblongata and in the spinal cord. The fibres supply the *sternocleidomastoid* and *trapezius muscles*. Branches join the vagus nerves and supply the *pharyngeal* and *laryngeal muscles*.

XII. Hypoglossal nerves (motor)

These nerves arise from nerve cells in the medulla oblongata. They supply the *muscles of the tongue* and *muscles surrounding the hyoid bone*.

A summary of the cranial nerves is given in Table 12:3.

AUTONOMIC NERVOUS SYSTEM

The autonomic or involuntary part of the nervous system controls the functions of the body carried out automatically, i.e., initiated in the brain below the level of the cerebrum. Although stimulation does not occur voluntarily the individual may be conscious of its effects, e.g., an increase in the heart rate.

The following list provides some examples of physiological activities controlled by the autonomic nervous system:

Table 12:3 Summary of the cranial nerves

Name and no.	Central connection	Peripheral connection	Function
I. Olfactory (sensory)	Smell area in temporal lobe of cerebrum through olfactory bulb	Mucous membrane in roof of nose	Sense of smell
II. Optic (sensory)	Sight area in occipital lobe of cerebrum Cerebellum	Retina of the eyes	Sense of sight Balance
III. Oculomotor (motor)	Nerve cells near floor of aqueduct of midbrain	Superior, inferior and medial rectus muscles of the eye Ciliary muscles of the eye Circular muscle fibres of the iris	Moving the eyeball Focusing Regulating the size of the pupil
IV. Trochlear (motor)	Nerve cell near floor of aqueduct of midbrain	Superior oblique muscles of the eyes	Movement of the eyeball
V. Trigeminal (mixed)	Motor fibres from the pons varolii Sensory fibres from the trigeminal ganglion	Muscles of mastication Sensory to gums, cheek, lower jaw, iris, cornea	Chewing Sensation from the face
VI. Abducent (motor)	Floor of fourth ventricle	Lateral rectus muscle of the eye	Movement of the eye
VII. Facial (mixed)	Pons varolii	Sensory fibres to the tongue Motor fibres to the muscles of the face	Sense of taste Movements of facial expression
VIII. Vestibulocochlear (sensory) (a) vestibular (b) cochlear	Cerebellum Hearing area of cerebrum	Semicircular canals in the inner ear Organ of Corti in cochlea	Maintenance of balance Sense of hearing
IX. Glossopharyngeal (mixed)	Medulla oblongata	Parotid glands Back of tongue and pharynx	Secretion of saliva Sense of taste Movement of pharynx
X. Vagus (mixed)	Medulla oblongata	Pharynx, larynx; organs, glands, ducts, blood vessels in the thorax and abdomen	Movement and secretion
XI. Accessory (motor)	Medulla oblongata	Sternocleidomastoid, trapezius, laryngeal and pharyngeal muscles	Movement of the head, shoulders, pharynx and larynx
XII. Hypoglossal (motor)	Medulla oblongata	Tongue	Movement of tongue

Rate and force of the heart beat
Secretion of the glands of the alimentary tract
Contraction of involuntary muscle
Size of the pupils of the eyes

The *efferent (motor) nerves* of the autonomic nervous system cause contraction of smooth muscle and glandular secretion. They arise from nerve cells in the brain and emerge at various levels between the midbrain and the sacral region of the spinal cord. Many of them travel within the same nerve sheath as the peripheral nerves of the central nervous system to reach the organs which they innervate.

For descriptive convenience the autonomic nervous system is divided into two parts: *sympathetic* (thoracolumbar outflow) and *parasympathetic* (craniosacral outflow).

SYMPATHETIC NERVOUS SYSTEM
(Thoracolumbar outflow)

Three neurones are involved in conveying impulses from their origin in the *hypothalamus* and *medulla oblongata* to effector organs and tissues (Fig. 12:40).

Neurone 1 has its cell in the *brain* and its fibre extends into the *spinal cord*.

Neurone 2 has its cell in the *lateral column of grey matter* in the spinal cord between the levels of the 1st thoracic and 2nd or 3rd lumbar vertebrae. The nerve fibre of this cell leaves the cord by the anterior root and terminates in one of the ganglia either in the *lateral chain of sympathetic ganglia* or passes through it to one of the *prevertebral ganglia*.

Neurone 3 has its cell in a ganglion and terminates in the organ or tissue supplied.

Sympathetic ganglia

The lateral chain of sympathetic ganglia. This is a chain of ganglia which extends from the upper cervical level to the sacrum, one chain lying on each side of the bodies of the vertebral. The ganglia are attached to each other by nerve fibres. *Preganglionic fibres* that emerge from the cord may synapse with the cell of the neurone 3 at the same level or they may pass up or down the chain through one or more ganglia before synapsing. For example, the nerve which dilates the pupil of the eye leaves the cord at

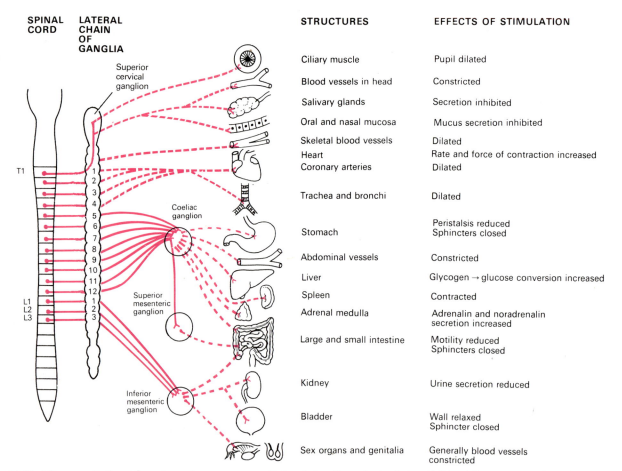

SPINAL CORD	LATERAL CHAIN OF GANGLIA	STRUCTURES	EFFECTS OF STIMULATION
	Superior cervical ganglion	Ciliary muscle	Pupil dilated
		Blood vessels in head	Constricted
		Salivary glands	Secretion inhibited
		Oral and nasal mucosa	Mucus secretion inhibited
T1	1 2 3 4	Skeletal blood vessels	Dilated
		Heart	Rate and force of contraction increased
		Coronary arteries	Dilated
	5 6 Coeliac ganglion	Trachea and bronchi	Dilated
	7 8 9	Stomach	Peristalsis reduced / Sphincters closed
	10 11	Abdominal vessels	Constricted
	12 Superior mesenteric ganglion	Liver	Glycogen → glucose conversion increased
L1 L2 L3	1 2 3	Spleen	Contracted
		Adrenal medulla	Adrenalin and noradrenalin secretion increased
		Large and small intestine	Motility reduced / Sphincters closed
	Inferior mesenteric ganglion	Kidney	Urine secretion reduced
		Bladder	Wall relaxed / Sphincter closed
		Sex organs and genitalia	Generally blood vessels constricted

Figure 12:40   The sympathetic outflow, the main structures supplied and the effects of stimulation. Solid red lines—preganglionic fibres, broken lines—postganglionic fibres.

the level of the 1st thoracic vertebra and passes up the chain to the superior cervical ganglion before it synapses with the cell of neurone 3. The *postganglionic fibres* then passes to the eye.

Prevertebral ganglia. There are three prevertebral ganglia situated in the abdominal cavity close to the origins of arteries of the same names:

Coeliac ganglion
Superior mesenteric ganglion
Inferior mesenteric ganglion

The ganglia consist of nerve cells rather diffusely distributed among a network of nerve fibres which form plexuses. Second neurone sympathetic fibres pass *through* the lateral chain to reach these ganglia.

PARASYMPATHETIC NERVOUS SYSTEM
(Craniosacral outflow)

Two neurones are involved in the transmission of impulses from their source to the effector organ (Fig. 12:41).

Neurone 1 has its cell either in the brain or in the spinal

cord. Those originating in the brain are the *cranial nerves* III, VII, IX and X, arising from nuclei in the midbrain and brain stem and their nerve fibres terminate outside the brain. The cells of the *sacral outflow* are in the lateral columns of grey matter at the distal end of the spinal cord. Their fibres leave the cord in sacral segments 2, 3 and 4 and synapse with second neurone cells in the walls of pelvic organs.

Neurone 2 has its cell either in a ganglion or in the wall of the organ supplied.

FUNCTIONS OF THE AUTONOMIC NERVOUS SYSTEM

The autonomic nervous system is involved in a complex of reflex activities which, like the reflexes described previously, depend on sensory input to the brain or spinal cord, and on motor output. In this case the reflex action is contraction of involuntary (smooth and cardiac) muscle or glandular secretion. These reflexes are co-ordinated in the brain below the level of consciousness, i.e., below the

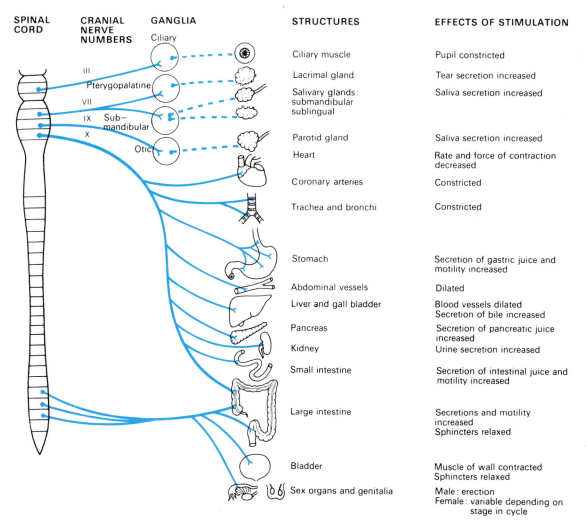

Figure 12:41 The parasympathetic outflow, the main structures supplied and the effects of stimulation. Solid blue lines—preganglionic fibres, broken lines—postganglionic fibres. Where there are no broken lines the 3rd neurone is in the wall of the structure.

level of the cerebrum. Some sensory input does reach consciousness and may result in temporary inhibition of the reflex action, e.g., reflex micturition can be inhibited temporarily.

The majority of the organs of the body are supplied by both sympathetic and parasympathetic nerves which have opposite effects that are finely balanced to ensure the optimum functioning of the organ.

Sympathetic stimulation as a whole has similar effects on the body as those produced by the hormones *adrenaline* and *noradrenaline* secreted by the medulla of the adrenal glands. It prepares the body to deal with excitement and stressful situations, e.g., strengthening its defences in danger and in extremes of environmental temperature. It is sometimes said that sympathetic stimulation mobilises the body for 'fight or flight'.

Parasympathetic stimulation has a tendency to slow down body processes except digestion and absorption of food and the functions of the genitourinary systems. Its general effect is that of a 'peace maker' allowing restoration processes to occur quietly and peacefully.

Normally the two systems function simultaneously producing a regular heart beat, normal temperature and an internal environment compatible with the immediate external surroundings.

EFFECTS OF AUTONOMIC STIMULATION ON THE VARIOUS BODY SYSTEMS

CARDIOVASCULAR SYSTEM

Sympathetic stimulation

1. Exerts an accelerating effect upon the sinuatrial node in the heart, increasing the rate and force of the heart beat.

2. Causes dilatation of the coronary arteries, increasing the blood supply to heart muscle.

3. Causes dilatation of the blood vessels supplying skeletal muscle, increasing the supply of oxygen and nutritional materials and the removal of metabolic waste products, thus increasing the capacity of the muscle to do work.

4. Causes sustained contraction of the spleen, increasing the volume of circulating blood.

5. Raises the blood pressure by constricting the small arteries and arterioles in the skin. In this way an increased blood supply is available for highly active tissue, such as skeletal muscle.

6. Constricts the blood vessels in the secretory glands of the digestive system, reducing the flow of digestive juices. This raises the volume of blood available for circulation in dilated blood vessels.

Parasympathetic stimulation

This has effects opposite to those of sympathetic stimulation on the heart, spleen and blood vessels.

RESPIRATORY SYSTEM

Sympathetic stimulation

This causes dilatation of the bronchi allowing a greater amount of air to enter the lungs at each inspiration. In conjunction with the increased heart rate, the oxygen intake and carbon dioxide output of the body are increased.

Parasympathetic stimulation

Produces constriction of the bronchi.

DIGESTIVE AND URINARY SYSTEMS

Sympathetic stimulation

1. *The liver* converts an increased amount of glycogen to glucose, making more carbohydrate immediately available to provide energy.

2. *The adrenal (suprarenal) glands* are stimulated to secrete adrenaline and noradrenaline which potentiate and sustain the effects of sympathetic stimulation.

3. *The stomach* and *small intestine*. Muscle contraction and secretion of digestive juices are inhibited, delaying digestion, onward movement and absorption of food.

4. *Urethral* and *anal sphincters*. The muscle tone of the sphincters is increased, inhibiting micturition and defaecation.

5. *The bladder wall* relaxes.

Parasympathetic stimulation

1. *The stomach* and *small intestine*. The rate of digestion and absorption of food is increased.

2. *The pancreas*. There is an increase in the secretion of pancreatic juice and the hormone insulin.

3. *Urethral* and *anal sphincters*. Relaxation of the internal urethral sphincter is accompanied by contraction of the muscle of the bladder wall and micturition occurs. Similar relaxation of the internal anal sphincter is accompanied by contraction of the muscle of the rectum and defaecation occurs. In both cases there is voluntary relaxation of the external sphincters.

EYE

Sympathetic stimulation

This causes contraction of the radiating muscle fibres of the iris, *dilating* the pupil. Retraction of the levator palpabral muscles occurs, opening the eyes wide and giving the appearance of alertness and excitement.

Parasympathetic stimulation

This causes contraction of the circular muscle fibres of the iris, constricting the pupil. The eyelids tend to close, giving the appearance of sleepiness.

SKIN

Sympathetic stimulation

1. Causes increased secretion of sweat, leading to increased heat loss from the body.

2. Produces contraction of the arrectores pilorum (the muscles in the skin), causing heat production and giving the appearance of 'goose flesh'.

3. Causes constriction of the blood vessels preventing heat loss.

There is no parasympathetic nerve supply to the skin, therefore, nerve supply which is anatomically sympathetic has the dual function of facilitating heat loss and increasing heat production.

GENITALIA

Sympathetic stimulation

This causes generalised vasoconstriction.

Parasympathetic stimulation

This causes generalised vasodilatation, with erection of the penis in the male.

AFFERENT IMPULSES FROM VISCERA

Sensory fibres from the viscera travel with autonomic fibres and are sometimes called *autonomic afferents*. The impulses they transmit are associated with:

1. Visceral reflexes, usually at an unconscious level
2. Sensation of, e.g., hunger, thirst, nausea, sexual sensation, rectal and bladder distension
3. Visceral pain

VISCERAL PAIN

Normally the viscera are insensitive to cutting, burning and crushing. However, a sensation of dull, poorly located pain is experienced when:

1. Visceral nerves are stretched
2. A large number of fibres are stimulated
3. There is ischaemia and the accumulation of metabolites
4. The pain threshold has been lowered by disease

If the cause of the pain, e.g., inflammation, affects the parietal layer of a serous membrane (pleura, peritoneum) the pain is acute and easily located. This is because the peripheral spinal (somatic) nerves supplying the superficial tissues also supply the parietal layer of serous membrane. They transmit the impulses to the cerebral cortex where *somatic pain* is perceived and accurately located. Appendicitis is an example of this type of pain. Initially it is dull and vaguely located around the midline of the abdomen. As the condition progresses the parietal peritoneum becomes involved and acute pain is clearly located in the lower right abdominal quadrant, i.e., over the appendix.

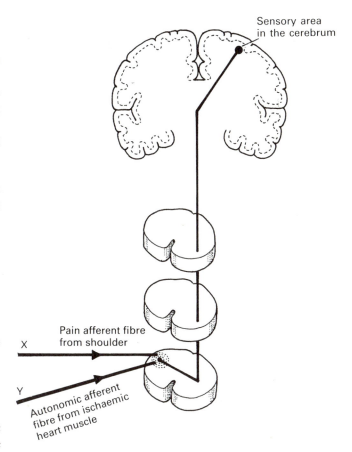

Figure 12:42   Referred pain. Pain perceived to originate from the tissues supplied by the damaged nerve. Y stimulates X and pain perceived in the shoulder. (See also Fig. 12:48B.)

REFERRED PAIN (Fig. 12:42)

In some cases of visceral disease pain may be perceived to occur in superficial tissues remote from the viscus, i.e., referred pain. This occurs when sensory fibres from the viscus enter the same segment of the spinal cord as somatic nerves, i.e., those from the superficial tissues. It is believed that the sensory nerve from the viscus stimulates the closely associated somatic nerve in the spinal cord and it transmits the impulses to the sensory area in the cerebral cortex where the pain is perceived as originating in the area supplied by the somatic nerve. Examples of referred pain are:

Tissue of origin of pain	Site of referred pain
Heart	Shoulder
Liver Biliary tract }	Right shoulder
Kidney } Ureter }	Loin and groin
Uterus	Low back
Male genitalia	Low abdomen
Prolapsed intervertebral disc	Leg

DISEASES OF THE NERVOUS SYSTEM

NEURONES AND NEUROGLIA

NEURONE DAMAGE

Damage to the nerve cells or their processes can lead to rapid necrosis with sudden acute functional failure, or to slow atrophy with gradually increasing dysfunction. These changes are associated with:

Anoxia	Trauma
Hypoglycaemia	Infections
Nutritional deficiencies	Ageing
Poisons, e.g., organic lead	

NEURONE REGENERATION (Fig. 12:43)

Neurones of the brain, spinal cord and ganglia reach maturity a few weeks after birth and are not replaced when they are damaged or die. The axons of *peripheral nerves* regenerate provided the cell remains intact. Distal to the damage the axon and myelin sheath disintegrate and are removed by macrophages, but the Schwann cells survive and proliferate within the neurilemma. The live proximal part of the axon grows along the original track (about 3 mm per day), provided the two parts of neurilemma are correctly positioned and in close apposition. Restoration of function depends on the re-establishment of satisfactory connections with the end organ. When the neurilemma is out of position or destroyed the sprouting axons and Schwann cells form a tumour-like cluster (*traumatic neuroma*), e.g., following some fractures and amputations of limbs.

NEUROGLIA DAMAGE

Astrocytes undergo necrosis and disintegrate when severely damaged. In less severe and chronic conditions there is proliferation of astrocyte processes and later cell atrophy (*gliosis*). This process occurs in many diseases and is analagous to fibrosis in other tissues.

Oligodendrocytes form and maintain myelin, having the same functions as Schwann cells in peripheral nerves. They increase in number around degenerating neurones and are destroyed in demyelinating diseases such as multiple sclerosis.

Microglia are derived from monocytes that migrate from the blood into the nervous system before birth, and are found mainly around blood vessels. Where there is inflammation and cell destruction the microglia increase in size and become phagocytic.

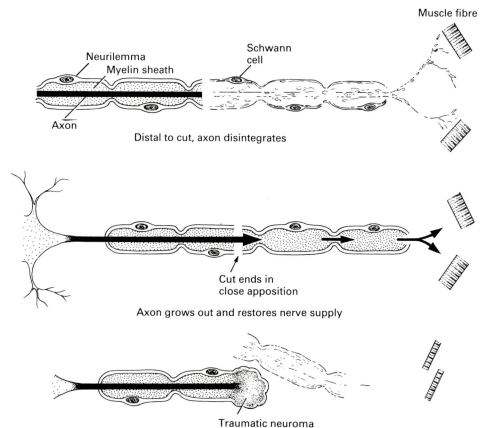

Figure 12:43 Regrowth of peripheral nerves following injury.

DISEASES OF THE BRAIN

INCREASED INTRACRANIAL PRESSURE

This is not a disease in itself but it is a very important complication of many conditions. Sometimes its effects are more serious than the condition causing it, e.g., by disrupting the blood supply or distorting the shape of the brain especially if the intracranial pressure (ICP) rises rapidly. A slow rise in ICP allows time for compensatory adjustment to be made, i.e., a slight reduction in the volume of circulating blood and of cerebral spinal fluid (CSF). The slower the rise in ICP, the more effective the compensation. As it reaches its limit a further small increase in pressure is followed by a sudden and possibly serious reduction in the cerebral blood flow, hypoxia and a rise in carbon dioxide level, causing arteriolar dilatation which further increases the ICP. This leads to progressive loss of functioning neurones, vasomotor paralysis and further cerebral hypoxia. The causes of increased ICP are:
1. Expanding lesions, e.g., haematoma, tumour
2. Cerebral oedema
3. Hydrocephalus, i.e., accumulation of excess CSF

EXPANDING LESIONS

These lesions cause brain damage in various ways (Fig. 12:44).

Displacement of the brain
Lesions causing displacement are usually one-sided but may affect both sides. Such lesions may cause:
1. Herniation of the cerebral hemisphere between the corpus callosum and the free border of the falx cerebri on the same side
2. Herniation of the midbrain between the pons and the free border of the tentorium cerebelli on the same side
3. Compression of the subarachnoid space and flattening of the cerebral convolutions
4. Distortion of the shape of the ventricles and their ducts
5. Herniation of the cerebellum through the foramen magnum
6. Protrusion of the medulla oblongata through the foramen magnum ('coning')

Obstruction of the flow of cerebrospinal fluid
The ventricles or their ducts may be pushed out of position or a duct obstructed. The effects depend on the position of the lesion, e.g., compression of the aqueduct of the midbrain causes dilatation of the lateral ventricles and the third ventricle, further increasing the ICP.

Vascular damage
There may be stretching or compression of blood vessels, causing:
1. Haemorrhage when stretched blood vessels break
2. Ischaemia and infarction due to compression of blood vessels
3. Papilloedema (oedema round the optic disc) due to compression of the retinal vein in the optic nerve sheath where it crosses the subarachnoid space

Neural damage
The vital centres in the medulla oblongata may be damaged when the increased ICP causes 'coning'. Stretching may damage cranial nerves, especially the oculomotor (III) and the abducent (VI), causing disturbances of eye movement and accommodation.

Bone changes
Prolonged increase of ICP causes bony changes, e.g.:
1. Erosion, especially of the sphenoid
2. Stretching and thinning before ossification is complete

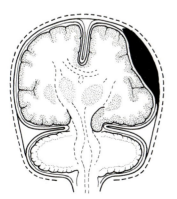 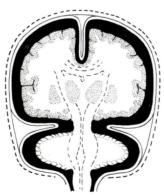 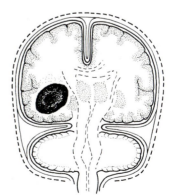

A. Subdural haematoma   B. Subarachnoid haemorrhage   C. Tumour

Figure 12:44 Effects of different types of expanding lesions inside the skull.

CEREBRAL OEDEMA

In this condition there is excess water in brain cells and/or in the interstitial spaces. It is associated with:

Traumatic injury
Hypoglycaemia
Haemorrhage
Infections, abcesses

Hypoxia
Local ischaemia, infarcts
Tumours
Inflammation

Cytotoxic oedema

The neurones and neuroglial cells contain excess water due to disturbances of their osmotic pressure caused by retention of excess electrolytes. The permeability of the capillary walls is normal.

Vasogenic oedema

Excess fluid collects in the interstitial spaces, especially round nerve fibres. The permeability of the capillary walls is increased and plasma proteins pass out of the capillaries, increasing the osmotic pressure in the interstitial spaces which causes oedema.

HYDROCEPHALUS

In this condition the volume of CSF is abnormally high and is usually accompanied by increased ICP. An obstruction is the most common cause. It is described as *communicating* when there is a subarachnoid obstruction, and *non-communicating* when the obstruction is in the system of ventricles, foramina or ducts.

PRIMARY HYDROCEPHALUS

This is usually caused by obstruction to the flow of CSF and may be communicating or non-communicating. Occasionally it is caused by malabsorption of CSF by the arachnoid villi.

Congenital primary hydrocephalus

This is due to malformation of the ventricles, foramina or ducts, usually at a narrow point.

Acquired primary hydrocephalus

This is caused by lesions that obstruct the circulation of the CSF, usually expanding lesions, e.g., tumours, haematomas or adhesions between arachnoid and pia maters, following meningitis.

SECONDARY HYDROCEPHALUS

Increases in the amount of CSF and ventricle capacity occur when there is atrophy of brain tissue, e.g., in senile dementia and following cerebral infarcts. There may not be a rise in ICP.

EFFECTS OF HYDROCEPHALUS

Enlargement of the head occurs in children when ossification of the cranial bones is incomplete but, in spite of this, the ventricles dilate and cause stretching and thinning of the brain. After ossification is complete, hydrocephalus leads to a marked increase in ICP and destruction of neural tissue.

HEAD INJURIES

The brain may be injured by a blow to the head or movement of the brain during sudden acceleration or deceleration of the head.

BLOW ON THE HEAD

At the site of injury there may be:
1. Scalp wound, with haemorrhage between scalp and skull bones
2. Damage to the underlying meninges and/or brain with local haemorrhage inside the skull
3. Depressed fracture of the skull, causing local damage to the underlying meninges and brain tissue
4. Temporal bone fracture, making an opening between the middle ear and the meninges
5. Fracture involving the air sinuses of the sphenoid, ethmoid or frontal bones, making an opening between the nose and the meninges

ACCELERATION—DECELERATION INJURIES

Because the brain floats relatively freely in 'a cushion' of CSF, sudden acceleration or deceleration has an inertia effect, i.e., there is delay between the movement of the head and the corresponding movement of the brain. During this period the brain is compressed and damaged at the site of impact. Other injuries include:
1. Nerve cell damage, usually to the frontal and parietal lobes, due to movement of the brain over the rough surface of bones of the base of the skull
2. Nerve fibre damage, especially following rotational movement
3. Haemorrhage due to rupture of blood vessels in the subarachnoid space on the side opposite the impact

COMPLICATIONS OF HEAD INJURY

If the individual survives the immediate effects, complications may develop hours or days later. Sometimes they are the first indication of serious damage caused by a seemingly trivial injury. Their effects may be to increase ICP, damage brain tissue or provide a route of entry for microbes.

INTRACRANIAL HAEMORRHAGE

Haemorrhage may occur at the site of injury or on the opposite side of the brain. If bleeding continues, the expanding haematoma damages the brain and increases the ICP.

Epidural haemorrhage

This usually occurs at the site of a fracture. In children it may occur without a fracture when the bones are still soft or the joints have not yet fused. Blood accumulates between the bone and the outer layer of dura mater (periosteum), stripping it from the bone. The epidural haematoma formed usually remains localised.

Acute subdural haemorrhage

This is due to haemorrhage from small veins in the dura mater or from larger veins between the layers of dura mater before they enter the venous sinuses. The blood may spread in the subdural space over one or both hemispheres. There may be concurrent subarachnoid haemorrhage, especially when there are extensive brain contusions and lacerations.

Chronic subdural haemorrhage

This may occur weeks or months after minor injuries and sometimes there is no history of injury. It occurs most commonly in people in whom there is some cerebral atrophy, e.g., the elderly and alcoholics. Evidence of increased ICP may be delayed if brain volume is reduced. The haematoma formed gradually increases in size due to repeated small haemorrhages and causes mild chronic inflammation and the accumulation of inflammatory exudate. In time it is isolated by a wall of fibrous tissue.

Intracerebral haemorrhage and cerebral oedema

These occur following contusions, lacerations and shearing injuries associated with acceleration and deceleration movements.

MENINGITIS

Microbes may spread to the meninges from the:
1. Skin, in compound fractures of the skull
2. Middle ear, in fractures of the temporal bone
3. Nose, in fractures of sphenoid, ethmoid or frontal bones when the air sinuses are involved

EPILEPSY

This may develop in the first week or several months after injury. Early development is most common after severe injuries, although in children the injury itself may have appeared trivial. After depressed fractures or large haematomas epilepsy tends to develop later.

CIRCULATORY DISTURBANCES AFFECTING THE BRAIN

There are a number of abnormalities of the circulation, not associated with head injuries, that affect the brain.

INTRACRANIAL HAEMORRHAGE

A distinction is made between intracerebral haemorrhage and subarachnoid haemorrhage. In both types the irritant effect of haemorrhage may cause arterial spasm, leading to ischaemia, infarction, fibrosis (gliosis) and localised brain damage.

INTRACEREBRAL HAEMORRHAGE

Prolonged hypertension leads to the formation of multiple microaneurysms in the walls of very small arteries in the brain. Rupture of one or more of these, due to continuing rise in blood pressure, is usually the cause of intracerebral haemorrhage. The most common sites are branches of the middle cerebral artery in the region of the internal capsule and the basal ganglia.

Severe haemorrhage

This causes destruction of tissue, a sudden increase in ICP and distortion and herniation of the brain. Death may follow in a few days, especially if the vital centres in the medulla oblongata are damaged by haemorrhage or if there is coning due to increased ICP.

Less severe haemorrhage

This causes paralysis and loss of sensation of varying severity, affecting the side of the body opposite the haemorrhage. If the bleeding stops and does not recur an *apoplectic cyst* develops, i.e., the haematoma is walled off by gliosis, the blood clot is gradually absorbed and the cavity filled with tissue exudate. When the ICP returns to normal some functions may be restored, e.g., speech and movement of limbs.

SUBARACHNOID HAEMORRHAGE

The most common cause is rupture of berry aneurysms which tend to form at congenitally weak areas of tunic media where arteries branch from the circle of Willis. The blood spreads in the subarachnoid space round the brain and spinal cord, causing a general increase in ICP without distortion of the brain. The irritant effect of the blood may cause arterial spasm, leading to ischaemia, infarction, gliosis and the effects of localised brain damage.

HAEMORRHAGE INTO TUMOURS

Within the confined space of the skull, haemorrhage in a tumour exacerbates the increased ICP, caused by the tumour.

HYPERTENSION

Sustained hypertension causes brain damage when:
1. Microaneurysms rupture
2. Pulsation of hard tortuous arteries damages brain tissue in their immediate vicinity
3. Cerebral oedema in malignant hypertension leads to hypertensive encephalopathy

HYPOXIA

Hypoxia may be due to:
1. Disturbances in the autoregulation of blood supply to the brain
2. Changes in cerebral blood vessels

When the systolic blood pressure falls below about 50 mmHg there is failure of the autoregulating mechanisms that control the blood flow to the brain by adjusting the calibre of the arterioles. The consequent rapid decrease in the cerebral blood supply leads to hypoxia and lack of glucose. If severe hypoxia is sustained for more than a few minutes there is irreversible brain damage. The neurones are affected first, then the neuroglial cells and later the meninges and blood vessels. Conditions in which autoregulation breaks down include:
a. Cardiorespiratory arrest
b. Sudden severe hypotension
c. Carbon monoxide poisoning
d. Hypercapnia (excess blood carbon dioxide)
e. Drug overdosage with, e.g., narcotics, hypnotics, analgesics

Changes in the cerebral circulation, leading to hypoxia, include:
a. Occlusion of a cerebral artery by, e.g., atheroma, thrombosis, rapidly expanding lesion, embolism
b. Arterial stenosis that occurs in arteritis, e.g. polyarteritis nodusum, tuberculous meningitis, syphilis, diabetes, in the elderly

If the individual survives the initial ischaemia, there is infarction, necrosis and loss of function of the affected area of brain.

INFECTIONS OF THE NERVOUS SYSTEM

The brain and spinal cord are relatively well protected from microbial infection by the blood–brain barrier. When infection does occur microbes may be:
1. Blood-borne from infection elsewhere in the body, e.g., lung abscess
2. Introduced through a skull fracture, e.g. infection from the nose through the air sinuses of the frontal, sphenoid or ethmoid bone, from the middle ear through the temporal bone, following a compound fracture
3. Spread through the skull bones from, e.g., middle ear infection, mastoiditis, skull bone infection
4. Introduced during a surgical procedure, e.g., lumbar puncture

The microbes usually involved are bacteria and viruses, occasionally protozoa and fungi. The infection may originate in the meninges (*meningitis*) or in the brain (*encephalitis*), then spread from one site to the other.

PYOGENIC INFECTION OF THE NERVOUS SYSTEM

MENINGITIS

Pyogenic infection of the meninges may remain localised in the extradural or subdural space. If an abscess forms it may rupture into the subarachnoid space and spread to the brain, causing encephalitis and further abscess formation. Infection of the arachnoid mater tends to spread diffusely in the subarachnoid space round the brain and spinal cord.

ENCEPHALITIS

Single or multiple sites may be involved, usually with abscess formation. The microbes are either blood-borne or spread from adjacent meningeal infection.

EFFECTS OF PYOGENIC ENCEPHALITIS

1. Inflammatory oedema and abscess formation lead to increased ICP, hypoxia and further brain damage.
2. When meningitis follows encephalitis, healing takes place with the formation of fibrous adhesions. These may interfere with CSF circulation and may cause hydrocephalus, or compress blood vessels, causing hypoxia.
3. Following encephalitis and brain abscess, healing is associated with gliosis. Destroyed nerve cells are not replaced so loss of function depends on the site and extent of brain damage.

VIRUS INFECTIONS OF THE NERVOUS SYSTEM

Virus infections of the nervous system are relatively rare even when caused by *neurotropic viruses*, i.e., those with an affinity for the nervous system. Viruses may cause meningitis, encephalitis or lesions of the neurones of spinal cord and peripheral nerves. Most viruses are blood-borne although a few travel along peripheral nerves, e.g., rabies virus and possibly polioviruses. They enter the body via:

1. Alimentary tract, e.g., poliomyelitis
2. Respiratory tract, e.g., shingles
3. Skin abrasions, e.g., rabies
4. Insect bites, e.g., encephalitis lethargica

The effects of virus infections vary according to the site and the amount of tissue destroyed. Viruses are believed to damage nerve cells by:

1. 'Taking over' their metabolism
2. Stimulating an immune reaction which may explain why signs of some infections do not appear until there is a high antibody titre, 1 to 2 weeks after infection

VIRUS MENINGITIS

This is sometimes called 'aseptic' meningitis because cultures for bacteria are sterile. It is usually a relatively mild infection followed by complete recovery.

VIRUS ENCEPHALITIS

The sites involved vary considerably and, as nerve cells are not replaced, loss of function reflects the extent of damage. In severe infection neurones and neuroglia may be affected, followed by necrosis and gliosis. Early senility may develop or, if vital centres in the medulla oblongata are involved, death may ensue. In mild cases recovery is usually complete with little loss of function.

HERPESVIRUSES

HERPES SIMPLEX ENCEPHALITIS

This is an acute condition, causing necrosis of areas of the cerebrum, usually the temporal lobes. If the patient survives the initial acute phase there may be residual dysfunction, e.g., behavioural disturbance and loss of memory.

HERPES ZOSTER NEURITIS (SHINGLES)

Herpes zoster viruses cause chickenpox (varicella) mainly in children and shingles (zoster) in adults. Susceptible children may contract chickenpox from a shingles sufferer.

Adults may be infected with the viruses but show no immediate signs of disease. The viruses may remain dormant in posterior root ganglia of the spinal nerves then become active years later, causing shingles. The factors believed to reactivate them include:

1. Local trauma involving the dermatome, i.e., the area supplied by the affected peripheral nerve
2. Exposure of the dermatome to irradiation, e.g., sun, X-rays
3. Depression of the immunological system, e.g., by drugs, old age

The posterior root gnaglion becomes acutely inflamed. From there the *viruses* pass along the sensory nerve to the skin. The skin becomes inflamed and vesicles containing serous fluid and viruses develop along the course of the nerve. This is accompanied by persistent pain and hypersensitivity to touch (*hyperaesthesia*). Recovery is usually slow and there may be some loss of sensation, depending on the severity of the disease. The infection is usually unilateral and the most common sites are:

1. Nerves supplying the trunk, sometimes two or three adjacent dermatomes
2. The ophthalmic division of the trigeminal nerve and, if vesicles form on the cornea, there may be ulceration, scarring and residual interference with vision

ENTEROVIRUSES

This group contains polioviruses, Coxsackie-viruses and echoviruses and they may cause a mild form of meningitis or paralytic disease.

ANTERIOR POLIOMYELITIS

This disease is usually caused by polioviruses and occasionally by other enteroviruses. The infection is spread by food contaminated by infected faecal matter and the initial virus multiplication occurs in the alimentary tract. The viruses are then blood-borne to the nervous system and invade anterior horn cells in the spinal cord. Usually there is a mild febrile illness with no indication of nerve cell damage. Paralytic disease is believed to be precipitated by muscular exercise during the early febrile stage. Irreversible damage to lower motor neurones causes muscle paralysis which, in the limbs, may lead to deformity because of the unopposed tonal contraction of antagonistic muscles. Death may occur due to paralysis of intercostal muscles or the diaphragm, or to destruction of the respiratory centre in the medulla oblongata.

RABIES VIRUS

All warm-blooded animals are susceptible to rabies which is endemic in many countries but not in Britain. The main reservoirs of virus are wild animals, some of which may be carriers. These may infect domestic pets which then become the main source of human infection. The viruses multiply in the salivary glands and are present in large numbers in saliva. They enter the body through skin abrasions and are believed to travel to the brain along the nerves or lymph vessels adjacent to nerves. The incubation period varies from about 2 weeks to several months, possibly reflecting the distance viruses travel between the site of entry and the brain. Extensive damage to the basal nuclei, midbrain, medulla oblongata and the posterior root ganglia of the peripheral nerves causes meningeal irritation, extreme hyperaesthesia, muscle spasm and convulsions. *Hydrophobia* is due to painful spasm of the throat muscles when swallowing.

In the advanced stages muscle spasm may alternate with flaccid paralysis and death is usually due to respiratory muscle spasm or paralysis.

DEMYELINATING DISEASES

These diseases are caused either by injury to axons or by disorders of cells that secrete myelin, i.e., oligodendrocytes and Schwann cells.

MULTIPLE OR DISSEMINATED SCLEROSIS (MS)

In this disease there are areas of demyelinated white matter irregularly distributed throughout the brain and spinal cord. Nerve cells in the brain and spinal cord may also be affected because of the arrangement of satellite oligodendrocytes round cell bodies. In the early stages there may be little damage to axons.

CAUSATIVE FACTORS

The causes of multiple sclerosis are not known. Several factors have been suggested and more than one of the following may be involved.

Environment before the age of 5 years
The disease is most prevalent in temperate climates. Susceptibility to MS appears to be acquired by children living in such climates until they are about 5 years of age. It has been found that if they move to another climate (colder or hotter) after that age they retain their susceptibility, whereas people moving into a temperate climate after 5 years of age appear not to be susceptible. There may be a link between environment and other factors but these are not yet known (Fig. 12:45).

Genetic factors and autoimmunity
Certain immunological characteristics of families of MS

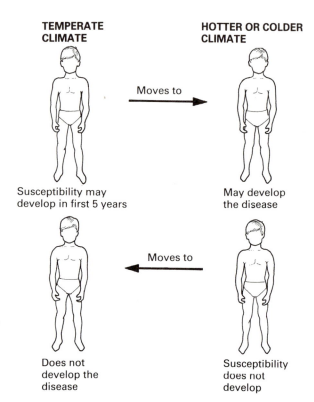

Figure 12:45   Climate and susceptibility to multiple sclerosis.

sufferers suggest an association between genetic factors and susceptibility.

EFFECTS OF MULTIPLE SCLEROSIS

Damage to grey matter leads to a variety of dysfunctions, depending on the sites and sizes of demyelinated plaques. Damage to white matter leads to muscle weakness, lack of co-ordination of movement and disturbed sensation. The optic nerves are commonly involved early in the disease, leading to visual disturbances.

The disease pattern is usually one of relapses and remissions of widely varying duration. Each relapse causes further loss of nervous tissue and progressive dysfunction.

ACUTE DISSEMINATING ENCEPHALOMYELITIS

This is a rare but serious condition that may occur:
1. During or very soon after a virus infection, e.g., measles, chickenpox, mumps, respiratory infection
2. Following primary immunisation against virus diseases, mainly in older children and adults

The cause of the acute diffuse demyelination is not known. It has been suggested that autoimmunity to myelin is triggered either by viruses during a virus infection or by those present in vaccines. The effects vary considerably, according to the distribution and degree of demyelination. The early febrile state may progress to paralysis and coma. Patients who survive the initial phase may recover with no residual dysfunction or have severe impairment of a wide variety of neurological functions.

PHENYLKETONURIA

Some genetic abnormalities cause deficiency in the production of enzymes required at various stages in metabolic chains. This causes accumulation of the intermediate metabolite that the enzyme should catalyse, and high concentrations of some metabolites are toxic to the nervous system. Phenylketonuria is the most common disease of this type. The enzyme *phenylalanine hydroxylase* converts *phenylalanine* to tyrosine in the liver. When the enzyme is absent phenylalanine accumulates in the liver cells, then overflows into the blood and impairs development of the brain, causing severe mental retardation (Fig. 12:46).

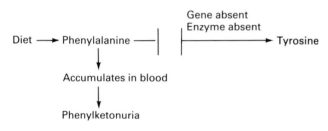

Figure 12:46 Phenylketonuria.

EFFECTS OF POISONS ON THE NERVOUS SYSTEM

Many chemical substances encountered either as drugs or in the environment may damage the nervous system. Neurone metabolism may be disturbed directly or result from damage to other organs, e.g., liver, kidneys. The outcome depends on the toxicity of the substance, the dose and the duration of exposure, ranging from short-term neurological disturbance to encephalopathy which may cause coma and death.

DEMENTIA

This name is given to a syndrome associated with a variety of diseases in which there is atrophy of neural tissue. It is not usually evident until after the age of 65. When it develops at an earlier age it is called *Alzheimer's disease*. There is usually loss of memory, altered behaviour, emotional instability and impaired intellect, judgement and

reasoning. The main cause is hypoxia due to cerebral atherosclerosis but others include:
1. Cerebral trauma
2. Infections, e.g., encephalitis, neurosyphilis
3. Multiple sclerosis
4. Chronic alcohol abuse
5. Vitamin B deficiency
6. Prolonged hypoglycaemia

DISEASES OF THE EXTRAPYRAMIDAL SYSTEM

PARKINSON'S DISEASE (Paralysis agitans)

In this disease there is gradual destruction of the neurones that produce dopamine, one of the chemical neurotransmitters. There is lack of control and co-ordination of muscle movement, resulting in:
1. Fixed muscle tone, expressionless features, stiff shuffling gait and stooping posture
2. Muscle tremor, e.g., 'pill rolling' movement of the fingers

The cause is unknown but some cases are associated with:
1. Previous encephalitis lethargica (an epidemic virus disease in the 1920s that died out)
2. Repeated trauma as in, e.g., 'punch drunk' boxers
3. Tumours, causing midbrain compression
4. Neuroleptic drugs, e.g., phenothiazines

There is progressive physical disability but the intellect is not impaired (Fig. 12:47).

Figure 12:47 Shuffling gait of Parkinson's disease.

CHOREA

HUNTINGTON'S CHOREA

This usually manifests itself between the ages of 30 and 40 years. It is caused by a genetic abnormality associated with deficient production of the neurotransmitter gamma aminobutyric acid (GABA). The extrapyramidal changes are accompanied by cortical atrophy. There are uncoordinated, involuntary movements, especially of the facial muscles and, as the disease progresses, dementia develops.

SYDENHAM'S CHOREA

This usually occurs betwen the ages of 5 and 15 years. The causes are unknown but it is commonly associated with rheumatic fever or endocarditis. There are rapid uncoordinated, involuntary muscle movements. In mild cases recovery takes place within about 4 weeks. In some cases the initial recovery may be followed by recurrences.

DISEASES OF THE SPINAL CORD AND PERIPHERAL NERVES

Because space in the neural canal and intervertebral foramina is limited any condition which distorts their shape or reduces the space may damage the spinal cord or peripheral nerve roots, or cause ischaemia by compressing blood vessels. Such conditions include:

1. Fracture and/or dislocation of vertebrae
2. Tumours of the meninges or vertebrae
3. Prolapsed intervertebral disc

The effects of disease or injury depend on the severity of the damage, the type and position of the neurones involved, i.e., motor, sensory, proprioceptor, autonomic, connector neurones in reflexes arcs in the spinal cord or in peripheral nerves.

MOTOR NEURONES

UPPER MOTOR NEURONE (UMN) LESIONS

Lesions of the UMNs above the level of the decussation of the pyramids affect the opposite side of the body, e.g., haemorrhage or infarction in the internal capsule of one hemisphere causes paralysis of the opposite side of the body. Lesions below the decussation level affect the same side of the body.

LOWER MOTOR NEURONE (LMN) LESIONS

The cells of LMNs are in the spinal cord and the axons are part of peripheral nerves. Lesions of LMNs lead to weakness or paralysis of the muscles they supply.

Table 12:4 gives a summary of the effects of damage to the motor neurones.

Table 12:4 Summary of effects of damage to motor neurones

Upper motor neurone	Lower motor neurone
Muscle weakness and spastic paralysis	Muscle weakness and flaccid paralysis
Exaggerated tendon reflexes	Absence of tendon reflexes
Muscle twitching	Muscle wasting
	Contracture of muscles
	Impaired circulation

The parts of the body affected depend on which neurones have been damaged and their site in the brain, spinal cord or peripheral nerve.

MOTOR NEURONE DISEASE

This is a chronic progressive degeneration of motor neurones, occurring mainly in men between 60 and 70 years of age. The cause is not known. Motor neurones in the cerebral cortex, brain stem and anterior horns of the spinal cord are destroyed and replaced by gliosis. Early effects are usually weakness and twitching of the small muscles of the hand, and muscles of the arm and shoulder girdle. The legs are affected later. Death is usually due to the involvement of the respiratory centre in the medulla oblongata.

SENSORY NEURONES

The sensory functions lost as a result of disease or injury depend on which neurones have been damaged and their position in the brain or spinal cord, or the peripheral nerve involved. In the brain, neurones connecting the thalamus and the cerebrum pass through the internal capsule. Damage in this area by, e.g., haemorrhage, usually from a berry aneurysm, may lead to loss of sensation but does not affect cerebellar function unless upper motor neurones have also been damaged. Spinal cord damage leads to loss of sensation and cerebellar function. Peripheral nerve damage leads to loss of reflex activity, loss of sensation and of cerebellar function.

MIXED MOTOR AND SENSORY CONDITIONS

SPINAL CORD

SUBACUTE COMBINED DEGENERATION OF THE SPINAL CORD

This occurs as a complication of pernicious anaemia, and occasionally of other chronic conditions, e.g., diabetes, leukaemia, carcinoma. The links between these diseases and the degenerative changes in the spinal cord are not known. There is degeneration of nerve fibre myelin in the posterior and lateral columns of white matter in the spinal cord, especially in the upper thoracic and lower cervical regions. Less frequently the changes occur in the posterior root ganglia and peripheral nerves. Demyelination of proprioceptor fibres leads to ataxia and involvement of upper motor neurones leads to increased muscle tone and spastic paralysis.

Although degeneration begins early in pernicious anaemia, sometimes before the anaemia is apparent, threatment with vitamin B_{12} may lead to complete recovery.

SYRINGOMYELIA

This dilatation of the central canal of the spinal cord occurs most commonly in the cervical region and is associated with congenital abnormality of the distal end of the fourth ventricle. As the central canal dilates, pressure causes progressive damage to sensory and motor neurones. Early effects include dissociated anaesthesia. i.e., insensibility to heat and pain. This is due to compression of the sensory fibres that cross the cord immediately they enter. In the long-term there is destruction of motor and sensory tracts, leading to spastic paralysis and loss of sensation and reflexes.

PROLAPSED INTERVERTEBRAL DISC (Fig. 12:48)

Prolapse of a disc is herniation of the nucleus polposus, causing the annulus fibrosus and the posterior longitudinal ligament to protrude into the neural canal. It is most common in the lumbar region. Bone disease or disc degeneration may cause progressive protrusion of nucleus polposus into the neural canal. The herniation disc may be:

1. One-sided, causing pressure damage to a nerve root
2. Midline, compressing the spinal cord, the anterior spinal artery and possibly bilateral nerve roots

The outcome depends upon the size of the hernia and the length of time the pressure is applied. Small herniations cause local pain due to pressure on the nerve endings in the posterior longitudinal ligament. Large herniations may cause:

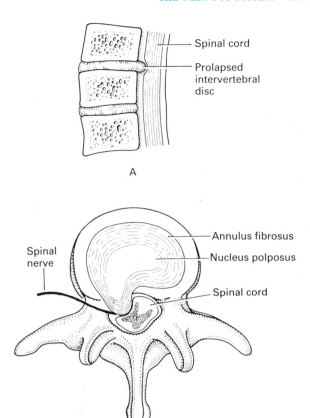

Figure 12:48   Prolapsed intervertebral disc. A, Viewed from the side. B. Viewed from above.

1. Unilateral or bilateral paralysis
2. Acute or chronic pain perceived to originate from the area supplied by the compressed sensory nerve, e.g., in the leg or foot.
3. Compression of the anterior spinal artery, causing ischaemia and possibly necrosis of the spinal cord
4. Local muscle spasm due to pressure on motor nerves

PERIPHERAL NERVES

NEUROPATHIES

This is a group of diseases of peripheral nerves not associated with inflammation. They are classified as:
1. Parenchymal, i.e., affecting neurones and/or myelin production and maintenance
2. Interstitial, i.e., affecting blood vessels and/or connective tissue adjacent to nerves

PARENCHYMAL NEUROPATHIES

Damage to a number of neurones and their myelin sheaths (polyneuropathy) occurs in, e.g.:
1. Nutritional deficiencies, e.g., folate, vitamins B_1, B_2, B_6, B_{12}

2. Metabolic disorders, e.g., diabetes
3. Chronic diseases, e.g., renal and hepatic failure, carcinoma
4. Toxic reactions to, e.g., lead, arsenic, mercury, carbon tetrachloride, aniline dyes, some drugs, such as phenytoin, chloroquine
5. Infections, e.g., influenza, measles, typhoid fever, diphtheria, leprosy

The long neurones are usually affected first, e.g., those supplying the feet and legs. The outcome depends upon the cause of the neuropathy.

INTERSTITIAL NEUROPATHY

Usually only one neurone is damaged and the most common cause is ischaemia, due to:
1. Pressure applied to cranial nerves in cranial bone foramina due to distortion of the brain by increased ICP
2. Compression of a nerve in a confined space caused by surrounding inflammation and oedema, e.g., the median nerve in the carpal tunnel
3. External pressure on a nerve, e.g., an unconscious person lying with an arm hanging over the side of a bed or trolley
4. Compression of the axillary (circumflex) nerve by ill-fitting crutches
5. Trapping a nerve between the broken ends of a bone
6. Ischaemia due to thrombosis of blood vessels supplying a nerve

The resultant dysfunction depends on the site and extent of the injury.

NEURITIS

ACUTE POST-INFECTIVE POLYNEURITIS
(Guillain-Barré syndrome)

This is a sudden, acute, progressive, bilateral ascending paralysis, beginning in the lower limbs and spreading to the arms, trunk and cranial nerves. It usually occurs 1 to 3 weeks after a viral infection and may be associated with an immunological reaction to myelin, precipitated by the infection. There is widespread inflammation accompanied by some demyelination of spinal, peripheral and cranial nerves and the spinal ganglia. Patients who survive the acute phase usually recover completely in weeks or months.

BELL'S PALSY

Compression of the facial nerve in the temporal bone foramen causes paralysis of facial muscles. The immediate cause is inflammation and oedema of the nerve but the underlying cause is unknown. Distortion of the features is due to muscle tone on the unaffected side, the affected side being expressionless. Recovery is usually complete within a few months.

DEVELOPMENTAL ABNORMALITIES OF THE NERVOUS SYSTEM

SPINA BIFIDA

This is a congenital malformation of the neural canal and spinal cord. The vertebral arches and dura mater are defective, most commonly in the lumbosacral region. The causes are not known. They may be of genetic origin or due to maternal upset, such as hypoxia, at a critical stage in development of the fetal vertebrae and spinal cord. The effects depend on the extent of the abnormality.

OCCULT SPINA BIFIDA

The skin over the defect is intact and excessive growth of hair over the site may be the only sign of abnormality.

MENINGOCELE

The skin over the defect is very thin and there is dilatation of the subarachnoid space posteriorly. The spinal cord is correctly positioned.

MENINGOMYELOCELE

The subarachnoid space is dilated posteriorly, the spinal cord is displaced backwards and is adherent to the posterior wall of the arachnoid dilatation. The skin is often deficient with leakage of CSF, and the meninges may become infected.

HYDROCEPHALUS see page 240

TUMOURS OF THE NERVOUS SYSTEM

Most tumours of the nervous system are of neuroglial origin. Neurones are rarely involved because they do not normally multiply. Metastases of nervous tissue tumours are rare. Because of this, the rate of growth of a tumour is more important than the likelihood of spread outside the nervous system. In this context, 'benign' means slow-growing and 'malignant' rapid-growing. Signs of raised ICP appear after the limits of compensation have been reached.

SLOW-GROWING TUMOURS

These allow time for adjustment to compensate for increasing intracranial pressure, so the tumour may be quite large before its effects are evident. Compensation involves gradual reduction in the volume of cerebrospinal fluid and circulating blood.

RAPID-GROWING TUMOURS

These do not allow time for adjustment to compensate for the rapidly increasing ICP, so the effects quickly become apparent. Complications include:
1. Neurological impairment, depending on tumour site and size
2. Effects of increased ICP (see p. 239)
3. Necrosis of the tumour, causing haemorrhage and oedema

Gliomas are usually astrocytomas, ranging from benign tumours to the highly malignant *glioblastoma multiforme*.

Meningiomas are usually benign, originating from arachnoid granulations.

Medulloblastomas are highly malignant neurone-cell tumours, occurring mainly in the cerebellum in children. They are believed to originate from primitive cells prior to differentiation into neurones and neuroglia.

Metastases in the brain
The most common primary sites are the breast, lungs and bone marrow (leukaemias). There are two forms:
1. Discrete multiple tumours, mainly in the cerebrum
2. Diffuse tumours in the arachnoid mater

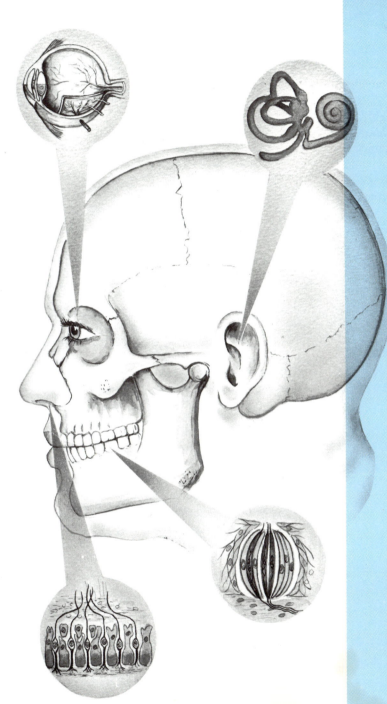

13

The Special Senses

13. *The Special Senses*

HEARING AND THE EAR

The ear is the organ of hearing. It is supplied by the eighth *cranial nerve*, i.e., the *cochlear part* of the *vestibulocochlear* nerve which is stimulated by vibrations caused by sound waves.

With the exception of the auricle (pinna), the structures that form the ear are encased within the petrous portion of the temporal bone.

STRUCTURE

The ear is divided into three distinct parts (Figs 13:1 and 13:7A):

External ear
Middle ear (tympanic cavity)
Internal ear

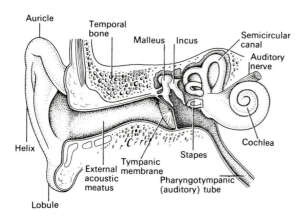

Figure 13:1   The parts of the ear.

EXTERNAL EAR

The external ear consists of the auricle (pinna) and the external acoustic meatus.

The auricle (Fig. 13:2)
The auricle is the expanded portion projecting from the side of the head. It is composed of *fibroelastic cartilage* covered with skin. It is deeply grooved and ridged and the most prominent outer ridge is the *helix*.

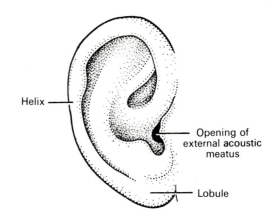

Figure 13:2   The auricle of the ear.

The lobule is the soft pliable part at the lower extremity, composed of fibrous and adipose tissue richly supplied with blood capillaries.

External acoustic meatus
This is a slightly 'S'-shaped tube about 2.5 cm long extending from the auricle to the *tympanic membrane* (ear drum). The lateral third is cartilaginous and the remainder is a canal in the temporal bone. The meatus is lined with a thin layer of skin, continuous with that of the auricle. There are numerous *ceruminous glands* in the skin of the lateral third. These are modified sweat glands that secrete *cerumin* (wax), a sticky material containing lysozyme and immunoglobulins. Foreign materials, e.g., dust, insects and microbes, are prevented from reaching the tympanic membrane by wax, hairs and the curvature of the meatus. Movements of the temporomandibular joint during chewing and speaking 'massage' the cartilaginous meatus, moving the wax towards the exterior.

The tympanic membrane (Fig. 13:3) completely separates the external acoustic meatus from the middle ear. It is oval-shaped with the slightly broader edge upwards and is formed by three types of tissue:

1. The outer covering of *hairless skin*
2. The middle layer of *fibrous tissue*
3. The inner lining of *mucous membrane* continuous with that of the middle ear

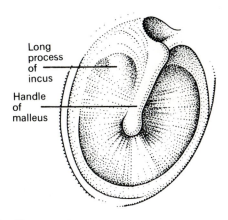

Figure 13:3 The tympanic membrane viewed from the outside showing the shadows cast by the malleus and the incus.

TYMPANIC CAVITY OR MIDDLE EAR

This is an irregular-shaped cavity within the petrous portion of the temporal bone. The cavity, its contents and the air sacs which open out of it are lined with mucous membrane. Air fills the cavity, reaching it through the *pharyngotympanic (auditory) tube* which extends from the nasopharynx. It is about 4 cm long and is lined with ciliated epithelium. The presence of air at atmospheric pressure on both sides of the tympanic membrane enables it to vibrate when sound waves strike it.

The lateral wall of the middle ear is formed by the tympanic membrane.

The roof and floor are formed by the temporal bone.

The posterior wall is formed by the temporal bone and has an opening leading to the *mastoid antrum* through which air passes to the mastoid air cells.

The medial wall is a thin layer of temporal bone in which there are two openings:

    Oval window (fenestra vestibuli)
    Round window (fenestra cochleae)

The oval window is occluded by part of a small bone called the *stapes* and the round window, by a fine sheet of *fibrous tissue*.

Auditory ossicles (Fig. 13:4)

These are three very small bones that extend across the cavity from the tympanic membrane to the oval window. They form a series of movable joints with each other and with the medial wall of the cavity at the oval window. They are: *the malleus, incus* and *stapes.*

The malleus is the lateral hammer-shaped bone. The handle is in contact with the tympanic membrane and the head forms a movable joint with the incus.

The incus is the middle anvil-shaped bone. Its body articulates with the malleus, the long process with the stapes, and it is stabilised by the short process, fixed by fibrous tissue to the posterior wall of the cavity.

The stapes is the medial stirrup-shaped bone. Its head articulates with the incus and its base fits into the oval window.

The three ossicles are held in position by fine ligaments.

INTERNAL EAR (Fig. 13:5)

The internal ear contains the organs of hearing and balance and is generally described in two parts, the *bony labyrinth* and the *membranous labyrinth*.

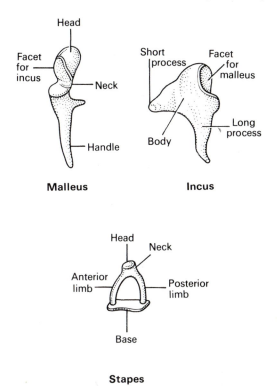

Figure 13:4 The auditory ossicles.

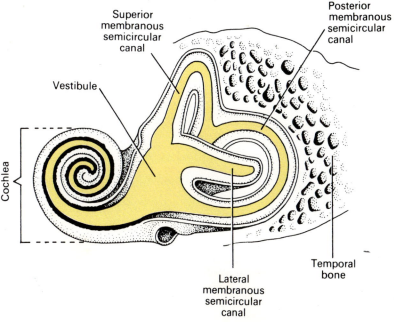

Figure 13:5 The internal ear. Membranous labyrinth coloured.

Bony labyrinth

This is a cavity within the temporal bone lined with periosteum. It is larger than the membranous labyrinth of the same shape which fits into it, like a tube within a tube. The space between the bony walls and the membranous tube is occupied by *perilymph*. The membranous labyrinth also contains fluid, the *endolymph*.

The bony labyrinth consists of:
1 vestibule
1 cochlea
3 semicircular canals

The vestibule is the expanded part nearest the middle ear. It contains the oval and round windows.

The cochlea resembles a snail's shell. It has a broad base where it is continuous with the vestibule and a narrow apex, and it spirals round a central bony column.

The semicircular canals are three tubes arranged so that one is situated in each of the three planes of space. They are continuous with the vestibule.

Membranous labyrinth

The membranous labyrinth is the same shape as its bony counterpart and is separated from it by perilymph. It contains endolymph. It is divided into the same parts: the vestibule which contains the *utricle* and *saccule*, the cochlea and three semicircular canals.

A cross-section (Fig. 13:6) shows the triangular shape of the membranous cochlea. Neuroepithelial cells and nerve fibres lie on the *basilar membrane* or base of the triangle. Many of the neuroepithelial cells are long and narrow and are arranged side by side. These cells and their nerve fibres form the true organ of hearing, the *organ of Corti*. The nerve fibres combine to form the *auditory part of the vestibulocochlear nerve* (eighth cranial nerve), which passes through a foramen in the temporal bone to reach the hearing area in the temporal lobe of the cerebrum.

PHYSIOLOGY OF HEARING (Figs 13:6 and 13:7)

Every sound produces *sound waves* or disturbances in the air, which travel at about 332 metres (1088) feet per second. The auricle, because of its shape, concentrates the waves and directs them along the auditory meatus causing the tympanic membrane to vibrate. Tympanic membrane vibrations are transmitted through the middle ear by movement of the ossicles. At their medial end the footplate of the stapes rocks to and fro in the oval window, setting up fluid waves in the perilymph. These indent the membranous labyrinth and the wave motion in the endolymph stimulates the neuroepithelial cells of the organ of Corti. The nerve impulses produced pass to the brain in the cochlear portion of the eighth cranial nerve (VIII). The fluid wave is finally expended into the middle ear by vibration of the membrane of the round window. This nerve, the *vestibulocochlear nerve*, transmits the impulses to various nuclei in the pons varolii and midbrain. Some of the nerve fibres pass to the hearing area in the cerebral cortex where sound is perceived.

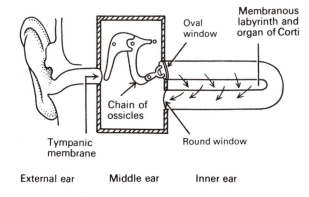

Figure 13:7A   Diagram of the ear showing the passage of sound waves. The cochlea has been uncoiled.

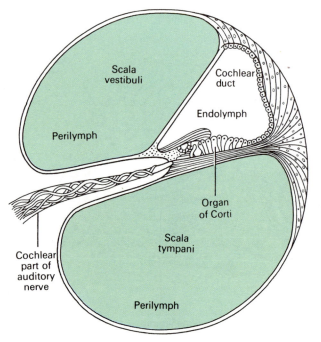

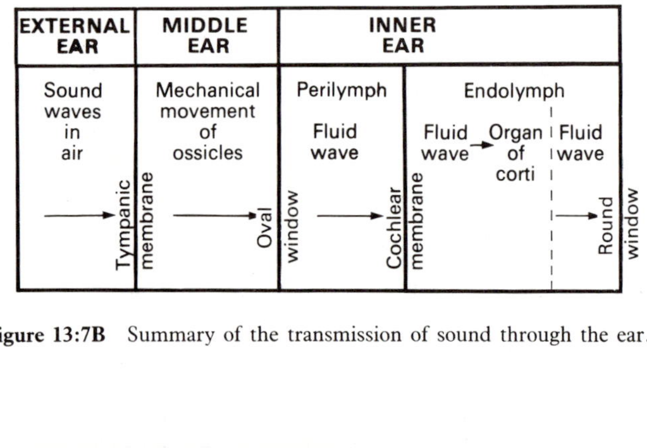

Figure 13:6   A section of the membranous cochlea showing the organ of Corti.

EXTERNAL EAR	MIDDLE EAR	INNER EAR		
Sound waves in air	Mechanical movement of ossicles	Perilymph Fluid wave	Endolymph	
			Fluid wave → Organ of corti	Fluid wave
→ (Tympanic membrane)	→ (Oval window)	→ (Cochlear membrane)		→ (Round window)

Figure 13:7B   Summary of the transmission of sound through the ear.

SEMICIRCULAR CANALS

The semicircular canals have no auditory function although they are closely associated with the cochlea. They provide information about the position of the head in space, contributing to maintenance of equilibrium and balance.

There are three semicircular canals, one lying in each of the three planes of space. They are situated above and behind the vestibule of the inner ear and open into it.

Structure of the semicircular canals (Fig. 13:8)

The semicircular canals, like the cochlea, are composed of an outer bony wall and inner membranous tubes or *ducts*. The membranous ducts contain endolymph and are separated from the bony wall by perilymph.

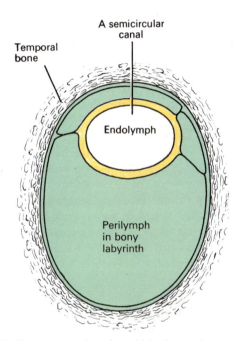

Figure 13:8   Transverse section of a semicircular canal.

The utricle is a membranous sac which is part of the vestibule and the three membranous ducts open into it at their dilated ends, the *ampullae*. The *saccule* is a part of the vestibule and communicates with the utricle and the cochlea.

In the walls of the utricle, saccule and ampullae there are fine specialised epithelial cells with minute projections, called *hair cells*. Amongst the hair cells there are the minute nerve endings of the *vestibular part* of the vestibulocochlear nerve.

Functions of the semicircular canal

The semicircular canals, utricle and saccule are concerned with balance. Any change of position of the head causes movement in the perilymph and endolymph which stimulates the nerve endings and the hair cells in the utricle, saccule and ampullae. The resultant nerve impulses are transmitted by the vestibular nerve which joins the cochlear nerve to form the *vestibulocochlear nerve*. The vestibular branch passes first to the *vestibular nucleus*, then to the *cerebellum*.

The cerebellum also receives nerve impulses from the eyes and the muscles and joints. Impulses from these three sources are coordinated and efferent nerve impulses pass to the cerebrum where position in space is perceived, and to muscles to maintain posture and balance.

DISEASES OF THE EAR

EXTERNAL OTITIS

Infection by *Staphylococcus aureus* is the usual cause of localised inflammation (boils) in the external auditory meatus. When more generalised, the inflammation may be caused by bacteria or fungi or by an allergic reaction to, e.g., dandruff, soaps, hair sprays, hair dyes.

ACUTE OTITIS MEDIA

This may be due to:
1. Spread of microbes from upper respiratory tract infection through the auditory (Eustachian) tube
2. Effusion of fluid into the middle ear cavity

Microbial infection leads to the accumulation of pus, rupture of the tympanic membrane and purulent discharge from the ear. The spread of infection may cause mastoiditis, meningitis, labyrinthitis, encephalitis and possibly brain abscess.

Effusion of fluid follows obstruction of the auditory tube by, e.g., pharyngeal inflammation or tumour. Air already present in the middle ear is absorbed. At first there is retraction of the tympanic membrane, then fluid is drawn into the low-pressure cavity from surrounding blood vessels. There may or may not be secondary infection.

CHRONIC OTITIS MEDIA

In this condition there is permanent perforation of the tympanic membrane following acute otitis media, mechanical or blast injuries. A serious complication occurs when the membrane has been ruptured for some time and squamous epithelial cells from the external ear grow into the middle ear, forming a *cholesteatoma*. This consists of desquamated epithelial cells and purulent material. Continued development of cholesteatoma may lead to:
1. Destruction of the ossicles and conduction deafness
2. Erosion of the roof of the middle ear and meningitis
3. Spread of infection to the inner ear, causing labyrinthitis

MÉNIÈRE'S DISEASE

In this condition there is generalised dilatation of the membranous labyrinth with destruction of the sensory cells in the ampulla and cochlea. It is usually unilateral at first but both ears may be affected later. The cause is not known. Ménière's disease is associated with recurrent episodes of incapacitating dizziness (vertigo), nausea and vomiting, lasting for several hours. Periods of remission vary from days to months. During and between attacks there may be continuous ringing in the affected ear (tinnitus). Loss of hearing is experienced during episodes, and permanent deafness gradually develops over a period of years as the organ of Corti is destroyed.

LABYRINTHITIS

This may be caused by spread of infection from the middle ear. In some generalised virus diseases the organ of Corti is destroyed, causing sudden total nerve deafness in the affected ear. The virus infections include mumps, measles, chicken-pox and influenza.

DEAFNESS IN YOUNG CHILDREN

This is usually nerve (perceptive) deafness and may be due to:
1. Genetic abnormality
2. Rubella in the mother in the first 3 months of pregnancy
3. Acute hypoxia at birth, or soon after

DEAFNESS IN OLDER CHILDREN AND ADULTS

CONDUCTIVE DEAFNESS

This is due to impaired transmission of sound waves from the outside to the oval window. It may be caused by:
1. Wax in the external auditory meatus
2. Acute or chronic otitis media
3. Otosclerosis in which the footplate of the stapes becomes fixed in the oval window
4. Traumatic rupture of the tympanic membrane and dislocation of the ossicles by, e.g., excessive force used when syringing an ear, attempting to remove a foreign body, a severe blow to the head, an explosion
5. Haemorrhage into the middle ear

SENSORY-NEURAL (PERCEPTIVE) DEAFNESS

This is the result of disease of the cochlea, auditory nerve or hearing area of the brain. The individual usually hears noise but cannot discriminate between sounds, i.e., hears but cannot understand. Causes include:
1. Presbycousis (the deafness of old age) due to degeneration of the sensory cells of the organ of Corti
2. Ménière's disease, progressive destruction of the organ of Corti
3. Trauma caused by:
   a. exposure to high-pitched loud noise for a prolonged period
   b. fracture of the base of the skull, involving the petrous portion of the temporal bone
4. Vascular changes in which the blood supply to the inner ear is interrupted
5. Labyrinthitis due to general virus diseases or to infection spread from the middle ear
6. Disease of, or injury to, the auditory nerve

SIGHT AND THE EYE

The eye is the organ of the sense of sight situated in the orbital cavity and it is supplied by the *optic nerve* (second cranial nerve).

It is almost spherical in shape and is about 2.5 cm in diameter. The space between the eye and the orbital cavity is occupied by fatty tissue. The bony walls of the orbit and the fat help to protect the eye from injury.

Structurally the two eyes are separate but, unlike the ear, some of their activities are co-ordinated so that they function as a pair. It is possible to see with only one eye but three-dimensional vision is impaired when only one eye is used, especially in relation to the judgement of distance.

STRUCTURE OF THE EYE (Fig. 13:9)

There are three layers of tissue in the walls of the eye. They are:
1. The outer fibrous layer: sclera and cornea
2. The middle vascular layer: choroid, ciliary body and iris
3. The inner nervous tissue layer: retina
Structures inside the eyeball are the lens, aqueous fluid (humour) and vitreous body.

SCLERA AND CORNEA

The sclera, or white of the eye, forms the outermost layer of tissue of the posterior and lateral aspects of the eyeball and is continuous anteriorly with the transparent *cornea*. It consists of a firm fibrous membrane that maintains the shape of the eye and gives attachment to the *extraocular* or *extrinsic muscles* of the eye (see p. 263).

Anteriorly the sclera continues as a clear transparent epithelial membrane, *the cornea*. Light rays pass through

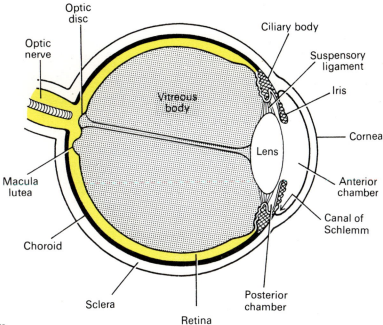

Figure 13:9 Section of the eye.

the cornea to reach the retina. The cornea is convex anteriorly and is involved in *refracting* or bending light rays to focus them on the retina.

CHOROID

The choroid lines the posterior five-sixths of the inner surface of the sclera. It is very rich in blood vessels and is a deep chocolate brown in colour. Light enters the eye through the pupil, stimulates the nerve endings in the retina then is absorbed by the choroid.

Ciliary body

The ciliary body is the anterior continuation of the choroid consisting of non-striated muscle fibres (*ciliary muscle*) and secretory epithelial cells. It gives attachment to the *suspensory ligament* which, at its other end, is attached to the capsule enclosing the lens. Contraction and relaxation of the ciliary muscle changes the thickness of the lens which refracts light rays entering the eye to focus them on the retina. The epithelial cells secrete *aqueous fluid* into the anterior segment of the eye, i.e., the space between the lens and the cornea. The ciliary body is supplied by parasympathetic branches of the oculomotor nerve (third cranial nerve). Stimulation causes contraction of the muscle and accommodation of the eye.

Iris

The iris extends anteriorly from the ciliary body and lies behind the cornea in front of the lens. It divides the anterior segment of the eye into *anterior* and *posterior chambers* which contain *aqueous fluid* secreted by the ciliary body. It is a circular body composed of pigment cells and two layers of muscle fibres, one circular and the other

radiating (Fig. 13:10). In the centre there is an aperture, the *pupil*.

The pupil varies in size depending upon the intensity of light. In bright light the circular muscle fibres contract and *constrict the pupil*. In dim light the radiating muscle fibres contract *dilating the pupil*.

The iris is supplied by parasympathetic and sympathetic nerves. Parasympathetic stimulation, supplied by the oculomotor nerve, constricts the pupil and sympathetic stimulation from the superior cervical ganglion, dilates the pupil.

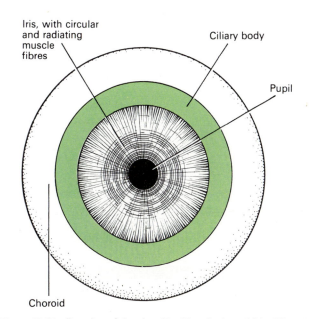

Figure 13:10 Drawing of the choroid, ciliary body and iris. Viewed from the front.

The colour of the iris depends on the number of pigment cells present. Albinos have no pigment cells and people with blue eyes have fewer than those with brown eyes.

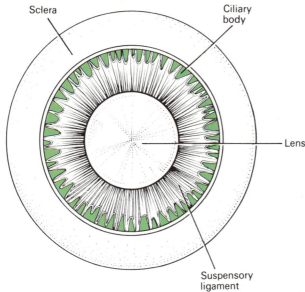

Figure 13:11 Drawing of the lens and suspensory ligament viewed from the front. The iris has been removed.

Lens (Fig. 13:11)

The lens is a *highly elastic* circular biconvex transparent body, lying immediately behind the pupil. It is suspended from the ciliary body by the *suspensory ligament* and enclosed within a transparent capsule. Its thickness is controlled by the ciliary muscle through the suspensory ligament. The lens bends light rays reflected by objects in front of the eye. It is the only structure in the eye that can vary its refractory power, achieved by changing its thickness. When the ciliary muscle contracts it *moves forward*, releasing its pull on the lens, increasing its thickness. The nearer the object the thicker the lens.

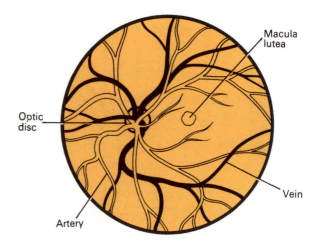

Figure 13:13 The retina as seen through the pupil.

RETINA (Figs 13:12 and 13:13)

The retina is the innermost layer of the wall of the eye. It is an extremely delicate membrane and is especially adapted to be stimulated by light rays. It is composed of several laycrs of nerve cells and nerve fibres, lying on a pigmented layer of epithelial cells which attach it to the choroid. The layer highly sensitive to light is the *layer of rods and cones*.

The retina lines about three-quarters of the eyeball and is thickest at the back and thins out anteriorly to end just behind the ciliary body. Near the centre of the posterior part there is an area that appears yellow in colour, hence the name, *macula lutea*. In the centre of the area there is a little depression called the *fovea centralis*, consisting of only *cone-shaped cells*. Towards the anterior part of the retina there are fewer cone- than rod-shaped cells.

The rods and cones contain photosensitive pigments involved in the conversion of light rays into nerve impulses. *Rhodopsin* (visual purple) is a pigment in rods that is so sensitive it is bleached by dim light. Vitamin A

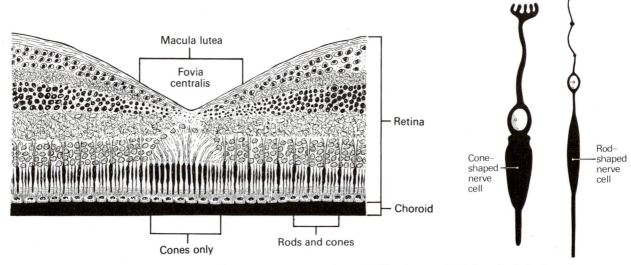

Figure 13:12A Magnified section of the retina.

Figure 13:12B Cone- and rod-shaped cells in the retina.

(retinal) is essential for its synthesis. Other pigments in cones are stimulated by colour.

About 0.5 cm to the nasal side of the macula lutea all the nerve fibres of the retina converge to form the *optic nerve* which passes through the sphenoid bone, eventually to reach the cerebral cortex in the occipital lobe of the cerebrum. The small area of retina where the optic nerve leaves the eye is the *optic disc* or *the blind spot*. It has no light-sensitive cells.

Blood supply to the eye

The eye is supplied with arterial blood by the *ciliary arteries* and the *central retinal artery*. These are branches of the ophthalmic artery, one of the branches of the internal carotid artery.

Venous drainage is by a number of veins, including the *central retinal vein* which eventually empty into the cavernous sinus.

The central retinal artery and vein are encased in the optic nerve, entering the eye at the optic disc

INTERIOR OF THE EYEBALL

The anterior segment of the eye, i.e., the space between the cornea and the lens, is incompletely divided into *anterior* and *posterior chambers* by the iris. Both chambers contain a clear *aqueous fluid (humour)* secreted into the posterior chamber by ciliary glands. It passes in front of the lens, through the pupil into the anterior chamber and returns to the circulation through the *canal of Schlemm* in the angle between the iris and cornea. There is continuous production and drainage but the intraocular pressure remains fairly constant between 1.3 and 2.6 kPa (10 to 20 mmHg).

Behind the lens and filling the cavity of the eyeball is the *vitreous body (humour)*. This is a soft, colourless, transparent, jelly-like substance composed of 99% water, some salts and mucoprotein. It maintains sufficient intraocular pressure to support the retina against the choroid and prevent the walls of the eyeball from collapsing.

The eye keeps its shape because of the intraocular pressure exerted by the vitreous body and the aqueous fluid. It remains fairly constant throughout life.

OPTIC NERVES (Fig. 13:14)

The fibres of the optic nerve originate in the retina of the eye. All the fibres converge to form the optic nerve about 0.5 cm to the nasal side of the macula lutea. The nerve pierces the choroid and sclera to pass backwards and medially through the orbital cavity. It then passes through the optic foramen of the sphenoid bone, backwards and medially to meet the nerve from the other eye at the *optic chiasma*.

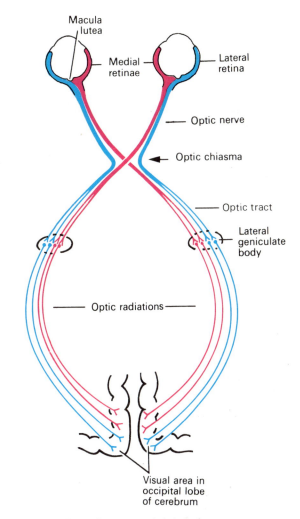

Figure 13:14   The optic nerves and their pathways.

Optic chiasma

This is situated immediately in front of and above the pituitary gland. In the optic chiasma the nerve fibres of the optic nerve from the *nasal side of each retina cross over* to the *opposite side*. The fibres from the *temporal side* do not cross but continue backwards on the same side.

Optic tracts

These are the pathways of the optic nerves, posterior to the optic chiasma. Each tract consists of the nasal fibres from the retina of one eye and the temporal fibres from the retina from the other. The optic tracts pass backwards through the cerebrum to synapse with nerve cells of the *lateral geniculate bodies*. From there the nerve fibres proceed backwards and medially as the *optic radiations* to terminate in the *visual area* of the cerebral cortex in the occipital lobes of the cerebrum. Other neurones originating in the lateral geniculate bodies convey impulses from the eyes to the *cerebellum* where they contribute to the maintenance of balance.

PHYSIOLOGY OF SIGHT

Light waves travel at a speed of 186 000 miles (300 000 kilometres) per second. Light is reflected into the eyes by objects within the field of vision. White light is a combination of all the colours of the visual spectrum (rainbow), i.e., red, orange, yellow, green, blue, indigo and violet. This can be demonstrated by passing white light through a glass prism which refracts or bends the rays of the different colours to a greater or lesser extent, depending on their wavelengths. Red light has the longest wavelength and violet the shortest (see Fig. 13:15). This range of colour is the *spectrum of visible light*. In a rainbow, white light from the sun is broken up by raindrops which act as prisms and reflectors.

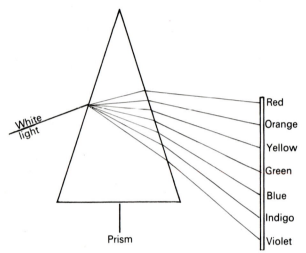

Figure 13:15   White light broken into the colours of the visible spectrum when passed through a prism.

The spectrum of light

The spectrum of light is broad but only a small part is visible to the human eye (Fig. 13:16). Beyond the long end there are infrared (heat), radar and radio waves. Beyond the short end there are ultraviolet (UV), X-ray and cosmic waves. UV light is not normally visible because it is absorbed by a yellow pigment in the lens. Following removal of the lens (cataract operation), UV light is visible and it has been suggested that long-term exposure may damage the retina.

A specific colour is perceived when only one wavelength is *reflected* by the object and all the others are *absorbed*, e.g., an object appears red when only the red wavelength is reflected. Objects appear white when all wavelengths are reflected, and black when they are all absorbed.

In order to achieve clear vision, light reflected from objects within the visual field is focused on to the retina of both eyes. The processes involved in producing a clear image are *refraction of the light rays* and *accommodation of the eyes*.

Although these may be considered as separate processes, effective vision is dependent upon their co-ordination.

REFRACTION OF THE LIGHT RAYS

When light rays pass from a medium of one density to a medium of a different density they are refracted or bent (Fig. 13:17). This principle is used in the eye to focus light on the retina. Before reaching the retina light rays pass successively through the conjunctiva, cornea, aqueous fluid, lens and vitreous body. They are all more dense than air and, with the exception of the lens, they have a constant refractory power, similar to that of water.

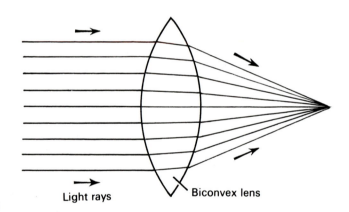

Figure 13:17   Refraction of light rays passing through media of different densities.

Lens

The lens is a biconvex elastic transparent body suspended behind the iris from the ciliary body by the suspensory ligament. It is the only structure in the eye that changes

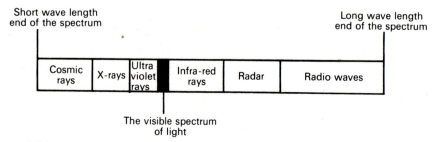

Figure 13:16   The spectrum of light.

its refractive power. All light rays entering the eye need to be bent (refracted) to focus them on the retina. Light from distant objects needs least refraction and as the object comes closer the amount needed is increased. To increase the refractive power the ciliary muscle contracts, releasing its pull on the suspensory ligament and the anterior surface of the lens bulges forward, increasing its convexity. When the ciliary muscle relaxes it slips backwards, increasing its pull on the suspensory ligament, making the lens thinner.

Looking at near objects 'tires' the eyes more quickly due to the continuous use of the ciliary muscle.

ACCOMMODATION OF THE EYES TO LIGHT
(Figs 13:18 and 13:19)

Three factors are involved in accommodation:
  Pupils
  Movement of the eyeballs (convergence)
  Lens

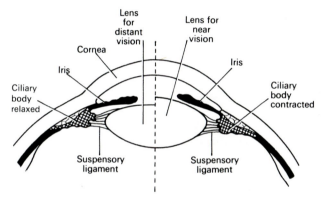

Figure 13:18 Diagram of the difference in the shape of the lens for near and distant vision.

Size of the pupils

Pupil size influences accommodation by controlling the amount of light entering the eye. In a bright light the pupils are constricted. In a dim light they are dilated.

If the pupils were dilated in a bright light, too much light would enter the eye and damage the retina. In a dim light, if the pupils were constricted, insufficient light would enter the eye to stimulate the nerve endings in the retina.

The iris consists of one layer of circular and one of radiating smooth muscle fibres. Contraction of the circular fibres constricts the pupil, and contraction of the radiating fibres dilates it. The size of the pupil is controlled by nerves of the autonomic nervous system. Sympathetic stimulation dilates the pupils and parasympathetic stimulation causes constriction.

Movements of the eyeballs — convergence

Light rays from objects enter the two eyes at different angles and for clear vision they must stimulate *corresponding areas* of the two retinae. Extraocular muscles move the eyes and to obtain a clear image they rotate the eyes so that they *converge* on the object viewed. This co-ordinated muscle activity is under autonomic control. When there is voluntary movement of the eyes both eyes move and convergence is maintained. The nearer an object is to the eyes the greater the eye rotation needed to achieve convergence. If convergence is not complete there is double vision, *diplopia*.

FUNCTION OF THE RETINA

The retina is the *photosensitive* part of the eye. The light-sensitive cells are the *rods* and *cones*. Light rays cause chemical changes in photosensitive pigments in these cells and they emit nerve impulses which pass to the occipital lobes of the cerebrum via the optic nerves.

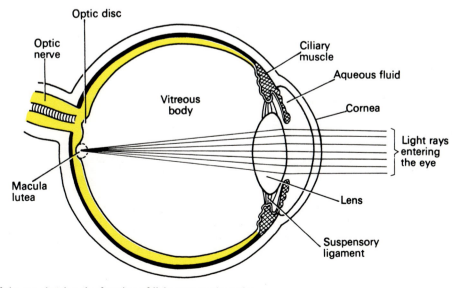

Figure 13:19 Section of the eye showing the focusing of light rays on the retina.

The rods are more sensitive than the cones. They are stimulated by *low intensity or dim light*, e.g., by the dim light in the interior of a darkened room.

The cones are sensitive to *bright light and colour*. The different wavelengths of light stimulate photosensitive pigments in the cones, resulting in the perception of different colours. In a bright light the light rays are focused on the macula lutea.

The rods are more numerous towards the periphery of the retina. *Visual purple (rhodopsin)* is a photosensitive pigment present only in the rods. It is bleached by bright light and when this occurs the rods cannot be stimulated. Rhodopsin is quickly reconstituted when an adequate supply of vitamin A is available. When the individual moves from an area of bright light to one of dim light there is a variable period of time when it is difficult to see. The rate at which *dark adaptation* takes place is dependent upon the reconstitution of rhodopsin. It is easier to see a dim star in the sky at night if the head is turned slightly away from it because light of low intensity is focused on the area of the retina where there is the greatest concentration of rods. If looked at directly the light intensity is not sufficient to stimulate the less sensitive cones in the area of the macula lutea. For the same reason, in dim evening light different colours cannot be distinguished.

BINOCULAR VISION (Fig. 13:20)

Binocular or stereoscopic vision has certain advantages. Each eye 'sees' a scene slightly differently. There is an overlap in the middle but the left eye sees more on the left than can be seen by the other eye and vice versa. The images from the two eyes are fused in the cerebrum so that only one image is perceived.

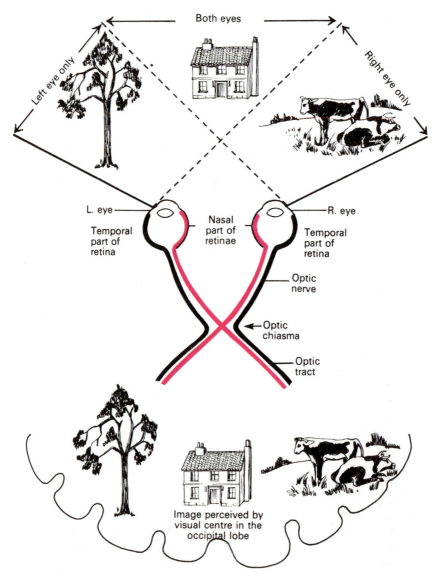

Figure 13:20   Diagram of the parts of the visual field — monocular and binocular.

Binocular vision provides a much more accurate assessment of one object relative to another, e.g., its distance, depth, height and width. Some people with monocular vision may find it difficult to judge the speed and distance of an approaching vehicle.

EXTRAOCULAR MUSCLES OF THE EYE

The eyeballs are moved by six extrinsic muscles, attached at one end to the eyeball and at the other to the walls of the orbital cavity. There are four *straight* and two *oblique* muscles (Fig. 13:21). They are:

Medial rectus   Superior oblique
Lateral rectus   Inferior oblique
Superior rectus
Inferior rectus

They consist of striated muscle fibres. Movement of the eyes to look in a particular direction is under voluntary control but co-ordination of movement, needed for convergence and accommodation to near or distant vision, is under autonomic control.

The medial rectus rotates the eyeball inwards.
The lateral rectus rotates the eyeball outwards.
The superior rectus rotates the eyeball upwards.
The inferior rectus rotates the eyeball downwards.
The superior oblique rotates the eyeball so that the cornea turns in a downwards and outwards direction.
The inferior oblique rotates the eyeball so that the cornea turns upwards and outwards.

NERVE SUPPLY TO THE MUSCLES OF THE EYE

The oculomotor (third cranial) nerve supplies the:

Superior rectus ⎫
Inferior rectus ⎪
Medial rectus   ⎬ extrinsic muscles
Inferior oblique ⎭

Iris   ⎫
Ciliary muscle ⎬ intrinsic muscles

The trochlear (IV cranial) nerve supplies the superior oblique.

The abducent (VII cranial) nerve supplies the lateral rectus.

ACCESSORY ORGANS OF THE EYE

The eye is a delicate organ which is protected by several structures (Figs 13:22 and 13:23):

Eyebrows
Eyelids and eyelashes
Lacrimal apparatus

EYEBROWS

These are two arched ridges of the supraorbital margins of the frontal bone. Numerous hairs (eyebrows) project obliquely from the surface of the skin. They protect the anterior aspect of the eyeball from sweat, dust and other foreign bodies.

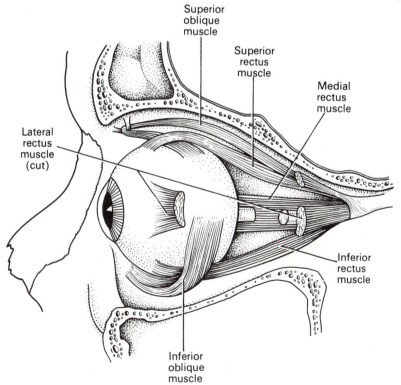

Figure 13:21 The extrinsic muscles of the eye.

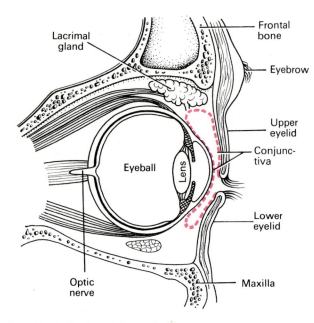

Figure 13:22 Section of the eye and its accessory structures.

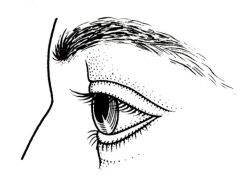

Figure 13:23 Side view of some structures which protect the eye.

EYELIDS

The eyelids are two movable folds of tissue situated above and below the front of each eye. On their free edges there are short curved hairs, the eyelashes. The layers of tissue which form the eyelids are:

A thin covering of skin

A thin sheet of areolar tissue

Two muscles — the *orbicularis oculi* and *levator palpebrae superioris*

A thin sheet of dense connective tissue, the *tarsal plate*, larger in the upper than in the lower eyelid, which supports the other structures

A lining of *conjunctiva*

Conjunctiva

This is a fine transparent membrane which lines the eyelids and is reflected on to the front of the eyeball. Where it lines the eyelids it consists of highly vascular columnar epithelium. Corneal conjunctiva consists of less vascular

stratified epithelium. When the eyelids are closed the conjunctiva becomes a closed sac. It protects the delicate cornea and the front of the eye. When drops of a drug are put into the eye they are placed in the lower conjunctival sac. The medial and lateral angles of the eye where the upper and lower lids come together are called respectively the *medial* and *lateral canthus*.

Eyelid margins

Along the edges of the lids there are numerous *sebaceous glands*, some with ducts opening into the hair follicles of the *eyelashes* and some onto the eyelid margins between the hairs. *Meibomian glands* (tarsal glands) are modified sebaceous glands embedded in the tarsal plates. Their ducts open onto the inside of the free margins of the eyelids. They secrete an oily material that is spread over the conjunctiva by blinking which delays evaporation of tears.

Function

The eyelids and *eyelashes* protect the eye from injury. If injury is feared or the conjunctiva touched very lightly the eyelids close. This is called the conjunctival or *corneal reflex*. Blinking spreads tears and Meibomian secretions over the cornea, preventing drying.

When the *orbicularis oculi* contracts the eyes close. When the *levator palpebrae* contract the eyelids open.

LACRIMAL APPARATUS (Fig. 13:24)

For each eye this consists of:

1 lacrimal gland and its ducts

2 lacrimal canaliculi

1 lacrimal sac

1 nasolacrimal duct

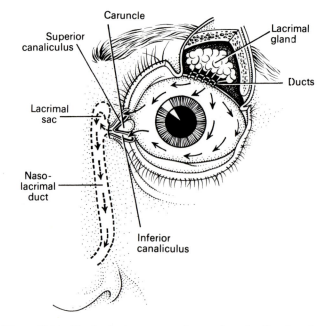

Figure 13:24 The lacrimal apparatus. Arrows show the direction of flow of tears.

The lacrimal glands are situated in recesses in the frontal bones on the lateral aspect of each eye just behind the supraorbital margin. Each gland is approximately the size and shape of an almond, and is composed of *secretory epithelial cells*. The glands secrete *tears* composed of water, salts and *lysozyme*, a bactericidal enzyme.

The tears leave the lacrimal gland by several small ducts and pass over the front of the eye under the lids towards the medial canthus where they drain into the *two lacrimal canaliculi*, the opening of each is called the *puncta*. One canaliculus lies above the other, separated by a small red body, the *caruncle*. The tears then drain into the *lacrimal sac* which is the upper expanded end of the *nasolacrimal duct*. This is a membranous canal approximately 2 cm long, extending from the lower part of the lacrimal sac to the nasal cavity, opening at the level of the inferior concha.

The fluid that fills the conjunctival sac, consisting of tears and the secretions of the Meibomian glands, is spread over the corneal conjunctiva by blinking. It washes away irritating materials, e.g., dust and grit, and the bactericidal lysozyme prevents microbial infection. The oiliness of this mixed fluid delays evaporation, preventing drying of the conjunctiva. In some emotional states the secretion of tears may be increased and, if the nasolacrimal duct cannot convey them all into the nasal cavity, they overflow.

DISEASES OF THE EYE

INFLAMMATION

STYE (HORDEOLA)

This is pyogenic infection of sebaceous or Meibomian glands of the eyelid margin. A 'crop' of styes may occur due to localised spread to adjacent glands. Infection of Meibomian (tarsal) glands may block their ducts, leading to cyst formation (*chalazion*) which may damage the cornea. The most common infecting organism is *Staphylococcus aureus*.

BLEPHARITIS

This is microbial or allergic inflammation of the eyelid margins. The most common causes are staphylococcal infection or allergy to dandruff. If ulceration occurs, healing by fibrosis may distort the eyelid margins, preventing complete closure of the eye. This may lead to drying of the eye, conjunctivitis and possibly corneal ulceration.

CONJUNCTIVITIS

This is microbial or allergic inflammation. Microbial infection is usually by staphylococci, streptococci, pneumococci, herpes zoster (shingles), herpes simplex or measles viruses. Gonococcal infection may be acquired by the baby at birth if the mother is infected. Infection may spread to the cornea, causing ulceration and patches of opacity when the ulcers heal. Allergic conjunctivitis may be a complication of other autoimmune diseases or be caused by a wide variety of antigens encountered in the environment, e.g., dust, pollen, fungus spores, animal dander, cosmetics, hair sprays, soaps.

TRACHOMA

This is a chronic inflammatory condition caused by *Chlamydia trachomatis* in which fibrous tissue forms in the conjunctiva and cornea, leading to eyelid deformity and possibly blindness. It occurs mostly where there is poor hygiene, especially in the Middle East, Africa and Southeast Asia. The microbes are usually spread by flies, communal use of contaminated washing water, cross infection between mother and child.

CORNEAL ULCER

This is local necrosis of corneal tissue, usually associated with corneal infection (*keratitis*) following trauma, or infection spread from the conjunctiva or eyelids. The most common infecting microbes are staphylococci, streptococci, pneumococci and herpes simplex viruses. Acute pain, photophobia and lacrimation interfere with sight during the acute phase. Extensive ulceration and healing by fibrosis cause opacity of the cornea and irreversible sight loss.

INFLAMMATION OF THE UVEAL TRACT (IRIS, CILIARY BODY, CHOROID)

ANTERIOR UVEITIS (Iritis, Iridocyclitis)

Iridocyclitis (inflammation of iris and ciliary body) is the more common and it may be acute or chronic. The infection may have spread from the outer eye but in most cases the cause is unknown. There is usually moderate to severe pain, redness, lacrimation and photophobia. In severe cases adhesions form between the iris and lens capsule, preventing the circulation of aqueous fluid in the posterior and anterior chambers. This may cause the lens to bulge and occlude the canal of Schlemm, raising intraocular pressure, i.e., a form of *glaucoma*. Acute infection usually resolves in several days or weeks while the chronic form may last for months or years.

POSTERIOR UVEITIS (Choroiditis, Chorioretinitis)

Chorioretinitis is the more common condition. It may be caused by spread of infection from the front of the eye or be secondary to a wide variety of systemic conditions, including rheumatoid arthritis, Reiter's disease, ulcerative colitis, brucellosis. Complications that may occur include retinal detachment due to accumulation of inflammatory exudate, secondary glaucoma, cataract.

GLAUCOMA

This is a group of conditions in which there is increased intraocular pressure due to defective drainage of aqueous fluid through the canal of Schlemm in the angle between the iris and cornea in the anterior chamber.

PRIMARY GLAUCOMAS

CHRONIC OPEN-ANGLE GLAUCOMA

There is a gradual painless rise in intraocular pressure with progressive loss of vision. Peripheral vision is lost first but may not be noticed until only central (tunnel) vision remains. As the condition progresses, atrophy of the optic disc occurs leading to irreversible blindness. It is commonly bilateral and occurs mostly in people over 40 years of age. The cause is not known but there is a familial tendency.

ACUTE CLOSED-ANGLE GLAUCOMA

This is most common in people over 60 years of age and usually affects one eye. During life the crystalline lens gradually increases in size, pushing the iris forward. In dim light when the pupil dilates the lax iris bulges still further forward, and may come into contact with the cornea and close the canal of Schlemm, raising the intraocular pressure. An acute attack may abort if the iris responds to bright light, constricting the pupil and releasing the pressure on the canal. After repeated attacks spontaneous recovery may be incomplete, leading to further increase in intraocular pressure accompanied by severe pain, headache, nausea and vomiting. The cornea becomes oedematous and opaque and vision is impaired.

CHRONIC CLOSED-ANGLE GLAUCOMA

Following repeated moderately severe attacks of acute glaucoma the angle between the iris and cornea is gradually reduced and the canal of Schlemm obstructed by adhesions. Although the intraocular pressure rises there is usually little pain. Peripheral vision deteriorates first followed by atrophy of the optic disc and blindness.

CONGENITAL GLAUCOMA

This abnormal development of the anterior chamber is often familial or due to maternal infection with rubella in early pregnancy.

SECONDARY GLAUCOMA

The most common primary disorder is uveitis with the formation of adhesions between the iris and lens capsule that obstruct the flow of aqueous fluid from the posterior to the anterior chamber. Other primary conditions include intraocular tumours, enlarged cataracts, central retinal vein occlusion, intraocular haemorrhage, trauma to the eye.

STRABISMUS (Squint, Cross-eye)

This is the inability of the eyes to move so that the same image falls on the corresponding parts of the retina in both eyes. It is caused by extraocular muscle weakness or defective nerve supply to the muscle, i.e., defective cranial nerves III, IV or VI. In most cases the image falling on the squinting eye is suppressed by the brain, otherwise there is double vision (diplopia).

CATARACT

This is opacity of the lens which may be degenerative or congenital, bilateral or unilateral. In degenerative cataract there is gradual development of lens opacity that may be due to senile degeneration, X-rays, infrared rays (heat), trauma, diabetes, uveitis and some drugs, e.g., long-term use of corticosteroids. Senile cataract usually develops in older age groups. Other degenerative forms may develop at an earlier age depending on the primary cause. Developmental cataract may be due to genetic abnormality, e.g., Down's syndrome, or maternal infection in early pregnancy, e.g., rubella.

RETINOPATHIES

VASCULAR RETINOPATHIES

Occlusion of the central retinal artery or vein causes sudden painless unilateral loss of vision. Arterial occlusion is usually due to embolism from, e.g., atheromatous plaques, endocarditis, retinal artery sclerosis. Venous occlusion is usually associated with arteriosclerosis in the elderly or with venous thrombosis elsewhere in the body. Predisposing factors include glaucoma, diabetes, hypertension, increased blood viscosity.

DIABETIC RETINOPATHY

Changes in retinal blood vessels are associated more with the duration of diabetes than with its severity. Capillary microaneurysms develop and later there may be proliferation of blood vessels. Haemorrhages, fibrosis and secondary retinal detachment may occur, leading to retinal degeneration and loss of vision.

RETINAL DETACHMENT

This occurs when a tear or hole in the retina allows fluid to accumulate between the layers of retinal cells or between the retina and choroid. It is usually localised at first but as fluid collects the detachment spreads. There are spots before the eyes, flashing lights due to abnormal stimulation of sensory cells, and progressive loss of vision. In many cases the cause is unknown but it may be associated with trauma to the eye or head, tumours, haemorrhage, cataract surgery when the pressure in the eye is reduced or diabetic retinopathy.

RETROLENTAL FIBROPLASIA

This may occur in premature infants who receive oxygen therapy. The high oxygen tension in the blood may cause constriction of immature retinal arterioles. After oxygen therapy ceases there is disordered development of retinal blood vessels and the formation of fibrovascular tissue in the vitreous body causing varying degrees of interference with light transmission. Blindness may be caused by haemorrhage in the vitreous humour, detachment of the retina or in very severe cases, by formation of an opaque membrane behind the lens.

RETINITIS PIGMENTOSA

This is a hereditary disease in which there is degeneration of the retina, mainly affecting the rods. Defective vision in dim light usually becomes apparent in early childhood, leading to tunnel vision and eventually, blindness.

KERATOMALACIA

In this condition there is corneal ulceration, usually with secondary infection. The lacrimal glands and conjunctiva may be involved. It is caused by chronic vitamin A and protein deficiency in the diet. There may be softening or even perforation of the cornea. Night blindness (defective adaptation to dim light) is usually an early sign of deficiency of vitamin A which is required for the reconstitution of rhodopsin (visual purple) after it has been exposed to light.

TUMOURS

CHOROIDAL MALIGNANT MELANOMA

This is the most common ocular malignancy and it occurs mostly in older people. Vision is not usually affected until the tumour causes retinal detachment, usually when well advanced. The tumour spreads locally in the choroid, and blood-borne metastases develop mainly in the liver, bones, lungs and brain.

RETINOBLASTOMA

This is a malignant tumour derived from embryonic retinal cells. It is usually evident before the age of 4 years and may be bilateral. it spreads locally to the vitreous and may grow along the optic nerve, invading the brain.

ACUTE DACRYOADENITIS

This is inflammation of the lacrimal gland, usually unilateral. It may be due to spread of infection from the eyelids or surrounding structures, or be associated with measles, mumps or influenza. The infection usually resolves but occasionally an abscess forms.

DACRYOCYSTITIS

This inflammation of the lacrimal sac is usually associated with partial or complete obstruction of the lacrimal duct. In infants there may be congenital stenosis of the duct. In adults the blockage may be due to nasal trauma, deviated nasal septum, nasal polyp or acute inflammatory nasal congestion.

SENSE OF SMELL

The nose has a dual function: respiration (see Ch. 7) and sense of smell (Fig. 13.25).

OLFACTORY NERVES (FIRST CRANIAL NERVES)

These are the sensory nerves of smell. They have their origins in special cells in the mucous membrane of the roof of the nose above the superior nasal conchae. On each side of the nasal septum nerve fibres from the cells pass through the *cribriform plate of the ethmoid bone* to the *olfactory bulb* where interconnections and synapses occur. From the bulb, bundles of nerve fibres form the *olfactory tract* which passes backwards to the olfactory area in the *temporal lobe* of the cerebral cortex in each hemisphere where the impulses are interpreted and odour perceived. (Fig. 13:26).

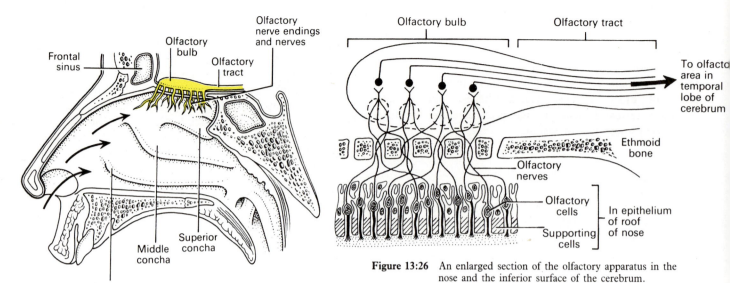

Figure 13:25 The olfactory structures.

Figure 13:26 An enlarged section of the olfactory apparatus in the nose and the inferior surface of the cerebrum.

PHYSIOLOGY OF SMELL

The sense of smell in human beings is generally less acute than in other animals. All odorous materials give off chemical particles which are carried into the nose with the inhaled air and stimulate the nerve cells of the olfactory region when in solution in mucus.

When an individual is continuously exposed to an odour, perception of the odour quickly decreases and eventually ceases. This loss of perception only affects that specific odour and adaptation probably occurs both in the cerebrum and in the nerve endings in the nose.

The air entering the nose is heated and convection currents carry eddies of inspired air from the main stream to the roof of the nose. 'Sniffing' concentrates more particles more quickly in the roof of the nose. This increases the number of special cells stimulated and thus the perception of the smell. The sense of smell may affect the appetite. If the odours are pleasant the appetite may improve and vice versa.

Inflammation of the nasal mucosa prevents odorous substances from reaching the olfactory area of the nose, causing loss of the sense of smell. The usual cause is the common cold.

SENSE OF TASTE

Taste buds are found in the papillae of the tongue and widely distributed in the epithelium of the tongue, soft palate, pharynx and epiglottis. They consist of small bundles of cells and nerve endings of the glossopharyngeal, facial and vagus nerves (cranial nerves VII, IX and X).

Some of the cells have hair-like microvilli on their free border, projecting towards tiny pores in the epithelium (Fig. 13:27). The nerve cells are stimulated by chemical substances in solution that enter the pores. The nerve impulses are transmitted to the thalamus then to the *taste area* in the cerebral cortex, one in each hemisphere, where taste is perceived. Four fundamental sensations of taste have been described — sweet, sour, bitter and salt. This is probably an oversimplification because perception varies widely and many 'tastes' cannot be easily classified. However, some tastes consistently stimulate taste buds in specific parts of the tongue.

Sweet and salty, mainly at the tip
Sour, at the sides
Bitter, at the back

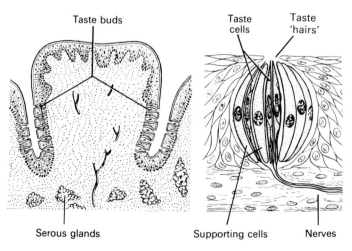

Figure 13:27 *Left*: A section of the vallate papilla. *Right*: A section of a taste bud—greatly magnified.

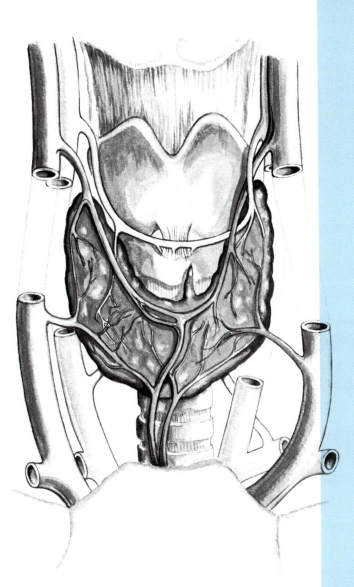

The Endocrine System

14. The Endocrine System

The endocrine system consists of glands widely separated from each other with no direct anatomical links. The glands are commonly referred to as the *ductless glands* because the *hormones* they secrete pass *directly* from the cells into the blood stream.

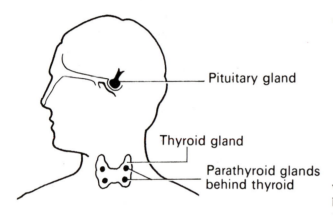

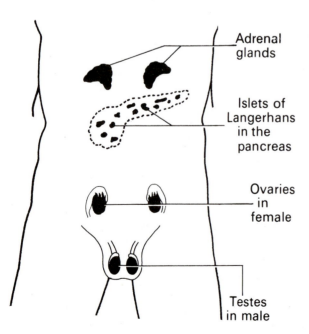

Figure 14:1 Diagram of the positions in the body of the endocrine glands.

A hormone is a chemical substance which, having been formed in one organ or gland, is carried in the blood to another organ (target organ) or tissue, probably quite distant, where it influences activity, growth, nutrition.

The internal environment of the body is regulated partly by the autonomic nervous system and partly by hormones.

The endocrine system consists of the following glands (Fig. 14:1):

1 pituitary gland
1 thyroid gland
4 parathyroid glands
2 adrenal (suprarenal) glands
Islets of Langerhans in the pancreas
1 pineal gland or body
2 ovaries in the female
2 testes in the male

The ovaries and the testes secrete hormones associated with the reproductive system, therefore their functions will be understood more readily if they are studied in Chapter 15.

PITUITARY GLAND AND HYPOTHALAMUS (Figs 14:2 and 14:3)

The pituitary gland and the hypothalamus act as a unit, regulating the activity of most of the other endocrine glands. The gland lies in the hypophyseal fossa of the sphenoid bone below the hypothalamus, to which it is attached by a *stalk*. It consists of two distinct parts that originate from different types of cells. The *adenohypophysis* (anterior lobe) is an upgrowth of tissue from the pharynx and the *neurohypophysis* (posterior lobe) is a downgrowth from the brain. There is a network of nerve fibres between the hypothalamus and the neurohypophysis and one of blood vessels between the hypothalamus and the adenohypophysis.

BLOOD SUPPLY

Arterial blood is supplied by branches from the internal carotid artery.

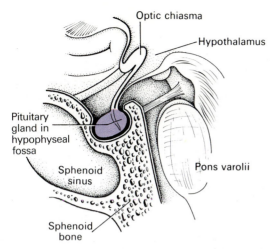

Figure 14:2   The position of the pituitary gland and its associated structures.

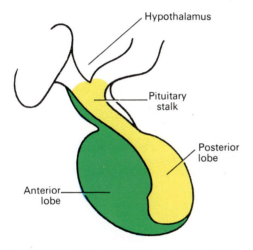

Figure 14:3   The parts of the pituitary gland and its relation to the hypothalamus.

The anterior lobe of the pituitary gland is supplied indirectly by blood that has already passed through a capillary bed in the hypothalamus. This network of blood vessels, *the pituitary portal system*, conveys blood from the hypothalamus to the anterior lobe where it enters thin-walled vascular sinusoids and is in very close contact with the cells. The blood pressure in the sinusoids is lower than in the portal system. In addition to providing oxygen and nutrition this blood conveys *releasing* and *inhibiting hormones* secreted by the *hypothalamus* that influence the secretion and release of hormones formed in the anterior lobe.

The posterior lobe is supplied directly by a branch from the carotid artery.

Venous blood from both lobes, containing hormones, leaves the gland in short veins that enter the venous sinuses between the layers of dura mater.

ADENOHYPOPHYSIS (Anterior lobe)

Some of the hormones secreted by the anterior lobe stimulate or inhibit secretion by other endocrine glands (target glands) while others have a direct effect on target tissues. Table 14:1 shows the main relationships between the hypothalamus, the adenohypophysis and the target glands or tissues.

The release of anterior pituitary hormones follows stimulation of the gland by *'releasing hormones'* produced by the hypothalamus and conveyed to the gland through the pituitary portal system of blood vessels. The whole system is controlled by a *negative feedback mechanism*. That is, when there is a low level of a hormone in the blood supplying the hypothalamus it produces the appropriate releasing hormone which stimulates release of hormone by the anterior pituitary and this in turn stimulates the target gland to produce and release its hormone. As a result the blood level of that hormone rises and inhibits the secretion of releasing factor by the hypothalamus (see Fig. 14:4).

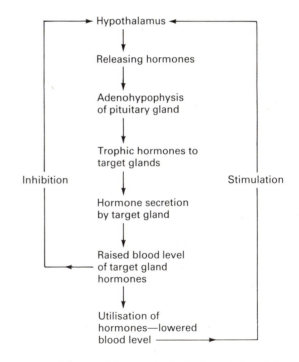

Figure 14:4   Diagram of the negative feedback regulation of the secretions of hormones by the anterior lobe of the pituitary gland.

ADENOHYPOPHYSEAL HORMONES (Table 14:1)

Growth hormone

Growth hormone (GH) is synthesised by the adenohypophysis. Its release is stimulated by GHRF (*somatotrophin*) and inhibited by GHRIH (*somatostatin*), both secreted by the hypothalamus. GH promotes protein synthesis in growth and repair of all tissues. Increased secretion is stimulated by exercise, anxiety, sleep and hypoglycaemia.

Thyroid stimulating hormone (TSH)

This hormone is synthesised by the adenohypophysis and its release is stimulated by TRH from the hypothalamus. It stimulates growth and activity of the thyroid gland which secretes the hormones *thyroxin* (T4) and *triiodothyronine* (T3). There is a circadian rhythmical release of TSH. It is highest between 9 p.m. and 6 a.m. and lowest between 4 p.m. and 7 p.m. Secretion is also regulated by a negative feedback mechanism. When the blood level of thyroid hormones is high secretion of TSH is reduced, and vice versa.

Adrenocorticotrophic hormone (ACTH)

Corticotrophin releasing factor (CRF) from the hypothalamus promotes the synthesis and release of ACTH by the adenohypophysis. This stimulates the adrenal cortex to secrete *cortisol*. Production of CRF is believed to be influenced by:

1. Nerve impulses from the higher centres
2. A low blood level of cortisol
3. Physical stress, especially exercise
4. Emotional stress
5. Hypoglycaemia

ACTH levels are highest at about 8 a.m. and fall to their lowest about midnight, although high levels sometimes occur at midday and 6 p.m. This circadian rhythm is maintained throughout life. It is associated with the sleep pattern and adjustment to changes takes several days, following, e.g., working-shift changes, travel to a different time-zone (jet-lag).

Prolactin

This hormone has a direct effect on the breasts immediately after parturition. The blood level of prolactin is not dependent on a hypothalamic releasing factor but it is lowered by the inhibiting factor, *dopamine*.

Prolactin together with oestrogens, corticosteroids, insulin and thyroxine is involved in initiating and maintaining lactation. The circadian rhythm of prolactin secretion is related to sleep, i.e., it is raised during any period of sleep, night and day. Emotional stress increases production.

Gonadotrophic hormones

The anterior lobe secretes *two gonadotrophic* or *sex hormones* in both the female and the male:

Follicle stimulating hormone (FSH)
Luteinising hormone (LH)

Female gonadotrophic hormones

The follicle stimulating hormone stimulates the development and ripening of the ovarian follicle (see Ch. 15). During its development the ovarian follicle secretes its own hormone, *oestrogen*. As the level of oestrogen increases in the blood so FSH secretion is reduced.

The luteinising hormone promotes the final maturation of the ovarian follicle and ovulation (discharge of the mature ovum). Its main function is to promote the formation of the *corpus luteum* which secretes the second ovarian hormone, *progesterone*. As the level of progesterone in the blood increases there is a gradual reduction in the production of the luteinising hormone.

Male gonadotrophic hormones

The follicle stimulating hormone stimulates the epithelial tissue of the *seminiferous tubules* in the testes to produce *spermatozoa* (male germ cells).

The luteinising hormone stimulates the *interstitial cells* in the testes to secrete the hormone *testosterone*.

Table 14:1   Hormones of the hypothalamus and adenohypophysis

Hypothalamus	Adenohypophysis	Target gland or tissue
GHRF	GH	many glands all tissues
GHRIF	GH inhibition	thyroid gland islets of Langerhans all tissues
TRH	TSH	thyroid
CRF	ACTH	adrenal cortex
none	PRL	breast
PIF	PRL inhibition	breast
LHRH	FSH	ovaries and testes
	LH	ovaries and testes

GHRF = growth hormone releasing factor
GH   = growth hormone (somatotrophin)
GHRIF = growth hormone release inhibiting factor (somatostatin)
TRH   = thyroid releasing factor
TSH   = thyroid stimulating hormone
CRF   = corticotrophin releasing factor
ACTH = adrenocorticotrophic hormone
PRL   = prolactin (lactogenic hormone)
PIF   = prolactin inhibiting factor
LH-RH = luteinising hormone releasing hormone
FSH   = follicle stimulating hormone
LH   = luteinising hormone

DISORDERS OF THE ADENOHYPOPHYSIS

These disorders result in hypersecretion or hyposecretion of one or more hormones. The abnormalities described here are those in which there is no specific target gland. The abnormalities of stimulating hormones are included with the descriptions of their target glands.

HYPERSECRETION OF ADENOHYPOPHYSEAL HORMONES

GIGANTISM AND ACROMEGALY

The most common cause is prolonged hypersecretion of growth hormone (GH) usually by a hormone-secreting tumour. The conditions are only occasionally due to excess growth hormone releasing factor (GHRF) secreted by the hypothalamus. As the tumour increases in size the pressure in the gland and surrounding tissues leads to:

1. Hyposecretion of other hormones of both the anterior and posterior lobes
2. Hyposecretion of hormone-releasing factors by the hypothalamus
3. Damage to the optic nerves, causing visual disturbances

Effects of excess GH

1. Excessive growth of bones
2. Enlargement of internal organs
3. Growth of excess connective tissue
4. Enlargement of the heart and a rise in blood pressure
5. Reduced glucose tolerance and a predisposition to diabetes mellitus

Gigantism occurs when there is excess GH while epiphyseal cartilages are still growing, i.e., before ossification of bones is complete. It is evident mainly in the bones of the limbs (Fig. 14:5).

Acromegaly occurs when there is excess GH after ossification is complete. The bones become abnormally thick, most noticeable as coarse facial features and excessively large hands and feet (Fig. 14:6).

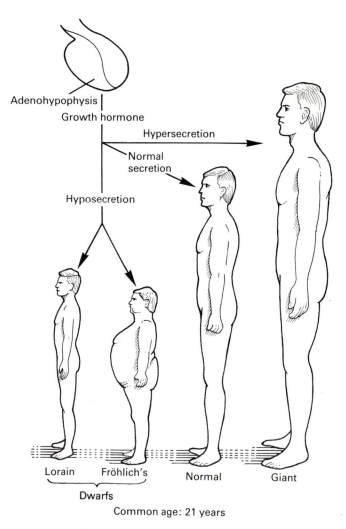

Figure 14:5   Effects of normal and abnormal growth hormone secretion.

HYPERPROLACTINAEMIA

This is caused by a hormone-secreting tumour. It causes sterility and amenorrhoea in women and sterility in men.

HYPOSECRETION OF ADENOHYPOPHYSEAL HORMONES

The number of hormones and the extent of hyposecretion varies. Panhypopituitarism is absence of all hormones. Causes of hyposecretion include:

1. Non-secretory pituitary tumours
2. Trauma, usually caused by fractured base of skull or surgery
3. Pressure caused by a tumour adjacent to the gland, e.g., glioma, meningioma
4. Ischaemic necrosis e.g., hypotension, following antepartum or postpartum haemorrhage
5. Ionising radiations or cytotoxic drugs

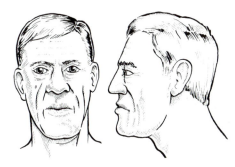

Figure 14:6   Acromegaly.

SIMMOND'S DISEASE (Sheehan's syndrome)

This panhypopituitarism occurs mostly in women following a difficult labour, especially if there has been haemorrhage and a substantial fall in blood pressure. Because of the arrangement of the blood supply, the gland is unusually susceptible to a fall in systemic blood pressure and becomes necrosed. The immediate effect may be failure of lactation and, later, deficient stimulation of target glands. The outcome depends on the extent of pituitary necrosis and hormone deficiency.

LORAIN–LEVI SYNDROME (Pituitary dwarfism)

This is caused by severe deficiency of GH, and possibly of other hormones, in childhood. Puberty is delayed and there may be episodes of hypoglycaemia. The individual is of small stature but is well proportioned and mental development is not affected. In the majority of cases the cause is unknown.

FRÖLICH'S SYNDROME

In this condition there is panhypopituitarism but the main features are associated with deficiency of GH, FSH and LH. In children the effects are diminished growth, lack of sexual development, obesity with female distribution of fat and retarded mental development. In a similar condition in adults, obesity and sterility are the main features. The hypothalamus may also be involved. It may be caused by a tumour but in most cases the cause is unknown.

NEUROHYPOPHYSIS (Posterior lobe)

The posterior lobe of the pituitary gland is composed of secretory cells called *pituicytes* and nerve fibres which arise from cells in the hypothalamus, the *supraoptic nucleus* and the *paraventricular nucleus*.

NEUROHYPOPHYSEAL HORMONES

The hormones released by the neurohypophysis are *oxytocin* and *antidiuretic hormone* (ADH or *vasopressin*). They are synthesised by the cells of the hypothalamus and migrate along nerve fibres to the posterior pituitary where they are stored in the nerve endings. Each hormone is released in response to a different stimulus (Fig. 14:7).

Oxytocin
Oxytocin promotes contraction of uterine muscle and contraction of myoepithelial cells of the lactating breast, squeezing milk into the large ducts behind the nipple. In late pregnancy the uterus becomes very sensitive to oxytocin. The amount secreted is increased just before and during labour and by sucking of the nipple.

Antidiuretic hormone (ADH) or vasopressin
This hormone has two main functions:
1. *Antidiuretic effect*. It increases the permeability to water of the distal convoluted and collecting tubules of the nephrons of the kidneys. As a result the reabsorption of water from the glomerular filtrate is increased. The amount of ADH secreted is influenced by the osmotic pressure of the blood circulating to the osmoreceptors in the hypothalamus. As the osmotic pressure rises the secretion of ADH increases and more water is reabsorbed. Conversely, when the osmotic pressure of the blood is low, the secretion of ADH is reduced, less water is reabsorbed and more urine is produced (see Fig. 10.10).

2. *Pressor effect*. In pharmacological doses, ADH stimulates contraction of smooth muscle, especially in blood vessel walls, raising the blood pressure. It is not clear whether physiological amounts have a significant pressor effect.

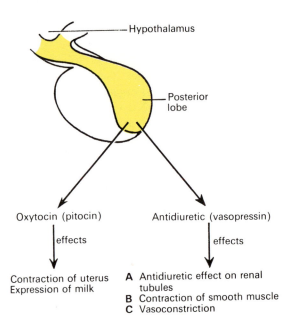

Figure 14:7 The hormones secreted by the posterior lobe of the pituitary gland (neurohypophysis) and their main functions.

DISORDERS OF THE NEUROHYPOPHYSIS

DIABETES INSIPIDUS

This is a relatively rare condition caused by hyposecretion of ADH. Water reabsorption by the kidneys is deficient, leading to excretion of excessive amounts of dilute urine, often more than 10 litres daily, causing dehydration, extreme thirst and polydipsia. Water balance is unlikely to be seriously disturbed unless intake is deficient.

THYROID GLAND (Fig. 14:8)

The thyroid gland is situated in the neck in front of with the larynx and trachea at the level of the 5th, 6th and 7th cervical and 1st thoracic vertebrae. It is a highly vascular gland surrounded by a fibrous capsule. It consists of *two lobes*, one on either side of the thyroid cartilage and upper cartilaginous rings of the trachea. The lobes are joined by a narrow *isthmus*, lying in front of the trachea.

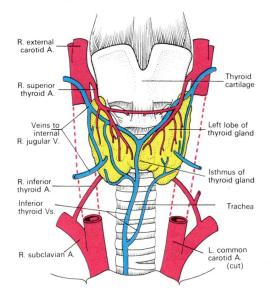

Figure 14:8   Position of thyroid gland and associated structures.

The lobes are roughly cone-shaped, about 5 cm long and 3 cm wide.

The arterial blood supply to the gland is through the superior and inferior thyroid arteries. The superior thyroid artery is a branch of the external carotid artery and the inferior thyroid artery is a branch of the subclavian artery.

The venous return is by the thyroid veins which drain into the internal jugular veins.

The gland is composed of epithelial cells which form closed spherical follicles, containing a thick, sticky, semi-fluid, structureless protein called *colloid* (Fig. 14:9).

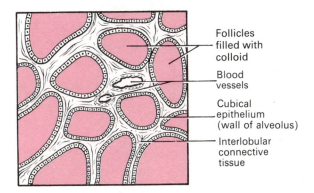

Follicles filled with colloid

Blood vessels

Cubical epithelium (wall of alveolus)

Interlobular connective tissue

Figure 14:9   The minute structure of the thyroid gland.

FUNCTIONS (Fig. 14:10)

Iodine is essential for the formation of the thyroid gland hormones, *thyroxine* (T4) and *triiodothyronine* (T3). It is ingested in food and most of it is taken up by the gland and used in hormone formation. After secretion, thyroid hormones combine with colloid and are stored in the follicles as *thyroglobulin*. Their release into the blood is regulated by *thyroid stimulating hormone* (TSH) from the adenohypophysis which is stimulated by *thyroid releasing hormone* (TRH) from the hypothalamus. Secretion of TRH is increased by a low blood level of T3 and T4 and decreased by a high level. Secretion of T3 and T4 begins about the 3rd month of fetal life and is increased at puberty and during pregnancy. Otherwise, it remains fairly constant throughout life.

The method of action of T3 and T4 is not known. It may act indirectly through cell enzymes and possibly by enhancing the effects of other hormones. Their influence is widespread and in general they promote:

1. Growth and differentiation of tissues
2. The provision of energy by oxidation

Calcitonin is a hormone produced by cells between the follicles of the thyroid gland. Its function is to reduce the blood level of calcium by inhibiting the reabsorption of calcium from bones. It has the opposite effect to *parathormone* secreted by the parathyroid glands. Its secretion is increased by a rise in blood calcium and by gastrointestinal hormones following a meal.

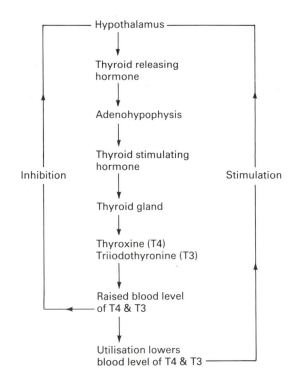

Figure 14:10   Regulation of secretion of thyroxine and triiodothyronine.

DISORDERS OF THE THYROID GLAND

Some disorders lead to hypersecretion, some to hyposecretion and some have little effect on secretion of thyroxin (T4) and triiodothyronine (T3). The disorders may occur in the hypothalamus, anterior pituitary or thyroid gland, or be due to insufficient intake of iodine in the diet. The main effects are caused by an abnormally high or low metabolic rate (MR).

HYPERTHYROIDISM

THYROTOXICOSIS (Toxic goitre)

The main effects are due to increased MR. A temporary increase may be tolerated but sustained increase causes:
1. Cardiac arrhythmias, because the heart becomes overstressed as it tries to supply extra oxygen and nutrition to the hyperactive body cells
2. Increased gluconeogenesis from body protein to provide extra energy, causing loss of weight, muscle wasting and weakness
3. Excess heat production as a by-product of the raised MR
4. Physical restlessness and mental excitability due to neurone hyperactivity

The gland may develop single or multiple hormone-secreting nodules or diffuse hyperplasia of secretory cells (Graves' disease).

GRAVES' DISEASE (Exophthalmic goitre)

There is diffuse hyperplasia of the thyroid gland with excess secretion of T3 and T4. It is believed to be caused by an autoimmune reaction to thyroid tissue and thyroglobulin, the antibodies acting as stimulating hormone (Fig. 14:11). Because of the feedback mechanism the abnormally high blood levels of T3 and T4 depress secretion of TRH from the hypothalamus and TSH from the anterior pituitary.

Exophthalmos (protrusion of the eyeballs), due to the deposition of excess fat and fibrous tissue behind the eyes, is often present in Graves' disease. It may be caused by autoimmunity different from that associated with hyperplasia of the gland. Effective treatment of thyrotoxicosis does not reduce the exophthalmos. In severe cases the eyelids may not completely cover the eyes during blinking and sleep, leading to drying of the conjunctiva and predisposing to infection. It does not occur in other forms of thyrotoxicosis.

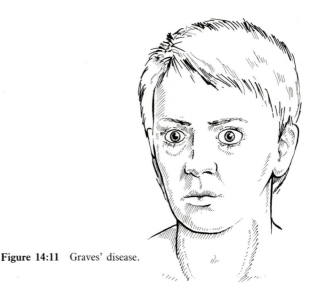

Figure 14:11   Graves' disease.

TOXIC NODULAR GOITRE

In this type of hyperthyroidism excess secretion of T3 and T4 produces the general effects of increased MR described above.

HYPOTHYROIDISM

Insufficient secretion of T3 and T4 causes cretinism in children and myxoedema in adults.

CRETINISM

The retarded mental and physical development is associated with hyposecretion of T3 and T4. In late intrauterine life and for the first few months after birth these hormones are essential for development of the brain and bones (Fig. 14:12)

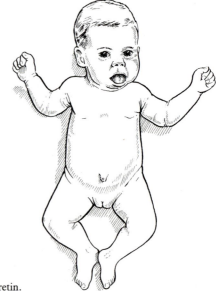

Figure 14:12   Cretin.

Endemic cretinism occurs in mountainous areas and areas far from the sea where soil and plant iodine content is very low. There is usually a family history of goitre extending over two or three generations. Iodine deficiency in the fetus causes developmental errors in early intrauterine life.

Sporadic cretinism is due to congenital underdevelopment of complete absence of the thyroid gland. The cause is unknown.

MYXOEDEMA

Deficiency of T3 and T4, occurring after normal mental and physical development is complete, results in an abnormally low MR and lack of response to demand for increased energy, e.g., by muscles during exercise (Fig. 14:13). Mental and physical processes become slower and there is reduced heat production with persistent feeling of cold even when the environmental temperature is high. Causes of myxoedema include:

1. Autoimmune thyroiditis, i.e., atrophy of the gland due to autoimmunity in which T3, T4 and thyroid gland tissue are the antigens
2. Severe and prolonged iodine deficiency
3. Deficiency of TRH and/or TSH

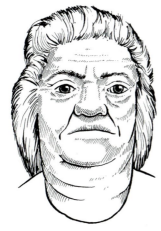

Figure 14:13   Myxoedema.

NON-TOXIC GOITRE

This is due to deficient secretion of T3 and T4 in spite of excess TRH and TSH. There is diffuse or nodular hyperplasia of the gland (Fig. 14:14). Sometimes the extra thyroid tissue is able to maintain normal hormone levels but if not, myxoedema develops. Underlying causes include:

1. Persistent iodine deficiency
2. Genetic abnormality
3. Chemicals that interfere with hormone synthesis, e.g., some drugs, including antithyrotoxicosis drugs, resorcinol, para-aminosalicylic acid, sulphonylureas

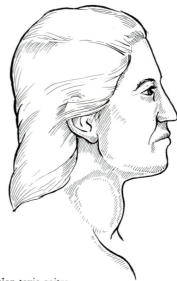

Figure 14:14   Non-toxic goitre.

The enlarged gland may cause pressure damage to adjacent tissues, especially if it lies in an abnormally low position, i.e., behind the sternum. The structures most commonly affected are the oesophagus, causing dysphagia; trachea, causing dyspnoea; recurrent laryngeal nerve, causing hoarseness of voice.

AUTOIMMUNE THYROIDITIS

The variants of this disease are Hashimoto's disease, primary myxoedema and focal thyroiditis. In all types there is autoimmunity to T3, T4, thyroglobulin and thyroid gland cells. The antibodies prevent synthesis and release of the hormones, causing myxoedema. The gland becomes swollen due to accumulation of lymphocytes and plasma cells. The causes are not known.

TUMOURS OF THE THYROID GLAND
BENIGN TUMOURS

Single and multiple adenomas are fairly common and many solitary nodules become cystic. If the adenoma secretes hormones, thyrotoxicosis may develop. The tumours have a tendency to become malignant, especially in the elderly.

MALIGNANT TUMOURS

These are relatively rare and the type of cell varies from well differentiated to anaplastic. The most common mode of spread is infiltration of the surrounding tissues, especially the trachea and recurrent laryngeal nerve.

PARATHYROID GLANDS (Fig. 14:15)

There are four small parathyroid glands, two embedded in the posterior surface of each lobe of the thyroid gland. They are surrounded by fine connective tissue capsules. The cells forming the glands are spherical in shape and are arranged in columns with channels containing blood between them.

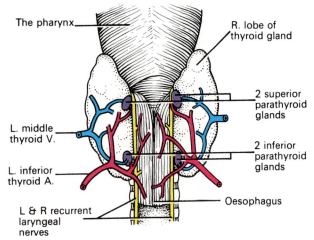

Figure 14:15 The positions of the parathyroid glands and their related structures viewed from behind.

FUNCTIONS

The function of the parathyroid glands is to secrete the hormone *parathormone* (PTH). Secretion is regulated by the blood level of *ionised calcium*. When this falls, secretion of PTH is increased and vice versa.

The main functions of parathormone are to maintain the blood concentration of calcium within normal limits. This is achieved by influencing the amount of calcium absorbed from the small intestine and reabsorbed from the renal tubules. If these sources provide inadequate supplies, parathormone stimulates osteoblasts and osteocytes to resorb calcium from bones.

Parathormone and calcitonin from the thyroid gland act together to maintain a normal blood calcium level under varying conditions.

DISORDERS OF THE PARATHYROID GLANDS

HYPERPARATHYROIDISM

Excess secretion of parathormone (PTH), usually by benign tumours of the gland, causes reabsorption of calcium from bones, raising the blood calcium level. The effects may be the formation of renal calculi, pyelonephritis, renal failure, calcification of soft tissues and tumours.

HYPOPARATHYROIDISM

PTH deficiency causes abnormally low blood calcium. This reduces absorption of calcium from the small intestine and resorption from bones and glomerular filtrate. Low blood calcium causes:

1. Increased skeletal muscle tone and, in severe cases, *tetany* (Fig. 14:16)
2. Development of cataract (opacity of the lens)
3. Behavioural disturbances and, in extreme cases, dementia

The causes of hypoparathyroidism include:

1. Damage to or removal of the glands during thyroidectomy
2. Ionising radiations, usually from radioactive iodine used to treat hyperthyroidism
3. Development of autoimmunity to parathyroid gland cells and PTH
4. Congenital abnormality of the glands

Tetany (carpopedal spasm) is caused by low blood levels of ionised calcium. There are very strong painful spasms of skeletal muscles, causing characteristic bending inwards of the hands, forearms and feet (Fig. 14:16). In children there may be laryngeal spasm and convulsions. It is associated with:

1. Hypoparathyroidism
2. Defective absorption from the intestine or dietary deficiency of calcium
3. Excretion of excess calcium in the urine in chronic renal failure
4. Alkalosis, due to persistent vomiting, ingestion of excess alkali to alleviate gastric disturbances or hyperventilation, which cause a fall in ionised calcium although the total calcium level is unchanged

Figure 14:16 Characteristic positions adopted during tetany spasms.

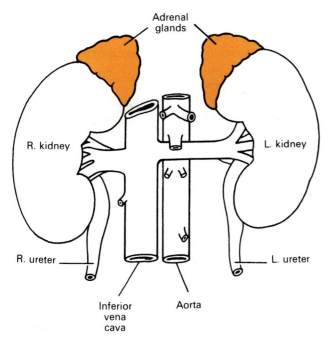

Figure 14:17   The positions of the adrenal glands and some of their associated structures.

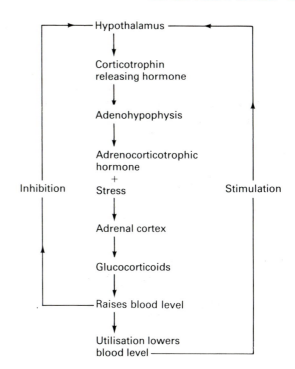

Figure 14:18A   Regulation of glucocorticoid secretion.

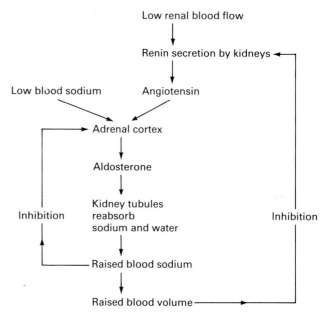

Figure 14:18B   Regulation of aldosterone secretion.

ADRENAL OR SUPRARENAL GLANDS (Fig. 14:17)

There are two adrenal glands, one situated on the upper pole of each kidney enclosed within the renal fascia. They are about 4 cm long and 3 cm thick.

The arterial blood supply to the glands is by branches from the abdominal aorta and renal arteries.

The venous return is by suprarenal veins. The right gland drains into the inferior vena cava and the left into the left renal vein.

The glands are composed of two parts which differ both anatomically and physiologically. The outer part is the *cortex* and the inner part, the *medulla*. The cortex is essential to life but the medulla is not.

ADRENAL CORTEX (Fig. 14:18A and B)

The adrenal cortex produces three groups of hormones:
1. Glucocorticoids
2. Mineralocorticoids
3. Androgens (sex hormones)

Glucocorticoids

Cortisol (hydrocortisone) and *corticosterone* are the main glucocorticoids. Secretion is stimulated by ACTH from the anterior pituitary and by stress. In non-stressful conditions secretion has marked circadian variations. The highest level of hormones occurs between 4 a.m. and 8 a.m. and the lowest, between midnight and 3 a.m.

Glucocorticoids have widespread effects on body systems. The main functions include:
1. Regulation of carbohydrate metabolism
2. Promotion of the formation of liver glycogen
3. Gluconeogenesis from protein, raising the blood glucose level
4. Depression of the body's inflammatory and allergic reactions
5. Promotion of sodium and water reabsorption from the renal tubules

Mineralocorticoids (Aldosterone)

Aldosterone is the main mineralo-corticoid. Its functions are associated with the maintenance of the electrolyte balance in the body. It stimulates the reabsorption of sodium by the renal tubules and when the amount of *sodium reabsorbed* is increased the amount of *potassium excreted* is increased. Indirectly this affects water excretion as water and electrolyte excretion are related.

The amount of aldosterone produced is influenced by the sodium level in the blood. If there is a fall in the sodium blood level, more aldosterone is secreted and more sodium reabsorbed.

Renin-angiotensin system. When renal blood flow is reduced the enzyme *renin* is secreted by kidney cells. This promotes the conversion of *angiotensinogen* to *angiotensin* which stimulates the production of aldosterone by the adrenal cortex. Aldosterone increases the reabsorption of sodium and water and the excretion of potassium by the kidneys. This raises the blood volume and the flow of blood through the kidneys, suppressing renin production and reducing aldosterone secretion.

Androgens

Sex hormones produced by the adrenal cortex are believed to have little significance compared to those produced by the gonads. They are associated with deposition of protein in muscles and retention of nitrogen, especially in males.

1. Painful adiposity of the face, neck and trunk
2. Excess protein catabolism, causing thinning of subcutaneous tissue and muscle wasting, especially of the legs
3. Diminished protein synthesis
4. Suppression of growth hormone, causing arrest of growth in children
5. Osteoporosis and kyphosis if vertebral bodies are involved
6. Excessive gluconeogenesis with hyperglycaemia and glycosuria
7. Atrophy of lymphoid tissue and depressed immune response
8. Susceptibility to infection due to reduced febrile response, depressed immune response and phagocytosis, impaired migration of phagocytes
9. Insomnia, excitability, euphoria, psychotic depression

HYPOSECRETION OF GLUCOCORTICOIDS

Inadequate secretion of cortisol causes diminished gluconeogenesis, low blood glucose, muscle weakness and pallor. It may be primary, i.e., due to adrenal cortex disease or secondary due to deficiency of ACTH from the anterior pituitary. In primary deficiency there is also hyposecretion of aldosterone but in secondary deficiency, aldosterone secretion is not usually affected.

HYPERSECRETION OF MINERALOCORTICOIDS

Excess aldosterone affects kidney function, causing:
1. Excessive reabsorption of sodium, chloride and water, causing hypertension
2. Excessive excretion of potassium, causing hypokalaemia which leads to cardiac arrhythmia, alkalosis, syncope and muscle weakness

PRIMARY ALDOSTERONISM (Conn's syndrome)

This is due to an excessive secretion of mineralocorticoids, independent of renin/angiotensin secretion. It may be caused by a tumour, usually affecting only one adrenal gland, or there may be bilateral hyperplasia of unknown origin.

SECONDARY ALDOSTERONISM

This is caused by overstimulation of normal glands by the excessively high blood levels of renin and angiotensin that result from low renal perfusion or low blood sodium chloride.

DISORDERS OF THE ADRENAL CORTEX

HYPERSECRETION OF GLUCOCORTICOIDS (Cushing's syndrome)

Cortisol is the main glucocorticoid hormone secreted by the adrenal cortex. Causes of hypersecretion include:
1. Hormone-secreting adrenal tumours, benign or malignant
2. Hypersecretion of adrenocorticotrophic hormone (ACTH) by the pituitary
3. Abnormal secretion of ACTH by a non-pituitary tumour, e.g., bronchial carcinoma, pancreatic tumour, carcinoid tumours
4. Prolonged therapeutic use of ACTH or glucocorticoids in high doses

Hypersecretion of cortisol has a wide variety of effects but they may not all be present. They include:

HYPOSECRETION OF MINERALOCORTICOIDS

Hypoaldosteronism results in failure of the kidneys to regulate sodium, potassium and water excretion, leading to:

1. Blood sodium deficiency and potassium excess
2. Dehydration and low blood volume, causing low blood pressure, especially if arteriolar constriction is defective due to deficiency of noradrenaline

There is usually hyposecretion of other cortical hormones, as in Addison's disease.

CHRONIC ADRENAL CORTEX INSUFFICIENCY (Addison's disease)

This is due to hyposecretion of all adrenal cortex hormones. The most common causes are autoimmunity to cortical cells, metastatic tumours and infections. Autoimmune disease of some other glands are associated with Addison's disease, e.g., thyroiditis, thyrotoxicosis, hypoparathyroidism. The outstanding effects are:

1. Muscle weakness and wasting
2. Gastrointestinal disturbances, e.g., vomiting, diarrhoea, anorexia
3. Increased pigmentation of the skin, especially of exposed areas, due to excess ACTH and the related melanin stimulating hormone secreted by the adenohypophysis
4. Listlessness and tiredness
5. Hypoglycaemia
6. Mental confusion
7. Menstrual disturbances and loss of body hair in women
8. Electrolyte imbalance, including low blood sodium and chloride and high potassium levels
9. Chronic dehydration, low blood volume and hypotension

The adrenal glands have a considerable reserve of tissue and Addison's disease is not usually severely debilitating unless more than 90% of cortical tissue is destroyed.

ADRENAL MEDULLA

The medulla is completely surrounded by the cortex. It is an outgrowth of tissue from the same source as the nervous system and its functions are closely allied to those of the sympathetic part of the autonomic nervous system. It is stimulated by its extensive sympathetic nerve supply to produce the *catecholamines, adrenaline* and *noradrenaline*. Noradrenaline is the chemical transmitter of the symp-

athetic nervous system. The adrenal medulla is not essential to life but its hormones greatly assist the body in its response to adverse environmental conditions. The main function of noradrenaline is maintenance of blood pressure by causing general vasoconstriction, except of the coronary arteries. Adrenaline is associated with potentiating the conditions needed for 'fight or flight' after the initial sympathetic stimulation, by, e.g.:

1. Constricting skin blood vessels
2. Dilatating muscle blood vessels
3. Converting glycogen to glucose

DISORDERS OF THE ADRENAL MEDULLA

TUMOURS

Hormone-secreting tumours are the main abnormality. The effects of excess adrenaline and noradrenaline include:

1. Hypertension, often associated with arteriosclerosis and cerebral haemorrhage
2. Raised blood glucose and glycosuria
3. Excessive sweating and alternate flushing and blanching
4. Raised metabolic rate

PHEOCHROMOCYTOMA

This is a *benign tumour*, occurring in one or both glands. The secretion of hormones may be at a steady high level or in intermittent bursts, usually triggered by injury in the region of the tumour.

NEUROBLASTOMA

This is a *malignant tumour*, occurring in infants and children under 15 years of age. Tumours that develop early tend to be highly malignant but there may be spontaneous regression. Haemorrhagic necrosis and calcification may destroy the gland. The tumour spreads rapidly to the thoracic lymph nodes, liver and bones.

ISLETS OF LANGERHANS

The cells which make up the islets of Langerhans are found in clusters irregularly distributed throughout the substance of the pancreas. Unlike the pancreatic tissue which produces a digestive juice there are no ducts leading from the clusters of islet cells. Their secretion passes directly into the pancreatic veins and circulates throughout the body.

There are three main types of cell in the islets of Langerhans: α cells that secrete *glucagon*, β cells that secrete *insulin* and δ cells that secrete *somatostatin* (GHIF) (see p. 272). Insulin and glucagon influence the level of glucose in the blood, each balancing the effects of the other. Glucagon tends to raise the blood glucose level and insulin reduces it.

The normal blood glucose level is between 3.5 and 5.5 mmol/litre (60 to 100 mg/100 ml). When there has been excessive exercise or an insufficient intake of carbohydrate foods it may fall to the low end of the normal range or even below. When this happens glucagon has the effect of *raising the blood glucose level* by mobilising the glycogen stores *in the liver*.

Insulin's functions are generally anticatabolic. Usually it acts to maintain homeostasis but when intake of nutrients is in excess of immediate needs it promotes storage. It acts on cell membranes, stimulating the uptake of glucose, amino acids and fats. In addition it is associated with:

1. Conversion of glucose to glycogen in liver and muscles
2. Synthesis of DNA and RNA
3. Storage of fat in adipose tissue
4. Prevention of protein and fat breakdown and gluconeogenesis

Secretion of insulin is stimulated by increased blood glucose and amino acids levels, gastrointestinal hormones, such as gastrin, secretin and pancreozymin. Secretion is inhibited by adrenaline and somatostatin (GHIF) secreted by δ cells of the islets of Langerhans and by adrenaline.

DISORDERS OF THE ISLETS OF LANGERHANS

DIABETES MELLITUS

This is due to deficiency or absence of insulin or to interference with insulin activity, causing varying degrees of failure to metabolise and store glucose.

TYPES OF DIABETES MELLITUS

INSULIN-DEPENDENT DIABETES (Juvenile, early-onset)

This occurs mainly in children and young adults and the onset is usually sudden. The deficiency or absence of insulin is due to the destruction of islet cells. The causes are unknown but there is a familial tendency, suggesting genetic involvement. In many cases antibodies to islet cells are present, probably to cells previously damaged by infection.

NON-INSULIN-DEPENDENT DIABETES (Late-onset)

This usually occurs in obese women over 65 years. The cause is unknown. Insulin secretion may be below or above normal. Deficiency of glucose inside body cells may occur when there is hyperglycaemia and a high insulin level. This may be due to changes in cell walls which block the insulin-assisted movement of glucose into cells.

SECONDARY DIABETES

This may develop as a complication of:

1. Other autoimmune endocrine diseases, e.g., acromegaly, thyrotoxicosis
2. Acute and chronic pancreatitis
3. Some drugs, e.g., corticosteroids, ACTH, thiazide diuretics

EFFECTS OF DIABETES

Raised blood glucose level
After the intake of a carbohydrate meal the blood glucose level remains high because:

1. Glucose metabolism by body cells is defective
2. Conversion of glucose to glycogen in the liver and muscles is diminished
3. There is gluconeogenesis from protein in response to the deficient intracellular glucose

Glycosuria and polyuria
The concentration of glucose in the glomerular filtrate is the same as in the blood and, although diabetes raises the renal threshold for glucose, it is not all reabsorbed by the tubules. The remaining glucose in the filtrate raises the osmotic pressure, water reabsorption is reduced and the volume of urine produced is increased. This causes electrolyte imbalance and excretion of urine of high specific gravity. Polyuria leads to hypovolaemia, extreme thirst and polydipsia.

Weight loss
In diabetes, cells fail to metabolise glucose, resulting in weight loss due to:

1. Gluconeogenesis from amino acids and body protein, causing tissue wasting and further increase in blood glucose
2. Catabolism of excess body fat, releasing some of its energy and excess production of ketoacids

Acidosis
This is due to the accumulation of the intermediate fat metabolite, acetyl coenzyme A, because it cannot enter the citric acid cycle without acetoacetic acid, a glucose intermediate metabolite. In diabetes the amount of available oxaloacetic acid is reduced because glucose metabolism is

reduced. As a result excess acetyl coenzyme A is converted to ketoacid which lowers the pH of the body fluids, causing acidosis. The effects are:

1. Hyperventilation and the excretion of excess bicarbonate
2. Acidification of urine and high filtrate osmotic pressure which leads to excessive loss of ammonia, sodium, potassium and water
3. Diabetic coma due to a combination of low blood pH, high plasma osmotic pressure and electrolyte imbalance

COMPLICATIONS OF DIABETES MELLITUS

DIABETIC COMA (Hyperglycaemic coma)

The factors that contribute to the development of hyperglycaemic coma in either type of diabetes include:

1. Hypovolaemia with severe dehydration due to persistent polyuria
2. High blood osmotic pressure due to excess blood glucose, leading to electrolyte imbalance
3. Dehydration due to polyuria
4. Acidosis due to accumulation of ketoacids

HYPOGLYCAEMIC COMA

This occurs in insulin-dependent diabetics when the insulin administered is in excess of that needed to balance the food intake and expenditure of energy. Because neurones are more dependent on glucose for their energy needs than are other cells, glucose deprivation causes disturbed neural function, leading to coma and, if prolonged, irreversible damage. Hypoglycaemia may be the result of:

1. Accidental overdose of insulin
2. Delay in taking food after insulin administration
3. Gastrointestinal disturbances in which carbohydrate absorption is diminished, e.g., vomiting, diarrhoea
4. Increased metabolic rate in, e.g., unexpected exercise, acute febrile illness
5. An insulin-secreting tumour, especially if it produces irregular bursts of secretion

CARDIOVASCULAR DISTURBANCES

Changes in blood vessels occur even when the disease is well controlled by diet and insulin.

Diabetic macroangiopathy
The most common lesions are atheroma and calcification of the tunica media of the large muscular arteries. In acute insulin-dependent diabetics these changes occur at a relatively early age. The most common sequelae are peripheral vascular disease, myocardial infarction, cerebral ischaemia and infarction.

Diabetic microangiopathy
There is thickening of the epithelial basement membrane of arterioles, capillaries and sometimes of venules. These changes may lead to:

1. Peripheral vascular disease, progressing to gangrene
2. Retinopathy, in which microaneurysms and small haemorrhages cause numerous small necrotic points in the retina, leading to loss of sight
3. Glomerulosclerosis, leading to the nephrotic syndrome and renal failure
4. Peripheral neuropathy, especially when myelination is defective

INFECTION

Diabetics are highly susceptibile to infection, especially by bacteria and fungi, possibly because phagocyte activity is depressed by insufficient intracellular glucose. Infection may cause:

1. Complications in areas affected by peripheral neuropathy and changes in blood vessels, e.g., in the feet when sensation and blood supply are impaired
2. Boils and carbuncles
3. Vaginal candidiasis (thrush)
4. Pyelonephritis

RENAL FAILURE

This is due to vascular changes and infection, and is a common cause of death in diabetics.

PINEAL GLAND OR BODY

The pineal gland is a small body situated in the brain below the corpus callosum and posterior to the third ventricle. It is reddish-grey in colour and is approximately 10 mm in length. It is surrounded by a fine capsule and is composed of epithelial cells arranged to form lobules which are surrounded by fine connective tissue.

Its functions are as yet unclear. It may be associated with the development of the gonads by influencing the release of gonadotrophic hormones from the anterior pituitary.

15 The Reproductive Systems

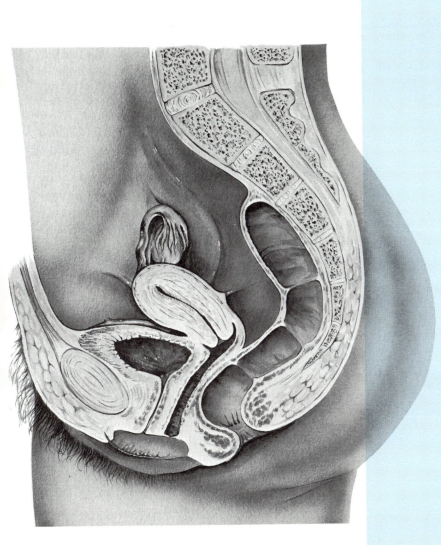

15. The Reproductive Systems

The ability to reproduce is one of the properties which distinguishes living from non-living matter. The more primitive the animal, the simpler the process of reproduction. In human beings the process is one of sexual reproduction.

The reproductive organs of the male and the female differ anatomically and physiologically. Both males and females produce specialised reproductive cells, called *gametes*, containing genetic material (*genes*) and *chromosomes*. Other body cells contain 46 chromosomes arranged in pairs. The gametes contains only one of each pair, i.e., 23 chromosomes. When an ovum is fertilised by a spermatozoon, the resultant *zygote* contains the full complement of 23 *pairs* of chromosomes, half obtained from the mother and half from the father. The zygote embeds in the wall of the uterus where it grows and develops during the 40-week *gestation period* before birth. The function of the female reproductive system is, therefore, to form the ovum and if it is fertilised, to nurture it until it is born then feed it with breast milk until it is able to take a mixed diet. The function of the male reproductive system is to form the spermatozoa and transmit it to the female.

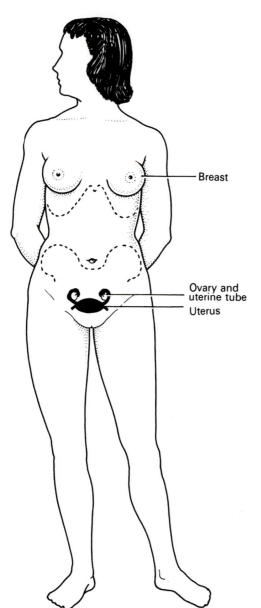

Figure 15:1 The female reproductive organs. Dotted lines denote the positions of the lower ribs and the pelvis.

FEMALE REPRODUCTIVE SYSTEM

The female reproductive organs, or genitalia, are divided into external and internal organs (Fig. 15:1).

EXTERNAL GENITALIA (Fig. 15:2)

The external genitalia are known collectively as the *vulva* which consists of:

Labia majora	Vestibule
Labia minora	Hymen
Clitoris	Greater vestibular glands

LABIA MAJORA

These are the two large folds which form the boundary of the *vulva*. They are composed of skin, fibrous tissue and fat and contain large numbers of sebaceous glands. anteriorly the folds join in front of the symphysis pubis, and posteriorly they merge with the skin of the perineum. At puberty hair grows on the mons pubis and on the lateral aspect of the labia majora.

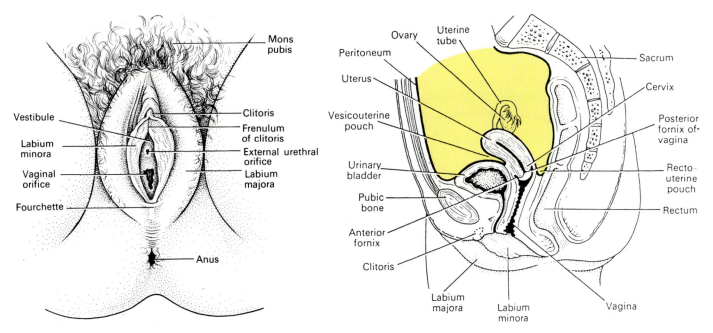

Figure 15:2 The external genitalia in the female.

Figure 15:3 The female reproductive organs in the pelvis and their associated structures. Lateral view.

LABIA MINORA

These are two smaller folds of skin between the labia majora, containing numerous sebaceous glands. Posteriorly they fuse to form the *fourchette*.

The cleft between the labia minora is the *vestibule*. The vagina, urethra and ducts of the greater vestibular glands open into the vestibule

CLITORIS

The clitoris corresponds to the penis in the male and contains erectile tissue.

HYMEN

The hymen is a thin layer of mucous membrane which partially occludes the opening of the vagina.

GREATER VESTIBULAR GLANDS

The greater vestibular glands (Bartholin's glands) lie in the labia majora, one on each side near the vaginal opening. They are about the size of a small pea and have ducts, opening into the vestibule. They secrete mucus that keeps the vulva moist.

Blood supply, lymph drainage and nerve supply

Arterial supply is by branches from the *internal pudendal arteries* that branch from the internal iliac arteries, and by the *external pudendal arteries* that branch from the internal iliac arteries, and by the *external pudendal arteries* that branch from the femoral arteries.

Veins form a large plexus which eventually drains into the internal iliac veins.

Lymph drainage is through the superficial inguinal glands.

Nerve supply is by branches from pudendal nerves.

PERINEUM

The perineum is the area extending from the fourchette to the anal canal. It is roughly triangular and consists of connective tissue, muscle and fat. It gives attachment to the muscles of the pelvic floor (see p. 342).

INTERNAL ORGANS (Figs 15:3 and 15:4)

The internal organs of the female reproductive system lie in the pelvic cavity and consist of the vagina, uterus, two uterine tubes and two ovaries.

VAGINA

The vagina is a fibromuscular tube lined with stratified epithelium, connecting the external and internal organs of reproduction. It runs obliquely upwards and backwards at an angle of about 45° between the bladder in front and rectum and anus behind. In the adult the anterior wall is about 7.5 cm (3 inches) long and the posterior wall about 9 cm long. The difference is due to the protrusion of the cervix through the anterior wall.

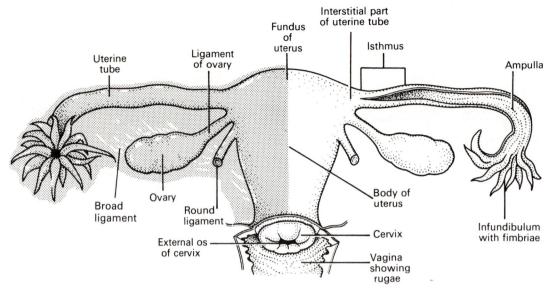

Figure 15:4 The female reproductive organs in the pelvis: posterior walls of the vagina and right uterine tube removed. Shaded area shows the arrangement of peritoneum.

Structure

The vagina has an outer covering of *areolar tissue*, a middle layer of *smooth muscle* and an inner lining of *stratified squamous epithelium*. It has no secretory glands but the surface is kept moist by cervical secretions. Between puberty and the menopause *Lactobacillus acidophilus* bacteria are normally present. These microbes secrete *lactic acid*, maintaining the pH between 4.9 and 3.5. The acidity inhibits the growth of most microbes that may enter the vagina from the perineum.

Blood supply, lymph drainage and nerve supply

An *arterial plexus* is formed round the vagina, derived from the uterine and vaginal arteries which are branches of the internal iliac arteries.

A *venous plexus*, situated in the muscular wall, drains into the internal iliac veins.

The lymph drainage is through the deep and superficial iliac glands.

The nerve supply consists of parasympathetic, and sympathetic autonomic fibres and somatic sensory fibres from the pudendal nerves.

UTERUS

The uterus is a hollow muscular pear-shaped organ, flattened anteroposteriorly. It lies in the pelvic cavity between the urinary bladder and the rectum in an antiverted antiflexed position.

Anteversion means that the uterus *leans forward*.

Anteflexion means that it is bent forward almost at right angles to the vagina with its anterior surface resting on the urinary bladder. As the bladder fills the degree of anteflexion is reduced slightly.

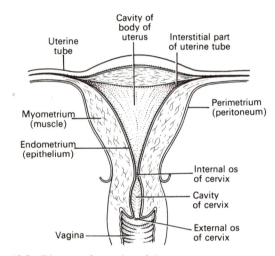

Figure 15:5 Diagram of a section of the uterus.

When the body is in the upright position the uterus lies in an almost horizontal position. It is about 7.5 cm long, 5 cm wide and its walls are about 2.5 cm thick. It weighs from 30 to 40 grams. The parts of the uterus are the fundus, body and cervix (Fig. 15:5).

The fundus is the dome-shaped part of the uterus above the openings of the uterine tubes.

The body is the main part. It is narrowest inferiorly at the *internal os* where it is continuous with the cervix.

The cervix protrudes through the anterior wall of the vagina, opening into it at the *external os*.

Structure

The walls of the uterus are composed of three layers of tissue: perimetrium, myometrium and endometrium.

Perimetrium consists of peritoneum, which is distributed differently on the various surfaces of the uterus.

Anteriorly it extends over the fundus and the body where it is reflected on to the upper surface of the urinary bladder. This fold of peritoneum forms the *vesicouterine pouch*.

Posteriorly the peritoneum extends over the fundus, the body and the cervix then it is reflected on to the rectum to form the *rectouterine pouch*.

Laterally only the fundus is covered because the peritoneum forms a double fold with the uterine tubes in the upper free border. This double fold is the *broad ligament* which, at its lateral ends, attaches the uterus to the sides of the pelvis.

Myometrium is the thickest layer of tissue in the uterine wall. It consists of a mass of smooth muscle fibres interlaced with areolar tissue, blood vessels and nerves.

Endometrium consists of columnar epithelium. It contains large number of mucus-secreting tubular glands. The upper two-thirds of the cervical canal is lined with mucous membrane. The lower third is lined with squamous epithelium, continuous with that of the vagina.

Blood supply, lymph drainage and nerve supply

The arterial supply is by the *uterine arteries* which are branches of the internal iliac arteries. They pass up the lateral aspects of the uterus between the two layers of the broad ligaments. They supply the uterus and uterine tubes and join with the ovarian arteries to supply the ovaries. Branches pass downwards to anastomose with the vaginal arteries to supply the vagina.

Venous drainage. The veins follow the same route as the arteries and eventually drain into the internal iliac veins.

Lymph drainage. There are deep and superficial lymph vessels which drain lymph from the uterus and the uterine tubes to the aortic lymph nodes and groups of nodes associated with the iliac blood vessels.

Nerve supply. The nerves supplying the uterus and the uterine tubes consist of parasympathetic fibres from the sacral outflow and sympathetic fibres from the lumber outflow.

SUPPORTS OF THE UTERUS

The uterus is supported in the pelvic cavity by: surrounding organs, muscles of the pelvic floor and ligaments that suspend it from the walls of the pelvis.

Supporting ligaments (Fig. 15:6)

Two broad ligaments are formed by a double fold of peritoneum, one on each side of the uterus. They hang down from the uterine tubes as though draped over them and at their lateral ends they are attached to the sides of the pelvis. The uterine tubes are enclosed in the upper free border and near this lateral ends they penetrate the posterior wall, opening into the peritoneal cavity. The ovaries are attached to the posterior wall, one on each side.

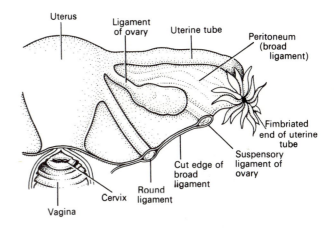

Figure 15:6 The main ligaments supporting the uterus. Only one side shown.

Blood and lymph vessels and nerves pass to the uterus and uterine tubes between the layers of the broad ligaments.

The round ligaments are bands of fibrous tissue between the two layers of broad ligament, one on each side of the uterus. They pass to the sides of the pelvis then through the *inguinal canal* to end by fusing with the labia majora.

Two uterosacral ligments originate from the posterior walls of the cervix and vagina and extend backwards, one on each side of the rectum, to the sacrum.

Two transverse cervical ligaments (cardinal ligaments) extend from the sides of the cervix and vagina to the side walls of the pelvis.

The pubocervical fascia extends forward from the transverse cervical ligaments on each side of the bladder and is attached to the posterior surface of the pubis.

Functions

After puberty the uterus goes through a regular cycle of changes, the *menstrual cycle*, which prepares it to receive, nourish and protect a fertilised ovum. It provides the environment for the growing fetus, during the 40-week gestation period, at the end of which the baby is born. The cycle is usually regular, lasting between 26 and 30 days. If the ovum is not fertilised the cycle ends with a short period of bleeding (menstruation).

If the ovum is fertilised the zygote embeds in the uterine wall which relaxes to accommodate the growing fetus. At the end of the gestation period *labour* begins and is concluded when the baby is born and the placenta extruded. During labour, the muscle of the fundus and body of the uterus contract intermittently and the cervix relaxes and dilates. As labour progresses the uterine contractions become stronger and more frequent. When the cervix is fully dilated the mother assists the birth of the baby by holding her breath and bearing down during the contractions.

UTERINE TUBES (Fallopian tubes)

The uterine tubes are about 10 cm long and extend from the sides of the uterus between the body and the fundus. They lie in the upper free border of the broad ligament and their trumpet-shaped lateral ends penetrate the posterior wall, opening into the peritoneal cavity close to the ovaries. The end of each tube has finger-like projections called *fimbrae*. The longest of these is the *ovarian fimbria* which is in close association with the ovary.

Structure

The uterine tubes have an outer covering of peritoneum (broad ligament), a middle layer of smooth muscle and are lined with ciliated epithelium.

Blood supply, lymph drainage and nerve supply

These are the same as for the uterus (see p. 289).

Function

The uterine tubes convey the ovum from the ovary to the uterus by peristalsis and ciliary movement. The mucus secreted by the lining membrane provides ideal conditions for movement of ova and spermatozoa. Fertilisation of the ovum usually takes place in the uterine tube then the zygote moves into the uterus.

OVARIES

The ovaries are the female gonads, or sex glands, and they lie in a shallow fossa on the lateral walls of the pelvis. Each is attached to the upper part of the uterus by *the ligament of the ovary* and to the back of the broad ligament by a broad band of tissue, *the mesovarium*. Blood vessels and nerves pass to the ovary in the mesovarium. They are 2.5 to 3.5 cm long, and 2 cm wide and 1 cm thick.

Structure and functions

The ovaries have two layers of tissue.

The medulla lies in the centre and consists of fibrous tissue, blood vessels and nerves.

The cortex surrounds the medulla. It has a framework of connective tissue, or *stroma*, covered by *germinal epithelium*. It contains *ovarian follicles*, each of which contains *an ovum*. Before puberty the ovaries are inactive but the stroma already contains immature (primordial) follicles. During the childbearing years one ovarian follicle matures, ruptures and releases its ovum into the peritoneal cavity during each menstrual cycle (Fig. 15:7).

Maturation of the follicle is stimulated by *follicle stimulating hormone* (FSH) from the anterior pituitary. While maturing, the follicle lining cells produce the hormone *oestrogen*. After ovulation the follicle lining cells develop into the *corpus luteum* (yellow body), under the influence of the *luteinising hormone* (LH) from the anterior pituitary. The corpus luteum produces the hormone *progesterone*. If the ovum is fertilised it embeds in the wall of the uterus where it grows and develops and produces *chorionic gonadotrophin* hormone which stimulates the corpus luteum to continue secreting progesterone for the first 3 months of the pregnancy (Figs 15:8 and 15:9). If the ovum is not fertilised the corpus luteum degenerates, menstruation occurs and the next cycle begins. Sometimes more than

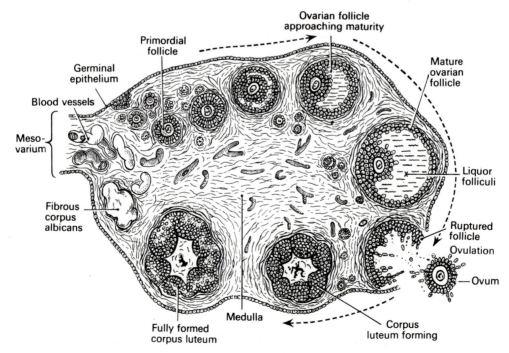

Figure 15:7   Diagram of a section of an ovary showing the stages of development of one ovarian follicle.

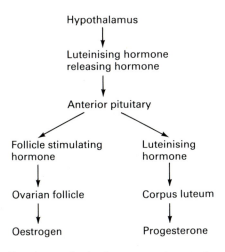

Figure 15:8   Female reproductive hormones and target tissues.

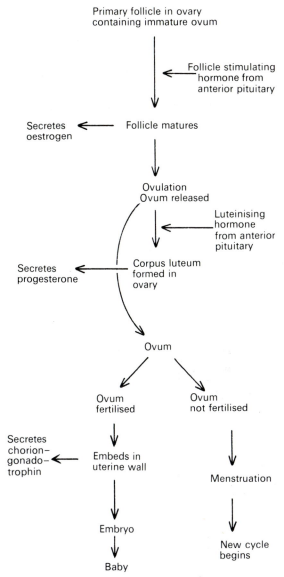

Figure 15:9   A summary of the stages of development of the ovum and the associated hormones.

one follicle matures at a time, releasing two or more ova in the same cycle. When this happens and the ova are fertilised the result is a multiple pregnancy.

Blood supply, lymph drainage and nerve supply

Arterial supply is by the *ovarian arteries* which branch from the abdominal aorta just below the renal arteries.

Venous drainage is into a plexus of veins behind the uterus from which the ovarian veins arise. The right ovarian vein opens into the inferior vena cava and the left into the left renal vein.

Lymph drainage is to the lateral aortic and pre-aortic lymph nodes. The lymph vessels follow the same route as the arteries.

Nerve supply. The ovaries are supplied by parasympathetic nerves from the sacral outflow and sympathetic nerves from the lumbar outflow. Their precise functions are not yet fully understood.

PUBERTY IN THE FEMALE

Puberty is the age at which the internal reproductive organs reach maturity. The ovaries are stimulated by the gonadotrophins from the anterior pituitary, the *follicle stimulating hormone* and the *luteinising hormone.*

The age of puberty varies between 10 and 14 years and a number of physical and psychological changes take place at this time:

1. The uterus, uterine tubes and the ovaries reach maturity
2. The menstrual cycle and ovulation begin
3. The breasts develop and enlarge
4. Pubic and axillary hair begins to grow
5. There is an increase in the rate of growth in height and widening of the pelvis
6. There is an increase in the amount of fat deposited in the subcutaneous tissue

MENSTRUAL CYCLE (Fig. 15:10)

This is a series of events, occurring regularly in females every 26 to 30 days throughout the childbearing period of about 36 years. The cycle consists of a series of changes that take place concurrently in the ovaries and uterine walls, stimulated by changes in the blood concentrations of hormones. Hormones secreted in the cycle are regulated by feedback mechanisms.

The *hypothalamus* secretes *luteinising hormone releasing hormone* (LH-RH). This stimulates the anterior pituitary to secrete:

1. *Follicle stimulating hormone* (FSH) which promotes the maturation of ovarian follicles and secretion of *oestrogen*, leading to ovulation
2. *Luteinising hormone* (LH) which stimulates the development of corpus luteum and secretion of *progesterone*

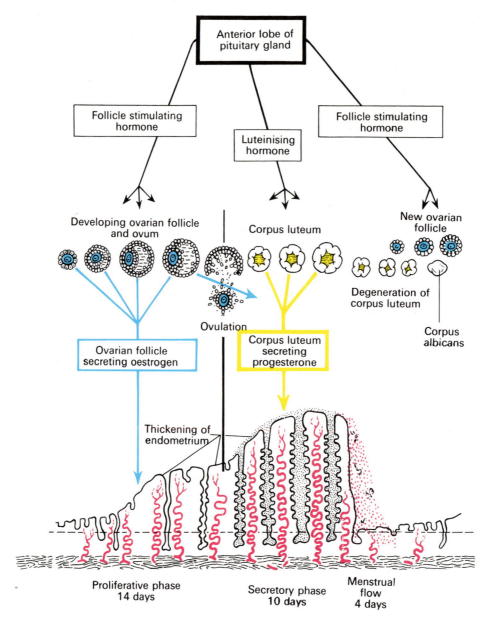

Figure 15:10   Diagram showing the endometrium at the various stages of the menstrual cycle and the associated hormones.

The hypothalamus responds to changes in the blood levels of oestrogen and progesterone. It is depressed by high levels and stimulated when they are low.

The phases of the menstrual cycle that denote changes in the uterine wall are:

Proliferative phase	10 days
Secretory phase	14 days
Menstrual phase	4 days

Proliferative phase

At this stage an ovarian follicle, stimulated by FSH, is growing towards maturity and is producing *oestrogen*. Oestrogen stimulates the proliferation of the endometrium in preparation for the reception of a fertilised ovum. The endometrium becomes thicker by rapid cell multiplication accompanied by an increase in the numbers of mucus-secreting glands and blood capillaries. This phase ends when *ovulation* occurs and oestrogen production stops.

Secretory phase

Immediately after ovulation, the lining cells of the ovarian follicle are stimulated by LH to develop the *corpus luteum*, which produces *progesterone*. Under the influence of progesterone the endometrium becomes oedematous and the secretory glands produce increased amounts of watery mucus. This is believed to assist the passage of the spermatozoa through the uterus to the uterine tubes where the ovum is usually fertilised. There is a similar increase in the secretion of watery mucus by the glands of the uterine tubes and into the vagina by cervical glands.

The ovum may survive in a fertilisable form for a very short time after ovulation, probably as little as 8 hours. The spermatozoa, deposited in the vagina during coitus, may be able to fertilise the ovum for only about 24 hours although they may survive for several days. This means that the period in each cycle during which fertilisation can occur is relatively short. However, the date of ovulation cannot be predicted with certainty, even when cycles are regular.

If fertilisation of the ovum does not occur the cycle goes into the third phase.

Menstrual phase

If the ovum is not fertilised, the high level of progesterone in the blood inhibits the activity of the pituitary gland and the production of luteinising hormone is considerably reduced. The withdrawal of this hormone causes degeneration of the corpus luteum and thus progesterone production is decreased. About 14 days, after ovulation the lining of the uterus degenerates and breaks down and menstruation begins. The menstrual flow consists of the extra secretions, endometrial cells, blood from the broken down capillaries and the unfertilised ovum.

When the amount of progesterone in the blood falls to a critical level another ovarian follicle is stimulated by the FSH and the next cycle begins.

If the ovum is fertilised there is no breakdown of the endometrium and no menstrual flow. The fertilised ovum (zygote) travels through the uterine tube to the uterus where it becomes embedded in the wall and produces the hormone *chorion gonadotrophin* which is similar to anterior pituitary luteinising hormone. This hormone keeps the corpus luteum intact enabling it to continue to secrete progesterone for the first 3 to 4 months of the pregnancy, inhibiting the maturation of ovarian follicles. During that time the *placenta* develops and produces oestrogen, progesterone and gonadotrophins. The placenta provides an indirect link between the circulation of the mother and that of the fetus. Through the placenta the fetus obtains nutritional materials, oxygen and antibodies and gets rid of carbon dioxide and other waste products.

MENOPAUSE (Climacteric)

The *menopause* usually occurs between the ages of 45 and 55 years, marking the end of the child bearing period. It may occur suddenly or over a period of years, sometimes as long as 10 years. It is caused by the changes in the concentration of the sex hormones. The ovaries gradually become less responsive to the FSH and LH and ovulation and the menstrual cycle become irregular, eventually ceasing. Several other phenomena may occur at the same time, including:

1. Short-term unpredictable vasodilatation with flushing,

sweating and palpitations, causing discomfort and disturbance of the normal sleep pattern
2. The breasts shrink
3. Axillary and pubic hair become sparse
4. The sex organs atrophy
5. Episodes of uncharacteristic behaviour sometimes occur

Similar changes occur after bilateral irradiation or surgical removal of the ovaries.

BREASTS OR MAMMARY GLANDS

The breasts or mammary glands are accessory glands of the female reproductive system. They also exist in the male but only in a rudimentary form.

In the female the breasts are quite small until puberty. Thereafter they grow and develop to their mature size under the influence of oestrogen and progesterone. During pregnancy these hormones stimulate further growth. After the baby is born the hormone *prolactin* from the anterior pituitary stimulates the production of milk, and *oxytocin* from the posterior pituitary stimulates the release of milk in response to the stimulation of the nipple by the sucking baby.

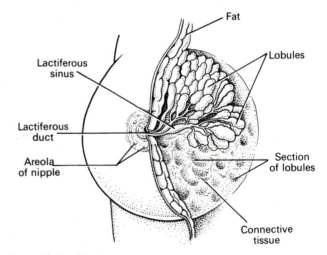

Figure 15:11   The breast.

Structure (Fig. 15:11)

The mammary glands consist of: glandular tissue, fibrous tissue and fatty tissue.

Each breast consists of about 20 lobes of *glandular tissue*, each lobe being made up of a number of lobules. The lobules consist of a cluster of alveoli which open into small ducts and these unite to form large excretory ducts, called *lactiferous ducts*. The lactiferous ducts converge towards the centre of the breast where they form dilatations or reservoirs for milk. Leading from these dilatations there are narrow ducts which open on to the surface at the nipple. Fibrous tissue supports the glandular tissue and ducts, and fat covers the surface of the gland and is found between the lobes.

The nipple is a small conical eminence at the centre of the breast surrounded by a pigmented area, *the areola*. On the surface of the areola there are numerous sebaceous glands (Montgomery's tubules) which lubricate the nipple in pregnancy.

Blood supply, lymph drainage and nerve supply

Arterial blood supply. The breasts are supplied with blood from the thoracic branches of the axillary arteries and from the internal mammary and intercostal arteries.

Venous drainage. This describes an anastomotic circle round the base of the nipple from which branches carry the venous blood to the circumference and end in the axillary and mammary veins.

Lymph drainage (Fig. 15:12). This is mainly into the axillary lymph vessels and nodes. Lymph may drain through the internal mammary nodes if the superficial route is obstructed .

Nerve supply. The breasts are supplied by branches from the fourth, fifth and sixth thoracic nerves which contain sympathetic fibres. There are numerous somatic *sensory nerve endings* in the breast especially around the nipple. When these *touch receptors* are stimulated by sucking, impulses pass to the hypothalamus and the flow of the hormone oxytocin is increased, promoting the release of milk.

Function

The mammary glands are active only during pregnancy and after the birth of a baby when they produce milk.

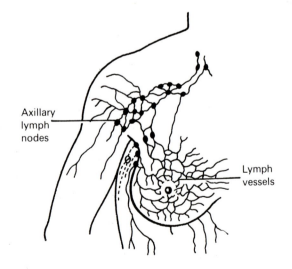

Figure 15:12 Lymph drainage from the breast.

DISEASES OF THE FEMALE REPRODUCTIVE SYSTEM

INFLAMMATION

Infection of the reproductive system may be classified as:
1. Non-specific, usually caused by a mixture of microbes, e.g., staphylococci, streptococci, coliform bacteria, *Clostridium perfringens* (*Cl. welchii*)
2. Specific, caused by sexually transmitted microbes, the most common of which is *Neisseria gonorrhoeae*

PELVIC INFLAMMATORY DISEASE (PID)

This infection may be specific or non-specific. It usually begins as vulvovaginitis, including the vulvar glands, then it may spread to the cervix, uterus, uterine tubes and ovaries. Upward spread is most common when microbes are present in the vagina before a surgical procedure, parturition or abortion, especially if some of the products of conception are retained.

Complications of PID include:
1. Infertility due to fibrous obstruction of uterine tubes
2. Peritonitis
3. Intestinal obstruction due to adhesions between the bowel and the uterus and/or uterine tubes
4. Bacteraemia which may lead to meningitis, endocarditis, suppurative arthritis
5. Bartholin's gland abscess or cyst formation if the duct is blocked

SPECIFIC INFECTIONS (Venereal diseases)

In general, the microbes that cause sexually transmitted diseases:
1. Are unable to survive outside the body for long periods
2. Have no intermediate host
3. Produce lesions in the genital area which discharge the infecting microbes

GONORRHOEA

This is the most commonly occurring venereal disease and affects men and women. It is caused by *Neisseria gonorrhoeae*. In the male, suppurative urethritis occurs and the infection may spread to the prostate gland, epididymis and testes. In the female the infection may spread from vulvar glands, vagina and cervix to the body of the uterus, uterine tubes, ovaries and peritoneum. Healing by fibrosis in the female may cause obstruction of the uterine tubes, leading to infertility. In the male it may cause urethral stricture.

Non-venereal transmission of gonorrhoea may cause *neonatal ophthalmia* in babies born to infected mothers.

SYPHILIS

This disease is caused by *Treponema pallidum*. There are three clearly marked stages although the third is now rarely seen in Britain. After an incubation period of several weeks, the *primary sore* (chancre) appears at the site of infection, e.g., the vulva, vagina, perineum, penis, round the mouth. After several weeks the chancre subsides spontaneously. *Secondary lesions* appear 3 to 4 months after infection. They consist of skin rashes and raised papules (condylomata lata) on the external genitalia and vaginal walls. These subside after several months and are followed by a latent period of a variable number of years. *Tertiary lesions* (gumma) develop in many organs and in a few cases the nervous system in involved, leading to general paralysis.

Sexual transmission occurs during the primary and secondary stages when discharge from lesions contains microbes. Congenital transmission occurs when microbes from an infected mother cross the placenta to the fetus. Accidental spread of infection may occur by blood transfusion if a donor's blood is taken during the incubation period after microbes have spread to the blood from the site of infection.

TRICHOMONAS VAGINALIS

These *protozoa* cause acute vulvovaginitis. It is usually sexually transmitted and is commonly present in women with gonorrhoea.

CANDIDIASIS

Candida albicans is the causative organism. It is a commensal in the vagina and causes infection in some circumstances, e.g., in diabetes, malnutrition and general debility.

VULVAR DYSTROPHIES

ATROPHIC DYSTROPHY

This is thinning of vulvar epithelium and the formation of fibrous tissue, occurring after the menopause due to oestrogen withdrawal. It predisposes to infection, especially in debilitated women, and to malignant epithelial neoplasia.

HYPERPLASTIC DYSTROPHY (Interepithelial neoplasia)

The extent of development of hyperplasia and dysplasia of cells of the skin of the vulva varies considerably. In the majority of cases the neoplasms are benign. In elderly women malignant tumours may develop which spread locally and there is early involvement of bilateral inguinal lymph nodes.

IMPERFORATE HYMEN

This is a congenital abnormality which may not be noticed until the onset of menstruation. Blood accumulates in the vagina, uterus and uterine tubes, and it may enter the peritoneal cavity and cause peritonitis. The uterine tubes may become obstructed by organised blood, leading to infertility.

DISORDERS OF THE CERVIX

INFLAMMATION (Cervicitis)

This occurs in most multiparous women and may be due to acute or chronic infection caused by specific or non-specific microbes. In non-specific infection there are several predisposing factors, e.g., trauma at childbirth, instruments used in gynaecological treatments, abnormal blood oestrogen levels, hypersecretion by cervical glands. In many cases the only indication of infection is excessive vaginal discharge (leukorrhoea). Chronic inflammation may follow acute attacks or develop gradually, and may predispose to malignancy.

CERVICAL INTRA-EPITHELIAL NEOPLASIA (CIN)

Dysplastic changes begin in the deepest layer of cervical epithelium, usually at the junction of the stratified squamous epithelium of the lower third of the cervical canal with the secretory epithelium of the upper two-thirds. The cell dysplasia may progress to involve the full thickness of epithelium, called *carcinoma-in-situ*. The cancer may develop further and spread locally to the vagina, uterine body and other pelvic structures. More widespread metastases occur late in the disease. All CIN cases do not develop to the cancerous stage but it is not possible to predict how far development will go, whether it will remain static or regress.

The disease takes 15 to 20 years to develop and it occurs mostly between 35 and 50 years of age. The carcinogens have not been identified but it has been suggested that herpes viruses and/or spermatazoal DNA may be involved. A high incidence is associated with some personal and social factors:

1. Early marriage and pregnancy
2. A high number of pregnancies
3. Coitus beginning at an early age
4. Frequent coitus
5. Promiscuity
6. Sexually transmitted disease
7. Low socioeconomic status

DISORDERS OF THE BODY OF UTERUS

ACUTE ENDOMETRITIS

This is usually caused by non-specific infection, following parturition, especially if some products of conception have been retained. The inflammation may subside after removal of retained products. The infection may spread to:

1. Myometrium, perimetrium and surrounding pelvic tissues which may lead to thrombosis of iliac veins
2. Uterine tubes, causing salpingitis, fibrosis, obstruction and infertility
3. Any of the above-mentioned areas, causing peritonitis and possibly adhesions

CHRONIC ENDOMETRITIS

This may follow an acute attack or be due to spread of pelvic inflammatory disease. It is also associated with the use of intrauterine contraceptive devices.

ENDOMETRIOSIS

This is the growth of endometrial tissue outside the uterus, most commonly in the ovaries, uterine tubes and other pelvic structures. The ectopic tissue reacts to sex hormones as does uterine endometrium, causing menstrual-type bleeding and formation of coloured cysts, 'chocolate cysts'. There is intermittent pain due to swelling, and recurrent haemorrhage causes fibrous tissue formation. Ovarian endometriosis may lead to extensive pelvic adhesions, involving the ovaries, uterus, uterine ligaments and the bowel. The cause is not clear but it has been suggested that there may have been:

1. Abnormal cell differentiation in the fetus
2. Regurgitation of menstrual material through the uterine tubes
3. Spread of endometrial cells in lymph and blood during menstruation

ADENOMYOSIS

This is growth of endometrium within the myometrium. The ectopic tissue may cause general or localised uterine enlargement. The lesions may cause dysmenorrhoea and irregular excessive bleeding, usually beginning between 40 and 50 years of age.

ENDOMETRIAL HYPERPLASIA

The hyperplasia may affect endometrial glands, causing cyst formation and/or focal hyperplasia of atypical cells.

The focal type frequently undergo malignant change. Both types are associated with a sustained high blood oestrogen level which may be due to:

1. Failure of ovarian follicles to mature and release their ova
2. Oestrogen-secreting ovarian tumours
3. Prolonged oestrogen therapy

LEIOMYOMA (Fibroid, Myoma)

These are very common, often multiple, benign tumours of myometrium. They are firm masses of smooth muscle encapsulated in compressed muscle fibres and they vary greatly in size. Large tumours may undergo degenerative changes if they outgrow their blood supply, leading to necrosis, fibrosis and calcification. They develop during the reproductive period and may be hormone dependent because they enlarge during pregnancy and when oral contraceptives are used. They tend to regress after the menopause. Large tumours may cause pelvic discomfort, frequency of micturition, menorrhagia, irregular bleeding, dysmenorrhoea and reduced fertility. Malignant change is rare.

ENDOMETRIAL CARCINOMA

This occurs mainly in nulliparous women between 50 and 60 years of age, especially if they are obese, hypertensive or diabetic. The tumour may develop as a diffuse mass, a localised plaque or a polyp and there is often ulceration and bleeding. As endometrium has no lymph vessels, local spread to other pelvic structures is usually advanced before distant metastases develop. Invasion of the ureters leads to hydronephrosis and uraemia which is commonly the cause of death.

DISORDERS OF THE UTERINE TUBES AND OVARIES

ACUTE SALPINGITIS

This infection usually spreads from the uterus, and only occasionally from the peritoneal cavity. The outcome may be:

1. Uneventful recovery
2. Chronic inflammation, leading to fibrous tubal obstruction and infertility
3. Pus formation (*pyosalpynx*) and further spread to the ovaries and peritoneal cavity, leading to fibrous tubal obstruction, infertility and/or pelvic adhesions.

ECTOPIC PREGNANCY

This is the implantation of a fertilised ovum outside the uterus, most commonly in a uterine tube. As the fetus grows the tube ruptures and its contents enter the peritoneal cavity, causing acute inflammation and possibly severe intraperitoneal haemorrhage.

OVARIAN TUMOURS

The majority of ovarian tumours are benign, usually occurring between 20 and 45 years of age. The remainder occur mostly between 45 and 65 years and are divided between borderline malignancy (low-grade cancer) and frank malignancy. There are three main types of cells involved: epithelial cells, germ cells and hormone-secreting cells (sex cord/stroma cells).

EPITHELIAL CELL TUMOURS

Most of these are borderline or malignant tumours. They vary greatly in size from very large to quite small and some are partly cystic. Large tumours may cause pressure, leading to gastrointestinal disturbances, frequency of micturition, dysuria and ascites. Those suspended by a pedicle may twist, causing ischaemia, necrosis, haemorrhage or rupture of a cyst. The principal methods of spread are invasion of local and peritoneal structures. Later lymph- and blood-spread metastases may develop.

GERM CELL OVARIAN TUMOURS

These occur mainly in children and young adults and only a few are malignant. Benign *dermoid cysts* are the most common type. These are thick-walled cysts containing a variety of uncharacteristic tissues, e.g., hair, skin, epithelium, teeth. They are usually small and have a tendency to twist on a pedicle, causing ischaemia and necrosis.

SEX CORD/STROMA CELL TUMOURS

These cells are the precursors of the ovarian follicle lining cells, luteal cells and fibrous supporting cells. Mixed tumours develop, some of which secrete hormones. Oestrogen-secreting tumours cause precocious sexual development in children. In adults the excess oestrogen may cause endometrial hyperlasia, endometrial carcinoma, cystic disease of the breast or breast cancer. Androgen-secreting tumours occasionally develop, causing atrophy of the breast and genitalia and the development of male sex characteristics.

METASTATIC OVARIAN TUMOURS

The ovaries are a common site of metastatic spread from primary malignant tumours in other pelvic organs, the stomach, pancreas and biliary tract.

DISORDERS OF THE BREASTS

INFLAMMATION

ACUTE NON-SUPPURATIVE INFLAMMATION

This occurs during lactation and is associated with painful congestion and oedema of the breast. It is of hormonal origin.

ACUTE SUPPURATIVE INFLAMMATION (Pyogenic mastitis)

The microbes enter through a nipple abrasion caused by the infant sucking. The most common causative microbes are *Staphylococcus aureus* and *Streptococcus pyogenes* usually acquired by the infant while in hospital. The infection spreads along the mammary ducts and, if it does not resolve, may become chronic or an abscess may form.

TUMOURS

BENIGN TUMOURS

Some are cystic and some solid and they usually occur in women nearing the menopause. They may originate from secretory cells, fibrous tissue or from ducts.

MALIGNANT TUMOURS

The incidence of breast cancer is highest in nulliparous women, especially those with early puberty and late menarche. The most common types of tumour are usually found in the upper outer quadrant of the breast. There is considerable fibrosis around the tumour that may cause retraction of the nipple and necrosis and ulceration of the overlying skin.

Early spread beyond the breast is via lymph to the axillary and internal mammary nodes. Local invasion involves the pectoral muscles and the pleura. Blood-spread metastases may occur later in many organs and bones, especially lumbar and thoracic vertebrae. The causes of breast cancer are not known but several predisposing factors have been suggested.

Oestrogen excess. Normally there is cyclic proliferation of cells in the mammary ducts consistent with the menstrual rise and fall of oestrogen secretion. Excess oestrogen may cause abnormal cell proliferation and malignant change.

Familial tendencies. There may be genetic factors associated with susceptibility.

Viruses. It has been suggested that viruses may be involved although none has been identified so far.

MALE REPRODUCTIVE SYSTEM

The male reproductive system consists of the following organs (Figs 15:13 and 15:14):

2 testes ⎫
2 epididymides ⎬ in the scrotum
2 deferent ducts (vas deferens)
2 spermatic cords
2 seminal vesicles
2 ejaculatory ducts
1 prostate gland
1 penis

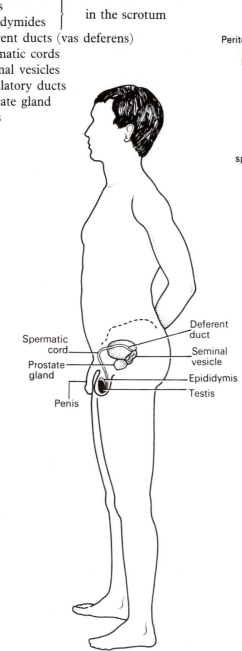

Figure 15:13   The male reproductive organs.

SCROTUM

The scrotum is a pouch of deeply pigmented skin divided into two compartments each of which contains one testis, one epididymis and the testicular end of a spermatic cord. It lies below the symphysis pubis, in front of the upper parts of the thighs and behind the penis.

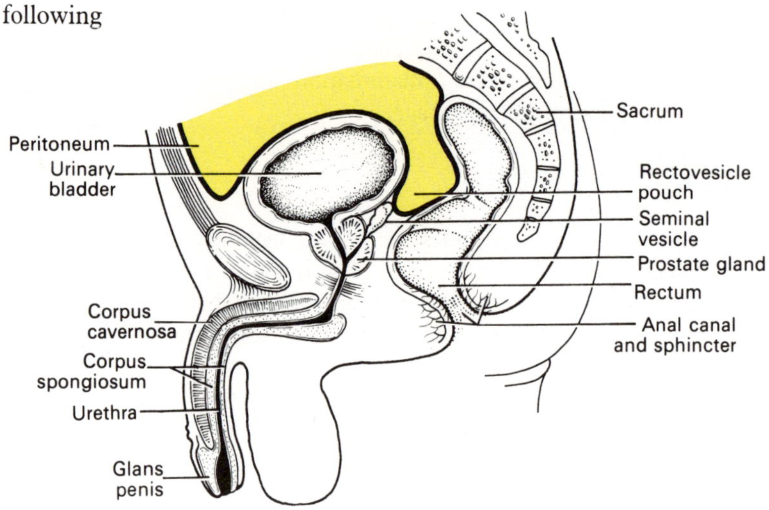

Figure 15:14   The male reproductive organs and their associated structures.

TESTES

The testes are the reproductive glands of the male and are the equivalent of the ovaries in the female. They are about 4.5 cm long, 2.5 cm wide and 3 cm thick and are suspended in the scrotum by the spermatic cords. They are surrounded by three layers of tissue.

The tunica vaginalis is the outer covering of the testes and is a downgrowth of the abdominal and pelvic peritoneum. During early fetal life the testes develop in the lumbar region of the abdominal cavity just below the kidneys. They then descend into the scrotum taking coverings of peritoneum with them. This peritoneum eventually surrounds the testes in the scrotum, becoming detached from the abdominal peritoneum. Descent of the testes into the scrotum should be complete by the 8th month of fetal life.

The tunica albuginea is a fibrous covering surrounding the testes situated under the tunica vaginalis. Ingrowths form septa dividing the glandular structure of the testes into *lobules*.

The tunica vasculosa consists of a network of capillaries supported by delicate connective tissue.

Structure of the testes

In each testis there are 200 to 300 lobules composed of *germinal epithelial cells*. Between the tubules there are groups of *interstitial cells* that secrete the hormone *testosterone*. At the upper pole of the testis the tubules combine to form a single tortuous tubule, the *epididymis*, which leaves the scrotum as the *deferent duct* (vas deferens) in the *spermatic cord* (see Figs 15:15 and 15:16). Blood and lymph vessels pass to the testes in the spermatic cords.

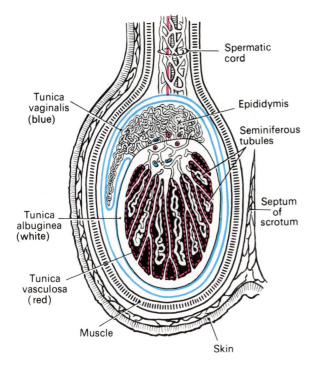

Figure 15:15 A longitudinal section of a testis and its coverings.

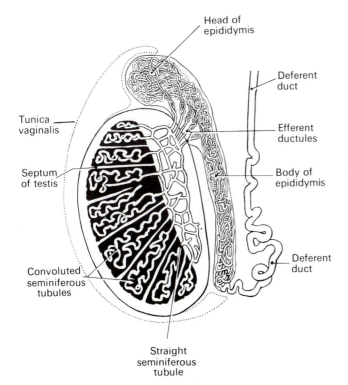

Figure 15:16 A longitudinal section of a testis and deferent duct.

SPERMATIC CORDS

There are *two spermatic cords*, one leading from each testis, consisting of:

1 testicular artery
1 testicular venous plexus
Lymph vessels
1 deferent duct (vas deferens)
Nerves

The spermatic cord suspends the testis in the scrotum. It is composed of fibrous tissue and smooth muscle, connective tissue surrounding the blood and lymph vessels, nerves and the deferent duct. It passes through the inguinal canal. At the deep inguinal ring the structures within the cord diverge.

The testicular artery branches from the abdominal aorta, just below the renal arteries.

The testicular vein passes into the abdominal cavity. The left vein opens into the left renal vein and the right into the inferior vena cava.

The lymph drainage is through lymph nodes around the aorta.

The deferent duct passes upwards from the testis through the inguinal canal and ascends medially towards the posterior wall of the bladder where it is joined by the duct from the *seminal vesicle* to form the *ejaculatory duct*.

The nerve supply is provided by branches from the tenth and eleventh thoracic nerves.

SEMINAL VESICLES

The seminal vesicles are two small fibromuscular pouches lined with columnar epithelium, lying on the posterior aspect of the bladder.

At its lower end each seminal vesicle opens into a short duct which joins with the corresponding deferent duct to form an ejaculatory duct.

EJACULATORY DUCTS

The ejaculatory ducts are two tubes about 2 cm long, each formed by the union of the duct from a seminal vesicle and a deferent duct. They pass through the prostate gland and join the prostatic urethra.

The ejaculatory ducts are composed of the same layers of tissue as the seminal vesicles

PROSTATE GLAND (Figs 15:17 and 15:18)

The prostate gland lies in the pelvic cavity in front of the rectum and behind the symphysis pubis, surrounding the first part of the urethra. It consists of an outer fibrous covering, a layer of smooth muscle and glandular

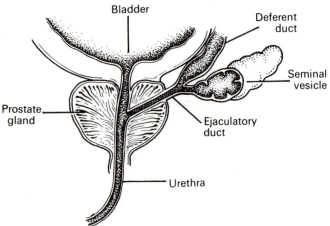

Figure 15:17   Section of the prostate gland and associated reproductive structures on one side.

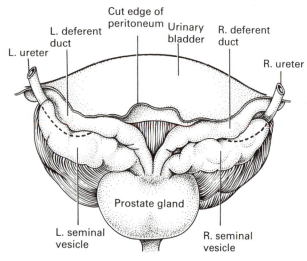

Figure 15:18   Structures associated with the prostate gland. Posterior view.

substance composed of columnar epithelial cell. It secretes a thin lubricating fluid that passes into the urethra through numerous ducts.

URETHRA AND PENIS

URETHRA

The male urethra provides a common pathway for the flow of urine and *semen*, the secretions of the male reproductive organs. It is about 19 to 20 cm long. It originates at the urethral orifice in the bladder where it is surrounded by the prostate gland, then passes through the perineum into the penis.

There are two urethral sphincters. The *internal sphincter* consists of smooth muscle fibres at the neck of the bladder above the prostate gland. The *external sphincter* consists of striated muscle fibres surrounding the membranous part.

PENIS (Fig. 15:19)

The penis has a *root* and a *body*. The root lies in the perineum and the body surrounds the urethra. It is formed by three elongated masses of *erectile tissue* and involuntary muscle. The erectile tissue is supported by fibrous tissue and covered with skin. It has a rich blood supply.

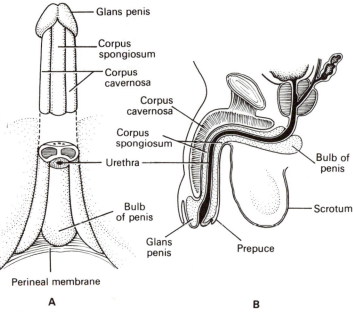

Figure 15:19   The penis. A. Viewed from below.
B. Viewed from the side.

The two lateral columns are called the *corpora cavernosa* and the column between them is the *corpus spongiosum*. At its tip it is expanded into a triangular structure known as the *glans penis*. Just above the glans the skin is folded upon itself and forms a movable double layers, the *foreskin* or *prepuce*. The penis is supplied by autonomic and somatic nerves. Parasympathetic stimulation leads to engorgement with blood and erection of the penis.

FUNCTIONS OF THE MALE REPRODUCTIVE SYSTEM

As in the female, the male reproductive organs are stimulated by the *gonadotrophic hormones* from the anterior lobe of the pituitary gland.

The follicle stimulating hormone stimulates the *seminiferous tubules* of the testes to produce male germ cells, the spermatozoa (Fig. 15:20).

The spermatozoa pass through the epididymis, deferent duct, seminal vesicle, ejaculatory duct and the urethra to be implanted in the female vagina during coitus.

In the epididymis and deferent duct the spermatozoa become more mature and are capable of independent movement through a liquid medium. An ejaculation usually consists of 2 to 5 ml of semen containing 40 to 100

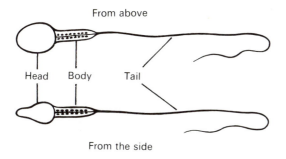

Figure 15:20 A spermatozoon.

million spermatozoa per ml. If they are not ejaculated, spermatozoa are reabsorbed by the tubules (Fig. 15:21).

Successful spermatogenesis takes place at a temperature about 3°C lower than normal body temperature. This lower temperature is possible because the testes in the scrotum are covered by only a thin layer of tissue containing very little fat.

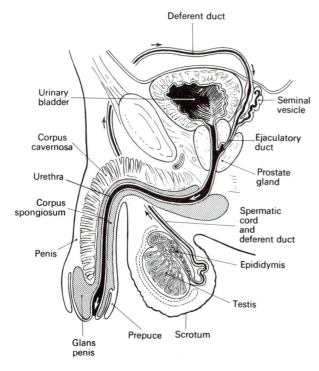

Figure 15:21 Section of the male reproductive organs. Arrows show the structures through which the spermatozoa pass.

Semen is the fluid ejaculated from the urethra during coitus. It consists of:

1. Spermatozoa
2. A viscid fluid which helps to nourish the spermatozoa, secreted by the seminal vesicles
3. A thin lubricating fluid produced by the prostate gland
4. Mucus secreted by glands in the lining membrane of the urethra

PUBERTY IN THE MALE

This occurs between the ages of 10 to 14 years. Luteinising hormone from the anterior lobe of the pituitary gland stimulates the *interstitial cells* of the testes to increase the production of *testosterone*. This hormone influences the development of the body to sexual maturity. The changes which occur at puberty are:

1. Growth of muscle and bone and a marked increase in height and weight
2. Enlargement of the larynx and deepening of the voice—it 'breaks'
3. Growth of hair on the face, axillae, chest, abdomen and pubis.
4. Enlargement of the penis, scrotum and prostate gland
5. Maturation of the seminiferous tubules and production of spermatozoa

In the male, fertility and sexual ability tend to decline gradually with ageing. There is no period comparable to the menopause in the female.

DISEASES OF THE MALE REPRODUCTION SYSTEM

Classification of infection in the male is the same as in the female, i.e., non-specific and specific (see p. 294). In addition to venereal diseases, specific infection of the testes (orchitis) may be caused by mumps viruses, blood-borne from the parotid glands.

PENIS

INFLAMMATION

Inflammation of the glans and prepuce may be caused by a specific or non-specific infection. In non-specific infections, or *balanitis*, lack of personal hygiene is an important predisposing factor, especially if *phimosis* is present, i.e., the orifice in the prepuce is too small to allow for its normal retraction. If the infection becomes chronic there may be fibrosis of the prepuce which increases the phimosis.

URETHRA

INFLAMMATION

Gonococcal urethritis is the most common specific infection. Non-specific infection may be spread from the bladder (cystitis) or be introduced during catheterisation, cystoscopy or surgery. Both types may spread throughout the system to the prostate, seminal vesicles, epididymis and testes. If infection becomes chronic, fibrosis may lead to urethral stricture or obstruction.

EPIDIDYMIS AND TESTES

INFLAMMATION

Non-specific epididymitis and orchitis are usually due to spread of infection from the urethra, commonly following prostatectomy. The microbes may spread either through the deferent duct (vas deferens) or via lymph. Specific epididymitis is usually caused by gonorrhoea spread from the urethra. Orchitis is more commonly caused by mumps viruses, blood-borne from the parotid glands. Acute inflammation with oedema occurs about a week after the appearance of parotid swelling. The infection is usually unilateral but, if bilateral, severe damage to germinal epithelium may result in sterility.

TUMOURS

Most testicular tumours are malignant. They occur in children and young adults in whom the affected testis has not descended or has been late in descending into the scrotum. The tumour tends to remain localised for a considerable time but eventually spreads in lymph to pelvic and abdominal lymph nodes, and more widely in blood. Occasionally hormone-secreting tumours develop and may cause precocious development in children.

PROSTATE GLAND

INFLAMMATION

Acute prostatitis is usually caused by non-specific infection, spread from the urethra or bladder, often following catheterisation, cystoscopy, urethral dilatation or partial resection of the gland. Chronic infection may follow an acute attack, but more commonly it develops insidiously and is not associated with known microbes. Fibrosis of the gland may occur during healing, causing urethral stricture or obstruction.

BENIGN PROSTATIC ENLARGEMENT

Hyperplastic nodules form around the urethra and in a few cases they may cause constriction or obstruction. Urethral stricture may prevent the bladder emptying completely during micturition, predisposing to infection which may spread upwards and cause pyelonephritis. Prostatic enlargement is common in men over 50 years. The cause is not clear, but it may be an acceleration of the ageing process associated with the decline in androgen secretion which changes the androgen/oestrogen balance.

MALIGNANT PROSTATIC TUMOURS

These are a relatively common cause of death in men over 50 years. The carcinogen is not known but changes in the androgen/oestrogen balance may be significant or viruses may be involved. Invasion of local tissues is widespread before lymph-spread metastases develop in pelvic and abdominal lymph nodes. Blood-spread metastases in bone are common and bone formation rather than bone destruction is a common feature. Lumbar vertebrae are a common site, possibly due to retrograde spread along the walls of veins. In many cases bone metastases are the first indication of malignant prostatic tumours.

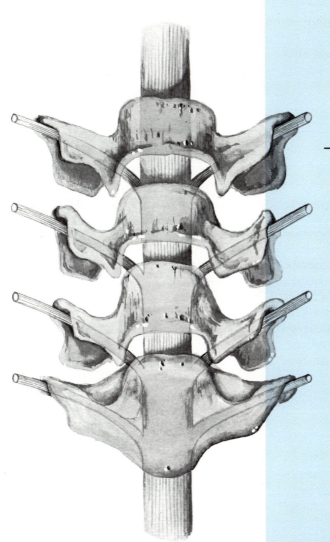

16

The Skeleton

16. *The Skeleton*

BONES

Bone is the hardest tissue in the body and when fully developed it is composed of:

Water	20%
Organic material	30 to 40%
Inorganic material	40 to 50%

There are two types of bone tissue, *compact* and *cancellous*.

COMPACT BONE (Fig. 16:1)

To the naked eye, compact bone appears to be solid but on microscopic examination large numbers of *Haversian systems* can be seen. These consist of a central Haversian canal, containing blood and lymph vessels and nerves, surrounded by concentric plates of bone (*lamellae*). Between these there are *lacunae* or spaces, containing lymph and *osteocytes*. *Canaliculi* link the lacunae with the lymph vessels in the Haversian canal and the osteocytes obtain nourishment from the lymph. This 'tubular' arrangement of lamellae gives bone greater strength than a solid structure of the same size would have.

CANCELLOUS BONE

To the naked eye, cancellous bone looks spongy. Microscopically, the Haversian canals are much larger than in compact bone and there are fewer lamellae, giving a honeycomb appearance. Red bone marrow is always present in cancellous bone.

PERIOSTEUM

Bones are almost completely covered by this vascular fibrous membrane. In the deeper layers of the periosteum there are *osteogenic cells* that increase the thickness of compact tissue and maintain the shape of the bones. It gives attachment to muscles and tendons and protects the bone from injury. Periosteum is replaced by hyaline cartilage on the articular surfaces of synovial joints and by dura mater on the inner surface of the cranial bones.

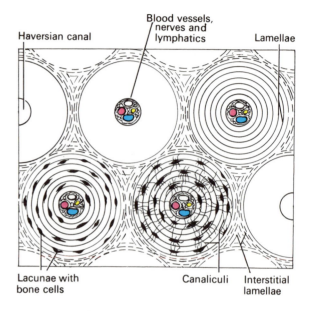

A. Cross section.

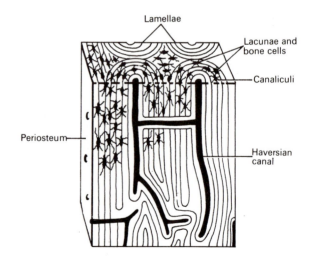

B. Longitudinal section.

Figure 16:1 Microscopic structure of bone.

TYPES OF BONES

These are classified as long, short, irregular, flat and sesamoid.

Long bones have a *diaphysis* or shaft and two *epiphyses* or extremities. The diaphysis is composed of compact bone with a central *medullary canal*, containing fatty *yellow bone marrow*. The epiphyses consist of an outer covering of compact bone with cancellous bone inside. The diaphysis and epiphyses are separated by *epiphyseal cartilages*. A bone grows in length by ossification of the diaphyseal surface of these cartilages and growth is complete when the cartilages stop growing and are completely ossified (Fig. 16:2). The focal points from which ossification begins are small concentrations of osteogenic cells, or *centres of ossification*. There is a primary centre in the diaphysis and one or more secondary centres in each epiphysis. Growth in thickness occurs by the deposition of bone under the periosteum.

Irregular, flat and sesamoid bones are composed of a relatively thin outer layer of compact bone with cancellous bone inside (Fig. 16:3).

DEVELOPMENT OF BONE TISSUE (Osteogenesis or ossification) (Fig. 16:4)

This begins before birth and is not complete until about the 25th year of life (Fig. 16:3). Long, short and irregular bones develop from cartilage models, flat bones from membrane and sesamoid bones from tendon. Bone development consists of two processes:

1. The secretion by osteoblasts of *osteoid*, i.e., collagen fibres in a mucopolysaccharide matrix which gradually replaces the cartilage and membrane models
2. Calcification of the osteoid immediately after its deposition

There are two types of arrangement of collagen in osteoid.

Non-lamellar bone (Woven bone)

The collagen fibres are deposited in bundles, then ossified. This occurs in:

1. Ossification of bones that originate as membrane models, e.g., skull bones
2. Healing of fractures, when collagen fibres are deposited in the granulation tissue that grows between the broken ends

Lamellar bone

The collagen fibres are deposited as in non-lamellar bone, organised into the characteristic Haversian systems then ossified. This occurs when cartilage models are replaced by bone and in fracture healing.

BONE CELLS (Osteogenic cells)

The cells responsible for bone formation are *osteoblasts* (these later develop into *osteocytes*) and *osteoclasts*. Osteoblasts, osteoclasts and *chondrocytes* (cartilage-forming cells) develop from the same parent fibrous tissue cells.

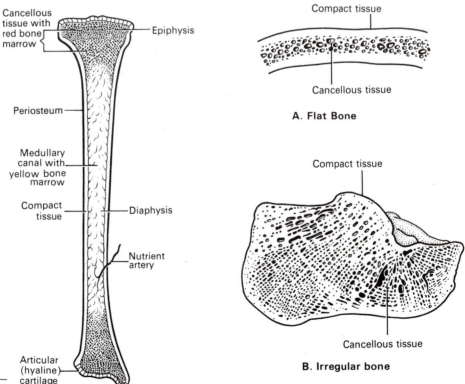

Figure 16:2 A mature long bone— longitudinal section.

Figure 16:3 Flat and irregular bones.

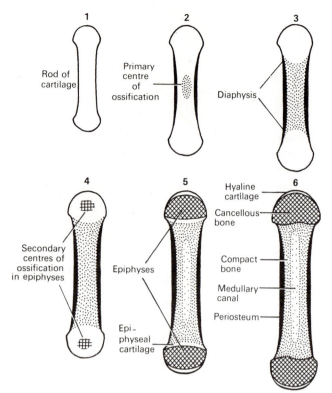

Figure 16:4 Diagram of the stages of development of long bones.

Differentiation into *osteogenic cells*, rather than *chondroblasts*, is believed to depend upon an adequate oxygen supply. This may be a factor affecting healing of fractures, i.e., if the oxygen supply is deficient there may be a preponderance of chondroblasts, resulting in a cartilaginous union of the fracture (see p. 323).

Osteoblasts

These are the bone forming cells and are present:
1. In the deeper layers of periosteum
2. In the centres of ossification of immature bone
3. At the ends of the diaphysis adjacent to the epiphyseal cartilages of long bones
4. At the site of a fracture

As bone develops, osteoblasts become trapped and remain isolated in lacunae. They stop forming new bone at this stage and are called *osteocytes*. Osteocytes receive nourishment from lymph in the canaliculi that radiate from the Haversian canals. Their functions are not clear but they may be associated with the movement of calcium between the bones and the blood.

Osteoclasts

These cells are derived from the same parent cells as osteoblasts. Their function is resorption of bone to maintain the optimum shape. This takes place at bone surfaces, i.e.:

1. Under the periosteum, to maintain the shape of bones during growth and to remove excess callus formed during healing of fractures
2. Round the walls of the medullary canal, to maintain the correct ratio of canal to lamellar bone during growth, and to canalise callus during healing

A fine balance of osteoblast and osteoclast activity maintains normal bone.

FUNCTIONS OF BONES

Bones have a variety of functions, they:
1. Provide the framework of the body
2. Give attachment to muscles and tendons
3. Permit movement of the body as a whole and of parts of the body, by forming joints that are moved by muscles
4. Form the boundaries of the cranial, thoracic and pelvic cavities, protecting the organs they contain
5. Contain red bone marrow in which blood cells develop
6. Provide a reservoir for calcium

Bones of the skeleton

Most bones have rough surfaces, raised protruberances and ridges which give attachment to muscle tendons and ligaments. These are not included in the following descriptions of individual bones unless they are of particular note, but many are marked on illustrations.

The bones of the skeleton are divided into two groups (Figs 16:5 and 16:6): *the axial skeleton* and *the appendicular skeleton*.

AXIAL SKELETON

This part consists of *the skull, vertebral column, ribs* and *sternum*. Together the bones forming these structures constitute the central bony core of the body, the axis.

SKULL (Figs 16:7 and 16:8)

The skull rests on the upper end of the vertebral column and its bony structure is divided into two parts: the cranium and the face.

CRANIUM

The cranium is formed by a number of flat irregular bones that provide a bony protection for the brain. It has a *base* upon which the brain rests and a *vault* that surrounds and covers it. The periosteum inside the skull bones consists of an outer layer of dura mater. The joints (*sutures*)

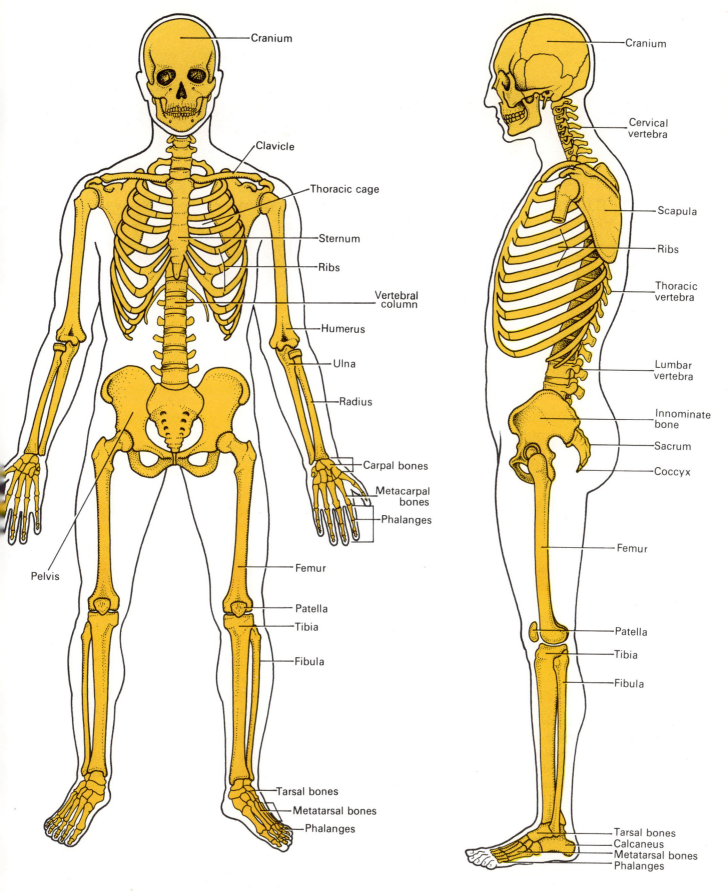

Figure 16:5 The skeleton. Anterior view.

Figure 16:6 The skeleton. Lateral view.

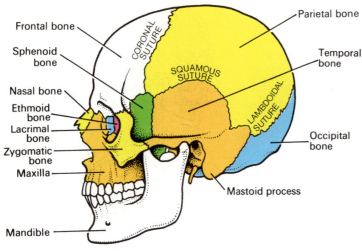

Figure 16:7   The bones of the skull and their joints or sutures.

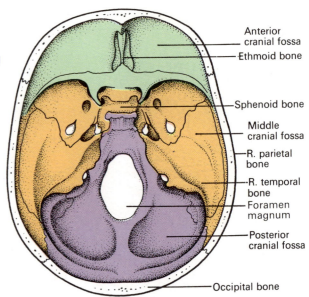

Figure 16:8   The bones forming the base of the skull and the cranial fossae. Viewed from above. Anterior cranial fossa — green. Middle cranial fossa — orange. Posterior cranial fossa — purple.

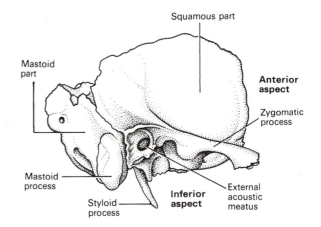

Figure 16:9   The right temporal bone. Lateral view.

between the bones are immovable (fibrous). The bones have numerous perforations through which nerves, blood and lymph vessels pass. The bones which form the cranium are:

1 frontal bone	1 occipital bone
2 parietal bones	1 sphenoid bone
2 temporal bones	1 ethmoid bone

Frontal bone

This is the bone of the forehead. It forms part of the orbital cavities and the prominent ridges above the eyes, the *supraorbital margins*. Just above the supraorbital margins, within the bone, there are two air-filled cavities or *sinuses* lined with ciliated mucous membrane which have openings into the nasal cavities.

The *coronal suture* joins the frontal and parietal bones and other fibrous joints are formed with the sphenoid, zygomatic, lacrimal, nasal, and ethmoid bones. The bone originates in two parts joined in the midline by the *frontal suture* (see Fig. 16:15).

Parietal bones

These bones form the sides and roof of the skull. They articulate with each other at the *sagittal suture*, with the frontal bone at the coronal suture, with the occipital bone at the *lambdoidal suture* and with the temporal bones at the *squamous sutures*. The inner surface is concave and is grooved by the brain and blood vessels (see Fig. 16:15).

Temporal bones (Fig. 16:9)

These bones lie one on each side of the head and form immovable joints with the parietal, occipital, sphenoid and zygomatic bones. Each temporal bone is divided into four parts.

The squamous part is the thin fan-shaped part that articulates with the parietal bone.

The mastoid process is a thickened part behind the ear. It contains a large number of very small air sinuses which communicate with the middle ear and are lined with squamous epithelium.

The petrous portion forms part of the base of the skull and contains the organ of hearing, the organ of Corti (p. 254).

The zygomatic process articulates with the zygomatic bone to form the zygomatic arch.

The temporal bone articulates with the mandible at the *temporomandibular joint*, the only movable joint of the skull. Immediately behind this articulating surface is the *auditory meatus* which passes inwards towards the petrous portion of the bone.

Occipital bone (Fig. 16:10)

This bone forms the back of the head and part of the base of the skull. It has immovable joints with the parietal, temporal and sphenoid bones. Its inner surface is deeply concave and the concavity is occupied by the occipital lobes of the cerebrum and by the cerebellum. The occiput has two articular condyles that form hinge joints with the first bone of the vertebral column, the *atlas*. Between the condyles there is the *foramen magnum* through which the spinal cord passes.

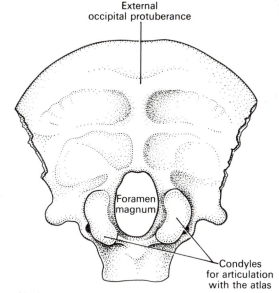

Figure 16:10   The occipital bone viewed from below.

Sphenoid bone (Fig. 16:11)

This bone occupies the middle portion of the base of the skull and it articulates with the temporal, parietal and frontal bones. On the superior surface of the middle of the bone there is a little saddle-shaped depression, the *hypophyseal fossa* (*sella turcica*) in which the *pituitary gland* rests. The body of the bone contains some fairly large air sinuses lined by ciliated mucous membrane with openings into the nasal cavities.

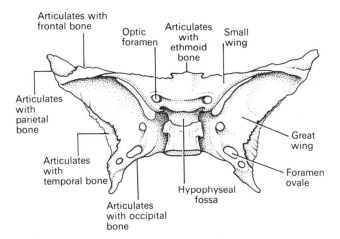

Figure 16:11   The sphenoid bone viewed from above.

Ethmoid bone (Fig. 16:12)

The ethmoid bone occupies the anterior part of the base of the skull and helps to form the orbital cavity, the nasal septum and the lateral walls of the nasal cavity. On each side there are two projections into the nasal cavities, the *upper* and *middle conchae* or *turbinated processes*. It is a very delicate bone containing many air sinuses lined with ciliated epithelium and with openings into the nasal cavities. The horizontal flattened part, the *cribriform plate*, forms the roof of the nasal cavities and has numerous small foramina through which nerve fibres of the *olfactory nerve* (the nerve of the sense of smell) pass upwards from the nasal cavities to the brain. There is also a very fine *perpendicular plate* of bone that forms the upper part of the *nasal septum*.

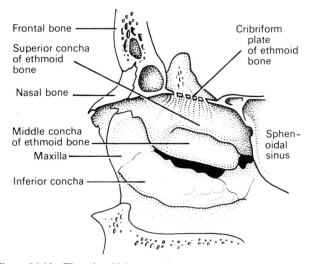

Figure 16:12   The ethmoid bone and its related structures.

FACE

The skeleton of the face is formed by 13 bones in addition to the frontal bone, already described. Figure 16:13 shows the relationships between the bones.

    2 zygomatic or cheek bones
    1 maxilla (originated as 2)
    2 nasal bones
    2 lacrimal bones
    1 vomer
    2 palatine bones
    2 inferior conchae or turbinated bones
    1 mandible (originated as 2)

Zygomatic or cheek bone

The zygomatic bones form the prominences of the cheeks and part of the floor and lateral walls of the orbital cavities.

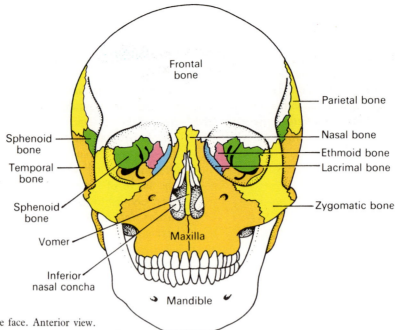

Figure 16:13   The bones of the face. Anterior view.

Maxilla or upper jaw bone

This originates as two bones but fusion takes place before birth. The maxilla forms the upper jaw, the anterior part of the roof of the mouth, the lateral walls of the nasal cavities and part of the floor of the orbital cavities. The *alveolar ridge* or *process* projects downwards and carries the upper teeth. On each side there is a large air sinus, the *maxillary sinus*, lined with ciliated mucous membrane and with openings into the nasal cavities.

Nasal bones

These are two small flat bones which form the greater part of the lateral and superior surfaces of the bridge of the nose.

Lacrimal bones

These two small bones are posterior and lateral to the nasal bones and form part of the medial walls of the orbital cavities. Each is pierced by a foramen for the passage of the *nasolacrimal duct* which carries the tears from the medial canthus of the eye to the nasal cavity.

Vomer

The vomer is a thin flat bone which extends upwards from the middle of the hard palate to form the main part of the nasal septum. Superiorly it articulates with the perpendicular plate of the ethmoid bone.

Palatine bones

These are two L-shaped bones. The horizontal parts unite to form the posterior part of the hard palate and the perpendicular parts project upwards to form part of the lateral walls of the nasal cavities. At their upper extremities they form part of the orbital cavities.

Inferior conchae or turbinated bones

Each concha is a scroll-shaped bone which forms part of the lateral wall of the nasal cavity and projects into it below the middle concha. The superior and middle conchae are parts of the ethmoid bone.

Mandible (Fig. 16:14)

This is the only movable bone of the skull. It originates as two parts which unite at the midline. Each half consists of two main parts: *a curved body* with the *alveolar ridge* containing the lower teeth and a *ramus* which projects upwards almost at right angles to the posterior end of the body.

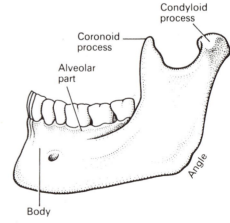

Figure 16:14   The mandible. Lateral view.

At the upper end the ramus divides into the *condyloid process* which articulates with the temporal bone to form the *temporomandibular* joint and the *coronoid process* that gives attachment to muscles and ligaments. The point where the ramus joins the body is the *angle* of the jaw.

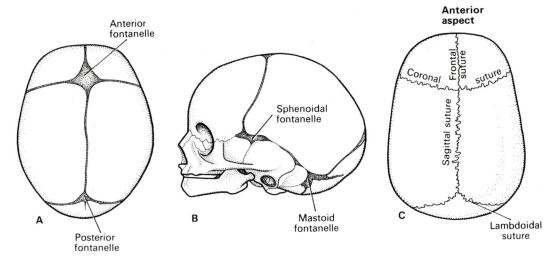

Figure 16:15   The skull showing the fontanelles and sutures. A. Fontanelles viewed from above. B. Fontanelles viewed from the side. C. Main sutures viewed from above when ossification is complete.

Hyoid bone

This is an isolated horse-shoe-shaped bone lying in the soft tissues of the neck just above the *larynx* and below the *mandible*. It does not articulate with any other bone but is attached to the styloid process of the temporal bone by ligaments. It gives attachment to the base of the tongue.

SINUSES

Sinuses containing air are present in the sphenoid, ethmoid, maxillary and frontal bones. They all communicate with the nose and are lined with ciliated mucous membrane. Their functions are:

1. To give resonance to the voice
2. To lighten the bones of the face and cranium, making it easier for the head to balance on top of the vertebral column

FONTANELLES OF THE SKULL (Fig. 16:15)

At birth, ossification of the cranial sutures is incomplete. Where three or more bones meet there are distinct membranous areas, or *fontanelles*. The two largest are: *the anterior fontanelle*, not fully ossified until the child is 12 to 18 months old and *the posterior fontanelle*, usually ossified 2 to 3 months after birth.

VERTEBRAL COLUMN (Fig. 16:16)

The vertebral column consists of 24 separate movable, irregular bones, the *sacrum* (five fused bones) and the *coccyx* (four fused bones). The 24 separate bones are in three groups: 7 cervical, 12 thoracic and 5 lumbar.

The movable vertebrae have many characteristics in common but some groups have distinguishing features.

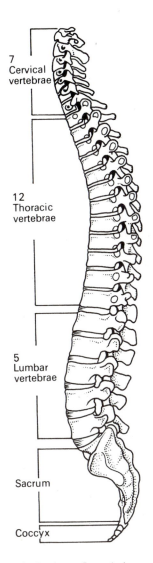

Figure 16:16   The vertebral column. Lateral view.

CHARACTERISTICS OF A TYPICAL VERTEBRA
(Fig. 16:17)

The body of each vertebra is situated anteriorly. The size varies with the site. They are smallest in the cervical region and become larger towards the lumbar region.

The neural arch encloses a large *vertebral foramen*. The ring of bone consists of *two pedicles* that project backwards from the body and *two laminae*. Where the pedicles and laminae unite, *transverse processes* project laterally and where the two laminae meet in the midline posteriorly they form a *spinous process*. The neural arch has four articular surfaces, two articulate with the vertebra above and two with the one below. The vertebral foramina form the neural canal that contains the spinal cord.

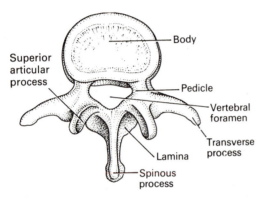

Figure 16:17   A lumbar vertebra showing the features of a typical vertebra — viewed from above.

SPECIAL FEATURES OF VERTEBRAE IN DIFFERENT PARTS OF VERTEBRAL COLUMN

Cervical vertebrae (Fig. 16:18)
The transverse processes have a foramen through which a vertebral artery passes upwards to the brain. The first two cervical vertebrae are atypical.

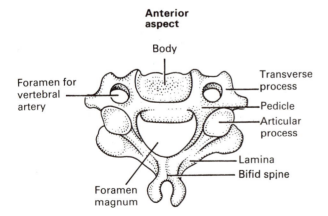

Figure 16:18   A cervical vertebra viewed from above.

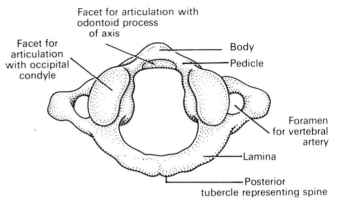

Figure 16:19A   The atlas viewed from above.

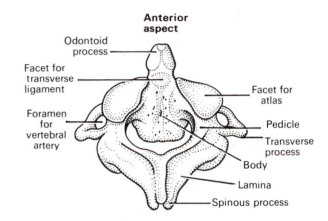

Figure 16:19B   The axis viewed from above.

The atlas (Fig. 16:19A) is the 1st cervical vertebra and it consists simply of a ring of bone with two short transverse processes. The anterior part of the large vertebral foramen is occupied by the *odontoid process* of the axis, which is held in position by a *transverse ligament* (Fig. 16:20).

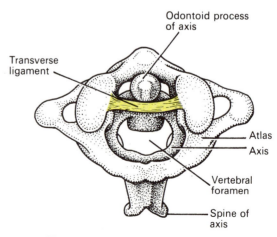

Figure 16:20   The atlas and axis in position showing the transverse ligament. Viewed from above.

Thus the odontoid process represents the body of the atlas. The posterior part is the true vertebral foramen and is occupied by the spinal cord. On its superior surface the bone has two articular facets which form joints with the condyles of the occipital bone of the skull. The nodding movement of the head takes place at these joints.

The axis (Fig. 16:19B) is the 2nd cervical vertebra. The body is small and has the upward projecting *odontoid process* or *dens* that articulates with the first cervical vertebra. The movement brought about at this joint is turning the head from side to side.

a *sacroiliac joint*, and at its inferior tip it articulates with the *coccyx*. The anterior edge of the base, *the promontory*, protrudes into the pelvic cavity. The vertebral foramina are present, and on each side of the bone there is a series of foramina for the passage of nerves.

Coccyx (Fig. 16:22)
This consists of the four terminal vertebrae fused to form a very small triangular bone, the broad base of which articulates with the tip of the sacrum.

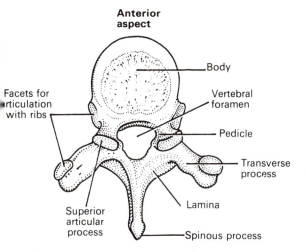

Figure 16:21A A thoracic vertebra. Viewed from above.

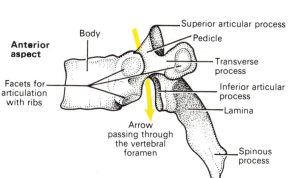

Figure 16:21B A thoracic vertebra. Viewed from the side.

Thoracic vertebrae (Fig. 16:21A and B)
The bodies and transverse processes have facets for articulation with the ribs.

Lumbar vertebrae
These have no special features.

Sacrum (Fig. 16:22)
This consists of five rudimentary vertebrae fused to form a wedge-shaped bone with a concave anterior surface. The upper part, or base, articulates with the 5th lumbar vertebra. On each side it articulates with the ilium to form

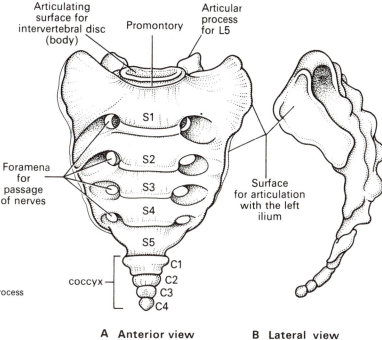

Figure 16:22 The sacrum and coccyx.

THE VERTEBRAL COLUMN AS A WHOLE

Intervertebral discs
The bodies of adjacent vertebrae are bound together by *intervertebral discs*, consisting of an outer rim of fibrocartilage (*annulus fibrosus*) and a central core of soft gelatinous material (*nucleus polposus*). They are thinnest in the cervical region and become progressively thicker towards the lumbar region. The posterior longitudinal ligament in the vertebral canal helps to keep them in place. They have a shock-absorbing function and the cartilaginous joints they form contribute to the flexibility of the vertebral column as a whole

Intervertebral foramina
When two adjacent vertebrae are viewed from the side, a foramen can be seen. Half of the wall is formed by the vertebra above and half by the one below (Fig. 16:23).

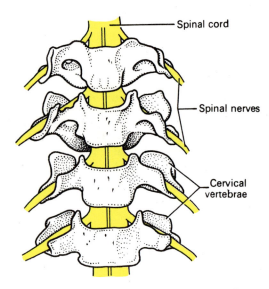

Figure 16:23 Cervical vertebra separated to show the spinal cord and spinal nerves emerging through the intervertebral foramina.

Throughout the length of the column there is an intervertebral foramen on each side between every pair of vertebrae, through which the spinal nerves, blood vessels and lymph vessels pass.

Curves of the vertebral column (Fig. 16:24)

When viewed from the side the vertebral column presents four curves, two *primary* and two *secondary*.

The fetus in the uterus lies curled up so that the head and the knees are more or less touching. This position shows the *primary curvature*. The secondary *cervical curve* develops when the child can hold up his head (after about 3 months) and the secondary *lumbar curve* develops when he stands upright (after 12 to 18 months). The thoracic and sacral primary curves are retained.

Ligaments of the vertebral column (Fig. 16:25)

These ligaments hold the vertebrae together and help to maintain the intervertebral discs in position.

The transverse ligament maintains the odontoid process of the axis in the correct position in relation to the atlas.

The anterior longitudinal ligament extends the whole length of the column and lies anteriorly to the vertebral bodies.

The posterior longitudinal ligament lies inside the vertebral canal and extends the whole length of the vertebral column in close contact with the posterior surface of the bodies of the bones.

The ligamenta flava connect the laminae of adjacent vertebrae.

The ligamentum nuchae and the *supraspinous ligament* connect the spinous processes, extending from the occiput to the sacrum.

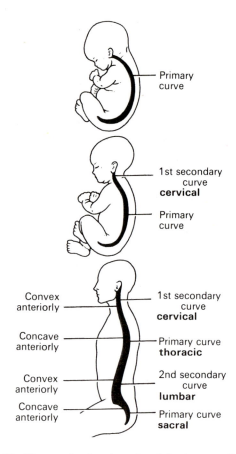

Figure 16:24 Diagram showing the order of development of the curves of the spine.

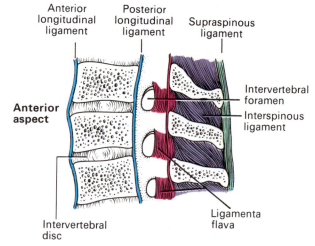

Figure 16:25 Section of the vertebral column showing the ligaments, intervertebral discs and intervertebral foramina.

Movements of the vertebral column

The movements between the individual bones of the vertebral column are very limited. However, the movements of the column as a whole are quite extensive and include *flexion* (bending forward), *extension* (bending backward), *lateral flexion* (bending to the side) and *rotation*. There is more movement in the cervical and lumbar regions than elsewhere.

Functions of the vertebral column

1. Collectively the vertebral foramina form the vertebral canal with provides a strong bony protection for the delicate spinal cord lying within it.

2. The pedicles of adjacent vertebrae form intervertebral foramina on each side through which spinal nerves, blood vessels and lymph vessels pass.

3. Because of the numerous individual bones, a certain amount of movement is possible.

4. It supports the skull.

5. The intervertebral discs act as shock absorbers, protecting the brain.

6. It forms the axis of the trunk, giving attachment to the ribs, shoulder girdle and upper limbs, and pelvic girdle and lower limbs.

THORACIC CAGE (Fig. 16:26)

The bones of the thorax or thoracic cage are: 1 sternum, 12 pairs of ribs and 12 thoracic vertebrae.

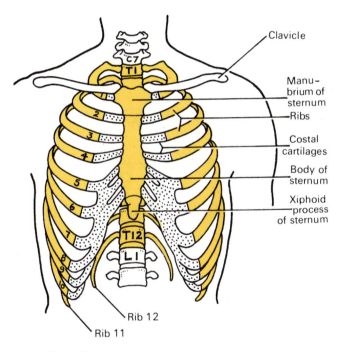

Figure 16:26   The thoracic cage. Anterior view.

Sternum or breast bone (Fig. 16:27)
This *flat bone* can be felt just under the skin in the middle of the front of the chest.

The manubrium is uppermost and articulates with the clavicles at the *sternoclavicular joints* and with the first two pairs of ribs.

The body or middle portion gives attachment of the ribs.

The xiphoid process is the tip of the bone. It gives attachment to the diaphragm, muscles of the anterior abdominal wall and the linea alba.

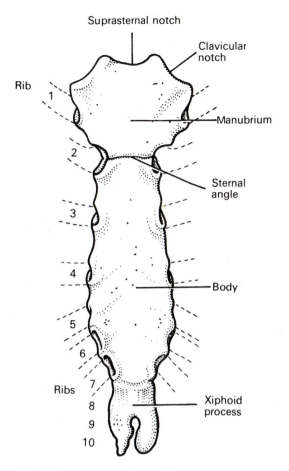

Figure 16:27   The sternum and its attachments.

Ribs
There are 12 pairs of ribs which form the bony lateral walls of the thoracic cage and articulate posteriorly with the thoracic vertebrae. The first 10 pairs are attached anteriorly to the sternum by *costal cartilages*, some directly and some indirectly (see Fig. 16:26)

The last two pairs (*floating ribs*) have no anterior attachment.

Characteristics of a rib (Fig. 16:28). The head articulates posteriorly with the bodies of two adjacent thoracic vertebrae and the tubercle with the transverse process of one. The sternal end is attached to the sternum by a costal cartilage, i.e., a band of hyaline cartilage. The superior border is rounded and smooth while the inferior border has a marked groove occupied by the intercostal blood vessels and nerves.

The first rib does not move during respiration.

The spaces between the ribs are occupied by the intercostal muscles. During inspiration, when these muscles contract, the ribs and sternum are lifted upwards and outwards increasing the capacity of the thoracic cavity.

Thoracic vertebrae
The 12 thoracic vertebrae have already been described.

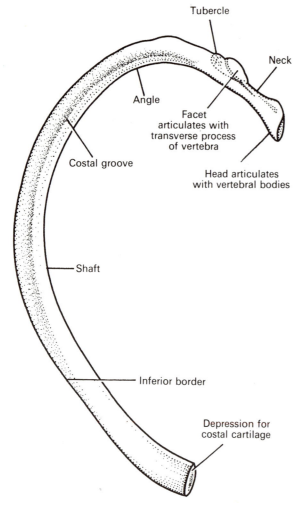

Figure 16:28   A typical rib viewed from below.

APPENDICULAR SKELETON

The appendicular skeleton consists of the shoulder girdle with the upper limbs and the pelvic girdle with the lower limbs (Fig. 16:31).

SHOULDER GIRDLE AND UPPER LIMB

Each shoulder girdle consists of the following bones:
   1 clavicle   1 scapula

Each upper extremity consists of the following bones:
   1 humerus   8 carpal bones
   1 radius   5 metacarpal bones
   1 ulna   14 phalanges

Clavicle or collar bone (Fig. 16:29)

The clavicle is a long bone which has a double curve. It articulates with the manubrium of the sternum at the *sternoclavicular joint* and forms the *acromioclavicular joint* with

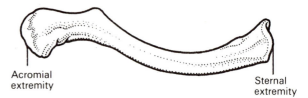

Figure 16:29   The right clavicle.

the *acromion process* of the scapula. The clavicle provides the only bony link between the upper extremity and the axial skeleton.

Scapula or shoulder blade (Fig. 16:30)

The scapula is a flat triangular-shaped bone, lying on the posterior chest wall superficial to the ribs and separated from them by muscles.

At the lateral angle there is a shallow articular surface, the *glenoid cavity* which, with the *head of the humerus*, forms the *shoulder joint*.

On the posterior surface there is a *spinous process* that projects beyond the lateral angle of the bone as the *acromion process* and overhangs the shoulder joint. It articulates with the clavicle at the *acromioclavicular joint*.

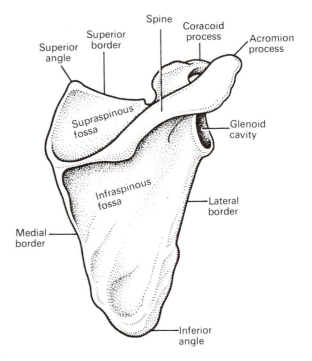

Figure 16:30   The right scapula. Posterior view.

Humerus (Fig. 16:32)

This is the bone of the upper arm. The head articulates with the glenoid cavity of the scapula, forming the shoulder joint. Distal to the head there are two roughened projections of bone, the *greater* and *lesser tubercles* and

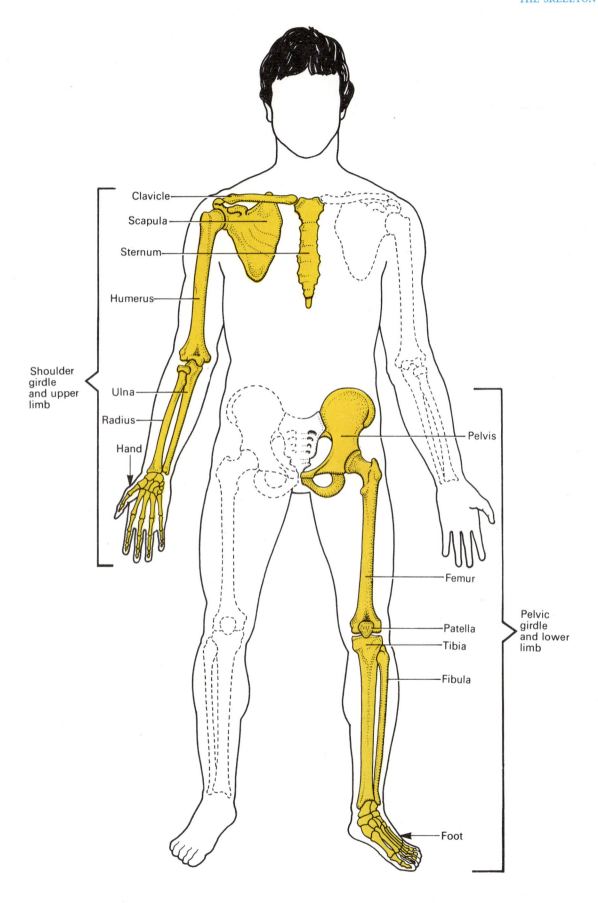

Clavicle
Scapula
Sternum
Humerus
Shoulder girdle and upper limb
Ulna
Radius
Hand

Pelvis
Femur
Patella
Tibia
Fibula
Pelvic girdle and lower limb
Foot

Figure 16:31 The bones of the appendicular skeleton.

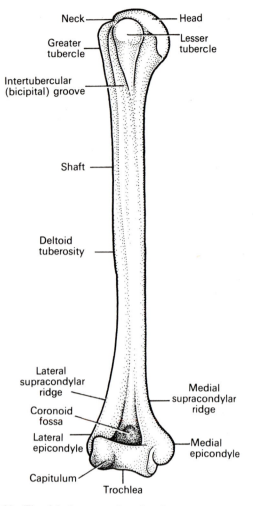

Figure 16:32 The right humerus. Anterior view.

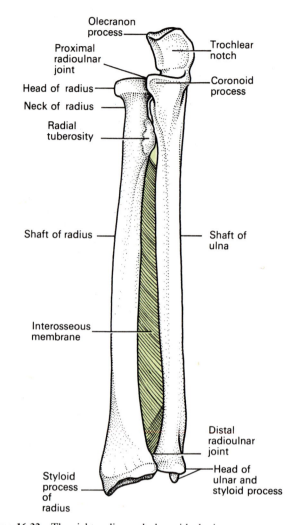

Figure 16:33 The right radius and ulna with the interosseous membrane. Anterior view.

between them there is a deep groove, the *bicipital groove* or *intertubercular sulcus*, occupied by one of the tendons of the biceps muscle.

The distal end of the bone presents two surfaces that articulate with the radius and ulna to form the elbow joint.

Ulna and radius (Fig. 16:33)

These are the two bones of the forearm. The ulna is medial to the radius and when the arm is in the anatomical position, i.e., with the palm of the hand facing forward, the two bones are parallel. They articulate with the humerus at the elbow, the carpal bones at the wrist and with each other at the *superior* and *inferior radioulnar* joints.

Carpal or wrist bones (Fig. 16:34)

There are eight carpal bones arranged in two rows of four. From without inwards they are:

Proximal row: scaphoid, lunate, triquetral, pisiform
Distal row: trapezium, trapezoid, capitate, hamate

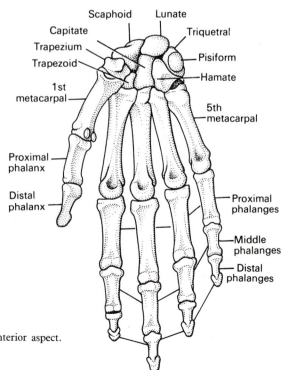

Figure 16:34 The bones of the wrist, hand and fingers. Anterior aspect.

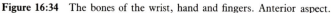

These bones are closely fitted together and held in position by ligaments which allow a certain amount of movement between them. The bones of the proximal row are associated with the wrist joint and those of the distal row form joints with the metacarpal bones. Tendons of muscles lying in the forearm cross the wrist and are held close to the bones by strong fibrous bands.

Metacarpal bones or the bones of the hand

These five bones form the palm of the hand. They are numbered from the thumb side inwards. The proximal ends articulate with the carpal bones and the distal ends with the phalanges.

Phalanges or finger bones

There are 14 phalanges, three in each finger and two in the thumb. They articulate with the metacarpal bones and with each other.

PELVIC GIRDLE AND LOWER LIMB

The bones of the pelvic girdle are:
    2 innominate bones   1 sacrum
The bones of the lower extremity are:
    1 femur   7 tarsal bones
    1 tibia   5 metatarsal bones
    1 fibula   14 phalanges
    1 patella

Innominate or hip bones (Fig. 16:35)

Each hip bone consists of three fused bones, the *ilium*, *ischium* and *pubis*. On its outer surface there is a deep depression, the *acetabulum*, which forms the hip joint with the almost-spherical head of femur.

The ilium is the upper flattened part of the bone and it presents the *iliac crest*, the anterior point of which is called the *anterior superior iliac spine*.

The pubis is the anterior part of the bone and it articulates with the pubis of the other hip bone at a cartilaginous joint, the *symphysis pubis*.

The ischium is the inferior and posterior part.

The union of the three parts takes place in the *acetabulum*.

The pelvis (Fig. 16:36)

The pelvis is formed by the two innominate bones which articulate anteriorly at the symphysis pubis and posteriorly with the sacrum at the sacroiliac joints. It is divided into two parts by the *brim of the pelvis*, consisting of the promontory of the sacrum and the *iliopectineal lines* of the innominate bones. The *greater or false pelvis* is above the brim and the *lesser or true pelvis* is below.

DIFFERENCES BETWEEN MALE AND FEMALE PELVES (Fig. 16:37)

The shape of the female pelvis allows for the passage of the baby during childbirth. In comparison with the male pelvis, the female pelvis has lighter bones, is more shallow and rounded and is generally more roomy.

Femur or thigh bone (Fig. 16:38)

The femur is the longest and strongest bone of the body. The head is almost spherical and fits into the *acetabulum* of the hip bone to form the *hip joint*. In the centre of the head there is a small depression for the attachment of the *ligament of the head of the femur*. This extends from the

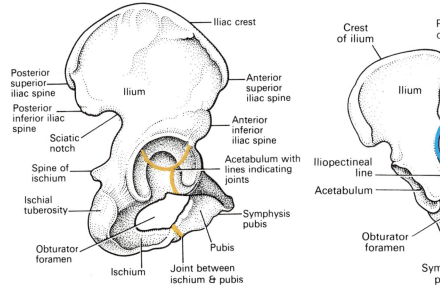

Figure 16:35 The right innominate bone. Lateral view.

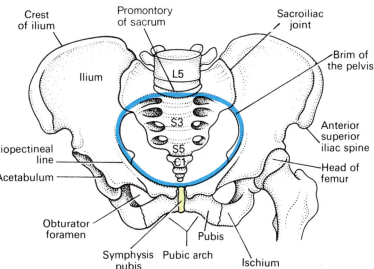

Figure 16:36 The bones of the pelvis and the upper part of the left femur.

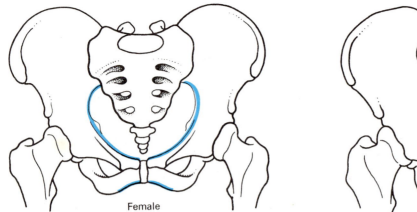
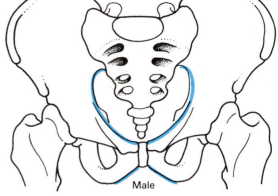

Female

Male

Figure 16:37   Diagram showing the difference in shape of the male and female pelves.

acetabulum to the femur and contains a blood vessel that supplies blood to an area of the head of the bone. The neck extends outwards and slightly downwards from the head to the shaft and most of it is within the capsule of the hip joint.

The posterior surface of the lower third forms a flat

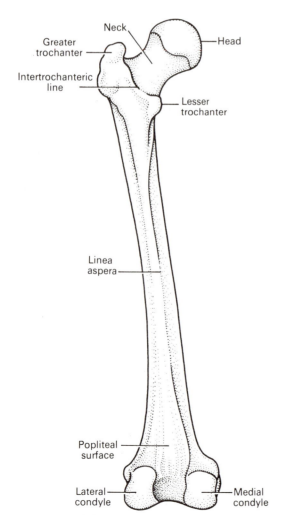

Neck

Greater trochanter

Head

Intertrochanteric line

Lesser trochanter

Linea aspera

Popliteal surface

Lateral condyle

Medial condyle

Figure 16:38   The left femur. Posterior aspect.

triangular area called the *popliteal surface*. The distal extremity has two articular *condyles* which, with the tibia and patella, form the knee joint.

Tibia or shin bone (Fig. 16:39)
The tibia is the medial of the two bones of the lower leg. The proximal extremity is broad and flat and presents two *condyles* for articulation with the femur at the knee joint. The head of the fibula articulates with the inferior aspect of the lateral condyle, forming the superior tibiofibular joint.

The distal extremity of the tibia forms the ankle joint with the *talus* and the fibula. The *medial malleolus* is a downward projection of bone medial to the ankle joint.

Fibula (Fig. 16:39)
The fibula is the long slender lateral bone in the leg. The head or upper extremity articulates with the lateral condyle of the tibia and the lower extremity articulates with the tibia then projects beyond it to form the *lateral malleolus*.

Patella or knee cap
This is a roughly triangular-shaped *sesamoid* bone associated with the knee joint. Its posterior surface articulates with the patellar surface of the femur in the knee joint and its anterior surface is in the *patellar tendon*, i.e., the tendon of the quadriceps femoris muscle.

Tarsal or ankle bones (Fig. 16:40)
There are seven tarsal bones which form the posterior part of the foot. They are:

1 talus	3 cuneiform
1 calcaneus	1 cuboid
1 navicular	

The talus articulates with the tibia and fibula at the ankle joint. The other bones articulate with each other and with the metatarsal bones.

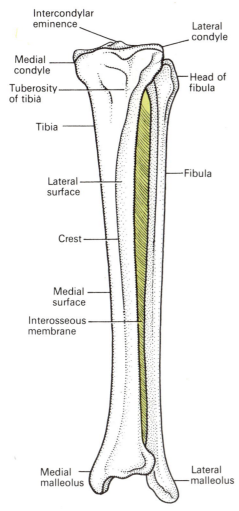

Figure 16:39   The left tibia and fibula with the interosseous membrane.

Metatarsal bones

These are five bones, numbered from within outwards, which form the greater part of the dorsum of the foot. At their proximal ends they articulate with the tarsal bones and at their distal ends, with the phalanges.

Phalanges

There are 14 phalanges arranged in a similar manner to those in the fingers, i.e., two in the great toe and three in each of the other toes.

ARCHES OF THE FOOT

The arrangement of the bones of the foot is such that it is not a rigid structure. This point is well illustrated by comparing a normal foot with a 'flat' foot. The bones have a bridge-like arrangement and are supported by muscles and ligaments so that four arches are formed, a medial and lateral longitudinal arch and two transverse arches.

Medial longitudinal arch

This is the highest of the arches and is formed by the calcaneus, navicular, three cuneiform and first three metatarsal bones. Only the calcaneus and the distal end of the metatarsal bones should touch the ground.

Lateral longitudinal arch

The lateral arch is much less marked than its medial counterpart. The bony components, are the calcaneus, cuboid and the two lateral metatarsal bones. Again only the calcaneus and metatarsal bones should touch the ground.

Transverse arches

These run across the foot and can be more easily seen by examining the skeleton than the live model. They are most marked at the level of the three cuneiform and cuboid bones.

MUSCLES AND LIGAMENTS WHICH SUPPORT THE ARCHES OF THE FOOT (Fig. 16:41)

As there are movable joints between all the bones of the foot, very strong muscles and ligaments are necessary to

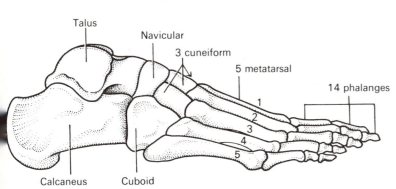

Figure 16:40   The bones of the foot. Lateral view.

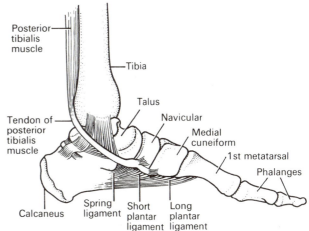

Figure 16:41   The tendons and ligaments supporting the arches of the foot. Medial view.

maintain the strength, resilience and stability of the foot during walking, running and jumping.

Posterior tibialis muscle

This is the most important muscular support of the medial longitudinal arch. It lies on the posterior aspect of the lower leg, originates from the middle third of the tibia and fibula and its tendon passes behind the medial malleolus to be inserted into the navicular, cuneiform, cuboid and metatarsal bones. It acts as a sling or 'suspension apparatus' for the arch.

Short muscles of the foot

This group of muscles is mainly concerned with the maintenance of the lateral longitudinal and transverse arches. They make up the fleshy part of the sole of the foot.

Plantar calcaneonavicular ligament or 'spring' ligament

This is a very strong thick ligament stretching from the calcaneus to the navicular bone. It plays an important part in supporting the medial longitudinal arch.

Plantar ligaments and interosseous membranes

These structures support the lateral and transverse arches.

HEALING OF BONES

Following a fracture, the broken ends of bone are joined by the deposition of new bone. This occurs in several stages (Fig. 16:42):

1. Formation of a haematoma between the ends of bone and in surrounding soft tissues
2. Development of acute traumatic inflammation
3. Accumulation of macrophages which phagocytose the haematoma, inflammatory exudate and small fragments of bone without blood supply (2 and 3 take about 5 days)
4. Growth of granulation tissue and new blood vessels
5. Development of large numbers of oesteoblasts that secrete non-lamellar osteoid which is quickly organised into lamellar osteoid and calcified, forming *callus* (in about 3 weeks)
6. Shaping of new bone by osteoclasts which remove excess callus and open up a medullary canal in the callus (in weeks or months)

FACTORS THAT DELAY HEALING OF FRACTURES

Infection

Microbes usually gain access through broken skin, although they may be blood-borne if healing is delayed. Inflammatory swelling increases pressure in the bone, leading to ischaemia and necrosis.

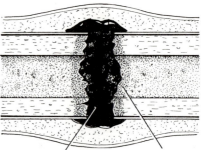

Haematoma and bone fragments   Inflamed area

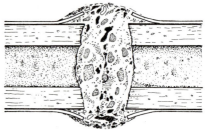

Phagocytosis of clot and debris. Granulation tissue formation begins

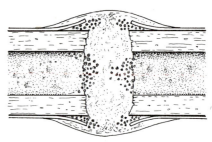

Osteoblasts begin to form new bone

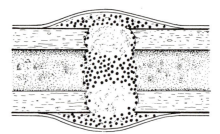

Gradual spread of new bone to bridge gap

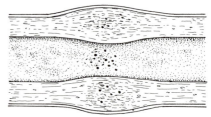

Bone healed. Osteoblasts reshape and canalise new bone

Figure 16:42   Stages in bone healing.

Fat embolism

Emboli, consisting of fat from the medullary canal, may enter torn veins. They are most likely to lodge in the lungs.

Tissue fragments between the ends of bone

Splinters of dead bone (*sequestra*) and soft tissue fragments not removed by phagocytosis delay osteogenesis.

Deficient blood supply

This delays growth of granulation tissue and new blood vessels. Hypoxia also reduces the number of osteoblasts and increases the number of chondroblasts that develop from their common parent cells. This may lead to cartilaginous union of the fracture. The most vulnerable sites, because of their normally poor blood supply, are the neck of femur, the scaphoid and the shaft of tibia.

Continued mobility

Continuous movement of the broken ends of bone tends to prevent osteogenesis. Fibrosis of the granulation tissue occurs, resulting in fibrous union of the fracture.

Old age

Healing of bones, like healing elsewhere, becomes slower with advancing age.

DISEASES OF BONES

OSTEOPOROSIS

In this condition the amount of bone tissue is reduced because its deposition does not keep pace with resorption. Cancellous bone is usually affected before compact bone, and the disease may be localised or general. The cause is not known.

OSTEOPOROSIS ASSOCIATED WITH IMMOBILITY

Localised osteoporosis

This occurs in limbs and is secondary to lack of movement. The bones return to normal when mobility is restored. Immobility may be due to:
1. Treatment of a fracture
2. Paralysis
3. Protracted severe pain that results in voluntary immobilisation, e.g., in arthritis

Generalised osteoporosis

This is usually secondary to prolonged unconsciousness. Excess calcium is resorbed from bone and raises the blood and urine calcium levels, predisposing to calculus formation in the kidneys which may lead to renal failure.

IDEOPATHIC OSTEOPOROSIS

This is believed to be due to acceleration of the normal generalised reduction in bone substance that occurs after middle age. It also occurs in some younger women, following oophorectomy. The cause is not entirely clear but one factor may be disturbance of the hormone balance between *anabolic steroids* (oestrogen and androgens) and *anti-anabolic steroids* (glucocorticoids).

PAGET'S DISEASE

In this disease there is hyperactivity of osteoblasts and osteoclasts and an abnormally rapid turnover of bone tissue. Bones become soft, thick and enlarged, and weight-bearing causes bowing. The bones most commonly affected are the pelvis, femur, tibia, lumbar vertebrae and the skull. The cause is not known but it mainly affects people over 40 years. Complications may include:
1. Compression of cranial nerves in the diminished cranial foramina due to thickening of the bones, e.g., compression of the auditory nerve, causing deafness
2. Fractures that may be spontaneous or follow minor trauma, occurring in spite of thickening of bones
3. Osteoarthritis, e.g., in the hip joint the acetabulum and head of the femur become soft and weight-bearing alters their shape; stresses in the mis-shapen joint damage the articular cartilage and, eventually, the underlying bone
4. Development of osteogenic sarcoma

RICKETS AND OSTEOMALACIA

Rickets occurs in children and osteomalacia in adults after ossification is complete. They are caused by deficiency of vitamin D which promotes calcification of bone and absorption of calcium in the small intestine. Vitamin D is present in some foods and is formed in the skin by the action of ultraviolet light on 7-dehydrocholesterol. Deficiency may be due to:
1. A diet deficient in foods containing the vitamin, e.g., dairy products, fish, meat
2. Malabsorption syndromes, e.g., in association with coeliac disease or following gastrointestinal surgery
3. Lack of exposure to sunlight, e.g., in housebound people or because of cultural modes of dress
4. Excretion of excess vitamin D or its precursors, e.g., in chronic renal disease or long continued renal dialysis
5. Long continued use of some anticonvulsant drugs that convert vitamin D to inactive substances

RICKETS

In this disease osteoid is deposited but calcification is incomplete. Although some growth of the epiphyseal cartilage continues, growth generally is stunted. The bones remain soft and those of the lower limbs become bowed by the weight of the body.

OSTEOMALACIA

In adults the normal processes of resorption and replacement of bone tissue are defective. As in rickets, osteoid is not calcified, the bones becoming soft and bowed.

INFECTION OF BONES

OSTEOMYELITIS

Microbes gain access to bones:
1. Through the skin in compound fractures
2. By spread from a local focus of infection, e.g., from a tooth abscess to the mandible or maxilla
3. Blood-borne from elsewhere in the body, commonly from a boil or paronychia
4. During a surgical procedure

The most common infecting organism is *Staphylococcus aureus* although mixed infection of the soft tissues of the feet, common in elderly diabetics, may spread to the bones. Infection of bone may lead to:
1. Necrosis, which prevents healing if pieces of dead bone (sequestra) are too large to be removed by phagocytosis
2. Pus formation in bone which causes considerable congestion in the confined space, and severe pain
3. Formation of a *subperiosteal abscess* that may rupture through the skin and drain through a sinus.

The outcome may be:
1. Resolution and complete repair
2. Delayed repair due to the presence of sequestrae
3. Chronic infection, often with acute exacerbations
4. Septicaemia and pyaemia if microbes or pus enter the blood stream, leading to abscess formation elsewhere in the body, most commonly in the lungs, kidneys, myocardium and endocardium
5. Spread of the infection to joints, especially if the part of the bone infected is inside the joint capsule, e.g., neck of femur

DEVELOPMENTAL ABNORMALITIES OF BONE

ACHONDROPLASIA

This is caused by a genetic abnormality. There is defective growth of cartilage, especially the epiphyseal cartilage of long bones, leading to dwarfism and under-development of the bones of the base of the skull.

OSTEOGENESIS IMPERFECTA

This is a congenital defect of osteoblasts, resulting in failure of ossification. The bones are brittle and fracture easily, either spontaneously or following very slight trauma.

TUMOURS OF BONE

BENIGN TUMOURS

Single or multiple tumours of bone (*exostoses*) or of cartilage (*endochondromas*) may develop for unknown reasons. They may cause pathological fractures or pressure damage to soft tissues, e.g., a vertebral exostosis may damage the spinal cord or a spinal nerve. Endochondromas have a tendency to undergo malignant change.

MALIGNANT TUMOURS

METASTATIC TUMOURS

The most common malignancies of bone are metastases of primary carcinomas of the breast, lungs and prostate gland. The usual sites are those with the best blood supply, i.e., cancellous bone, especially the bodies of the lumbar vertebrae and epiphyses of humerus and femur. Tumour fragments are spread in blood, and possibly along the walls of the veins from pelvic tumours to vertebrae. The effects of the malignancy may be:
1. Destruction of bone, leading to pathological fractures
2. Fibrosis of bone
3. Anaemia, leukopenia and thrombocytopenia but in most cases the link is not known

PRIMARY TUMOURS

Osteosarcoma
This is a rapidly growing tumour believed to develop from the precursors of osteogenic cells. In young people between 10 and 25 years of age the tumour develops most commonly in the medullary canal of long bones, especially the femur. It is usually well advanced before it becomes evident. In older people, usually over 60 years of age, it is often associated with Paget's disease and the bones most commonly affected are the vertebrae, skull and pelvis.

Chondrosarcoma
These relatively slow-growing tumours are usually the result of malignant change in benign tumours of cartilage cells (endochondromas). They occur mainly between the ages of 40 and 70 years.

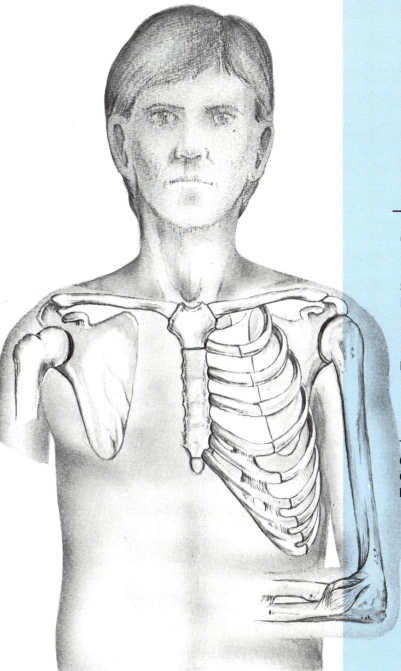

17

The Joints

17. The Joints

A joint is the site at which any two or more bones come together. Some joints have no movement (*fibrosis*), some only slight movement (*cartilaginous*) and some are freely movable (*synovial*).

FIBROUS OR FIXED JOINTS (Fig. 17:1)

These immovable joints have fibrous tissue between the bones, e.g., joints between the bones of the skull (sutures) and those between the teeth and the maxilla and mandible.

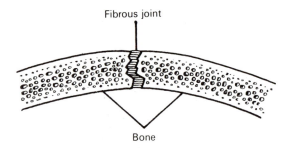

Figure 17:1 A fibrous or fixed joint, e.g., the sutures of the skull.

CARTILAGINOUS OR SLIGHTLY MOVABLE JOINTS (Fig. 17:2)

There is a pad of *white fibrocartilage* between the ends of the bones making up the joint which allows for very slight movement caused by compression of the pad of cartilage. Examples include the symphysis pubis and the joints between the bodies of the vertebrae.

SYNOVIAL OR FREELY MOVABLE JOINTS

Movements possible at synovial joints are:

Flexion or bending, usually forward but occasionally backward, e.g., knee joint

Extension means straightening or bending backward

Abduction is movement away from the midline of the body

Adduction is movement towards the midline of the body

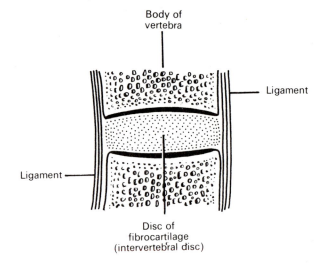

Figure 17:2 A cartilaginous or slightly movable joint, e.g., between the bodies of the vertebrae.

Circumduction is the combination of flexion, extension, abduction and adduction

Rotation is movement round the long axis of a bone
Pronation means turning the palm of the hand down
Supination means turning the palm of the hand up
Inversion is turning the sole of the foot inwards
Eversion is turning the sole of the foot outwards

Synovial joints are classified according to the range of movement possible or to the shape of the articular parts of the bones involved.

Ball and socket. The shape of the bones allows for a wide range of movement. Those possible are flexion, extension, abduction, adduction, rotation and circumduction. The joints are the shoulder and hip.

Hinge joints. These allow the movements of flexion and extension only. They are the elbow, knee, ankle, the joints between the atlas and the occipital bone, and the interphalangeal joints of the fingers and toes.

Gliding joints. The articular surfaces glide over each other, e.g., sternoclavicular joints, acromioclavicular joints and joints between the carpal bones and those between the tarsal bones.

Pivot joints. Movement is round one axis, (rotation), e.g., proximal and distal radioulnar joints and the joint between the atlas and the odontoid process of the axis.

Condyloid and saddle joints. Movements take place round two axes, permitting flexion, extension, abduction, adduction and circumduction, e.g., the wrist, temporomandibular, metacarpophalangeal and metatarsophalangeal joints.

CHARACTERISTICS OF A SYNOVIAL JOINT

All synovial joints have certain characteristics in common.

1. *Articular or hyaline cartilage.* The parts of the bones in contact are always covered with hyaline cartilage. It provides a smooth articular surface and is strong enough to bear the weight of the body.

2. *Capsular ligament.* The joint is surrounded and enclosed by a sleeve of fibrous tissue which holds the bones together. It is sufficiently loose to allow freedom of movement but strong enough to protect it from injury.

3. *Intracapsular structures.* Some joints have structures within the capsule, but outside the synovial membrane, which assist in the maintenance of stability. When these structures do not bear weight they are covered by synovial membrane.

4. *Synovial membrane.* This is composed of epithelial cells which secrete a thick sticky fluid, of egg-white consistency (*synovial fluid*). It acts as a lubricant, provides nutrient materials for the structures within the joint cavity and helps to maintain its stability. It prevents the ends of the bones from being separated as does a little water between two glass surfaces. Synovial membrane is found:

   a. Lining the capsular ligament
   b. Covering those parts of the bones within the joint capsule not covered with hyaline cartilage
   c. Covering all intracapsular structures that do not bear weight

Little sacs of synovial fluid or *bursae* are present in some joints. They act as cushions to prevent friction between a bone and a ligament or tendon, or skin where a bone in a joint is near the surface.

5. *Extracapsular structures.* Most joints have ligaments that blend with the capsule and provide additional stability.

6. *Muscles and movements.* Muscles or their tendons stretch across the joints they move. When the muscle contracts it shortens, pulling one bone towards the other.

7. *Nerve and blood supply.* Nerves and blood vessels crossing a joint usually supply the muscles that move it, and the joint structures.

MAIN SYNOVIAL JOINTS OF THE LIMBS

Individual synovial joints have the characteristics described above so only their distinctive features are included in this section.

SHOULDER JOINT (Fig. 17:3)

This *ball and socket joint* is formed by the glenoid cavity of the scapula and the head of the humerus. The capsular

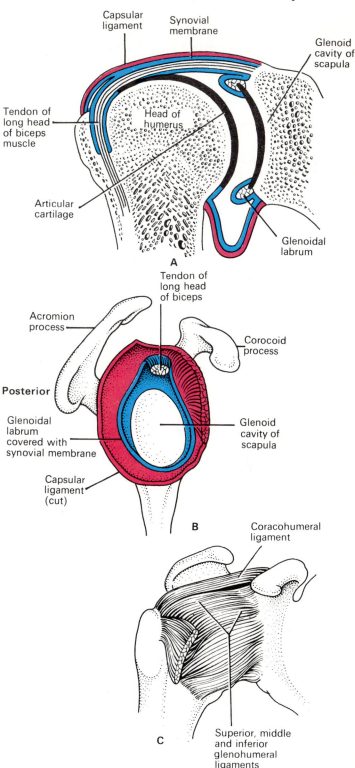

Figure 17:3 The shoulder joint. A. Section viewed from the front. B. The position of glenoidal labrum with the humerus removed, viewed from the side. C. The supporting ligaments viewed from the front.

ligament is very loose inferiorly to allow for the free movement normally possible at this joint. The glenoid cavity is deepened by a rim of fibrocartilage, the *glenoidal labrum*, which provides additional stability without limiting movement. The tendon of the long head of the *biceps muscle*, lying in the bicipital (intertubercular) groove of the humerus, extends through the joint cavity and is attached to the upper rim of the glenoid cavity. It has an important stabilising effect on the joint.

Synovial membrane forms a sleeve round the part of the tendon of the long head of the biceps muscles within the capsular ligament and covers the glenoid labrum.

Extracapsular structures consist of:

1. The *coracohumeral ligament*, extending from the coracoid process of the scapula to the humerus
2. The *transverse humeral ligament*, retaining the biceps tendon in the intertubercular groove

The stability of the joint may be reduced if these structures, together with the tendon of the biceps muscle, are stretched by repeated dislocations of the joint.

MUSCLES AND MOVEMENTS (Fig. 17:6)

Flexion: coracobrachialis, anterior fibres of deltoid and pectoralis major.
Extension: teres major, latissimus dorsi and posterior fibres of deltoid.
Abduction: deltoid.
Adduction: combined action of flexors and extensors.
Circumduction: flexors, extensors, abductors and adductors acting in series.
Medial rotation: pectoralis major, latissimus dorsi, teres major and anterior fibres of deltoid.
Lateral rotation: posterior fibres of deltoid.

Coracobrachialis muscle lies on the upper medial aspect of the arm. It arises from the coracoid process of the scapula, stretches across in front of the shoulder joint and is inserted into the middle third of the humerus. It flexes the shoulder joint.

Deltoid muscle fibres originate from the clavicle, acromion process and spine of scapula and radiate over the shoulder joint to be inserted into the deltoid tuberosity of the humerus. The anterior fibres cause flexion, the middle or main part, abduction and the posterior fibres extend the shoulder joint.

Pectoralis major lies on the anterior thoracic wall. The fibres originate from the middle third of the clavicle and from the sternum and are inserted into the lip of the bicipital groove of the humerus. It draws the arm forward and towards the body, i.e., flexes and adducts.

Latissimus dorsi arises from the posterior part of the iliac crest and the spinous processes of the lumbar and lower thoracic vertebrae. It passes upwards across the back then under the arm to be inserted into the bicipital groove

of the humerus, it adducts, medially rotates and extends the arm.

Teres major originates from the inferior angle of the scapula and is inserted into the humerus just below the shoulder joint. It extends and adducts the arm.

ELBOW JOINT (Fig. 17:4)

This *hinge* joint is formed by the trochlea and the capitulum of the humerus and the trochlear notch of the ulna and the head of the radius.

Extracapsular structures consist of anterior, posterior, medial and lateral strengthening ligaments.

MUSCLES AND MOVEMENTS (Fig. 17:6)

Flexion: biceps and brachialis.
Extension: triceps.

Biceps muscle lies on the anterior aspect of the upper arm. At its proximal end it is divided into two parts (heads) each of which has its own tendon. The short head rises from the coracoid process of the scapula and passes in front of the shoulder joint to the arm. The long head originates from the rim of the glenoid cavity and its tendon passes through the joint cavity and the bicipital groove to the arm. It is retained in the bicipital groove by a *transverse ligament* which stretches across the groove. The distal tendon crosses the elbow joint and is inserted into the radial tuberosity. It helps to stabilise and flex the shoulder joint and at the elbow joint it assists with flexion and supination.

Brachialis lies on the anterior aspect of the upper arm deep to the biceps. It originates from the shaft of the humerus, extends across the elbow joint and is inserted into the ulna just distal to the joint capsule. It is the main flexor of the elbow joint.

Triceps muscle lies on the posterior aspect of the humerus. It arises from three heads, one from the scapula and two from the posterior surface of the humerus. The insertion is by a single tendon to the olecranon process of the ulna. It extends the elbow joint.

PROXIMAL AND DISTAL RADIOULNAR JOINTS

The *proximal radioulnar joint*, formed by the rim of the head of the radius rotating in the radial notch of the ulna, is in the same capsule as the elbow joint. The *annular ligament* is a strong extracapsular ligament which encircles the head of the radius and keeps it in contact with the radial notch of the ulna (Fig. 17:4).

The *distal radioulnar joint* is a pivot joint between the distal end of the radius and the head of the ulna (Fig. 17:5).

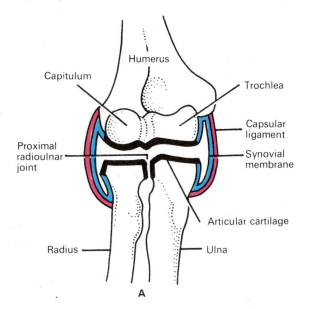

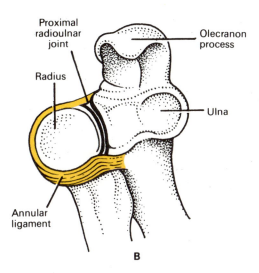

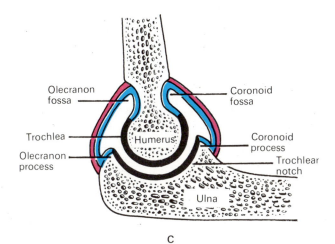

Figure 17:4   The elbow and proximal radioulnar joints. A. Section viewed from the front. B. The proximal radioulnar joint, viewed from above. C. Section of the elbow joint, partly flexed, viewed from the side.

MUSCLES AND MOVEMENTS (Fig. 17:6)

Pronation: pronator teres.
Supination: supinator and biceps.

Pronator teres lies obliquely across the front of the forearm. It arises from the medial epicondyle of the humerus and the coronoid process of the ulna, passes obliquely across the forearm to be inserted into the lateral surface of the shaft of the radius. It rotates the radioulnar joints, changing the hand from the anatomical to the writing position, i.e., pronation.

Supinator muscle lies obliquely across the posterior and lateral aspects of the forearm. Its fibres arise from the lateral epicondyle of the humerus and the upper part of the ulna and are inserted into the lateral surface of the upper third of the radius. It rotates the radioulnar joints, changing the hand from the writing to the anatomical position, i.e. supination.

WRIST JOINT (Fig. 17:5)

This is a *condyloid* joint between the distal end of the radius and the proximal ends of the scaphoid, lunate and triquetral. A disc of *white fibrocartilage* separates the ulna from the joint cavity and articulates with the carpal bones. It also separates the inferior radioulnar joint from the wrist joint.

Extracapsular structures consist of medial and lateral ligaments and anterior and posterior radiocarpal ligament.

MUSCLES AND MOVEMENTS (Fig. 17:6)

Flexion: flexor carpi radialis and the flexor carpi ulnaris.
Extension: extensors carpi radialis (longus and brevis) and the extensor carpi ulnaris.
Abduction: flexor and extensors carpi radialis.
Adduction: flexor and extensor carpi ulnaris.

Flexor carpi radialis lies on the anterior surface of the forearm. It originates from the medial epicondyle of the humerus and is inserted into the second and third metacarpal bones. It flexes the wrist joint, and when acting with the extensor carpi radialis, abducts the joint.

Flexor carpi ulnaris lies on the medial aspect of the forearm. It originates from the medial epicondyle of the humerus and the upper parts of the ulna and is inserted into the pisiform, the hamate and the fifth metacarpal bones. It flexes the wrist, and when acting with the extensor carpi ulnaris, adducts the joint.

Extensor carpi radialis longus and brevis lie on the posterior aspect of the forearm. The fibres originate from the lateral epicondyle of the humerus and are inserted by a long tendon into the second and third metacarpal bones. They extend and abduct the wrist.

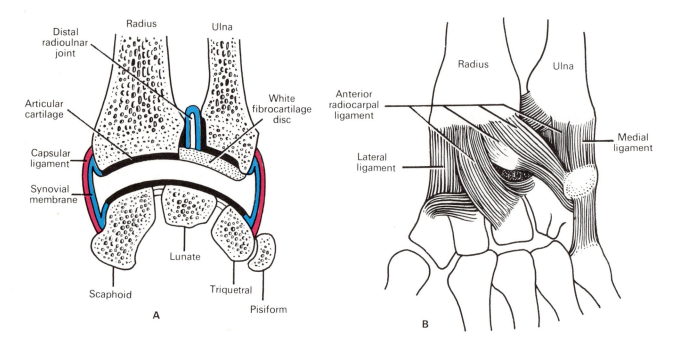

Figure 17:5   The wrist and distal radioulnar joints. Anterior view. A. Section. B. Supporting ligaments.

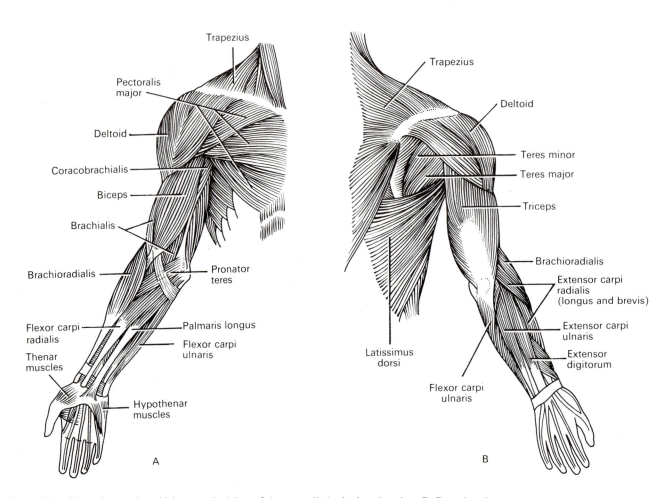

Figure 17:6   The main muscles which move the joints of the upper limb. A. Anterior view. B. Posterior view.

Extensor carpi ulnaris lies on the posterior surface of the forearm. It originates from the lateral epicondyle of the humerus and is inserted into the fifth metacarpal bone. It extends and adducts the wrist.

JOINTS OF THE HANDS AND FINGERS

There are synovial joints between the carpal bones between the carpal and metacarpal bones, between the metacarpal bones and proximal phalanges and between the phalanges. The powerful movements that occur at these joints are produced by muscles in the forearm which have tendons extending into the hand. These tendons are encased in sleeves of synovial membrane and they are held close to the wrist bones by strong transverse ligaments. The tendons move smoothly within the synovial sheath as the joints move. Many of the finer movements of the fingers are produced by numerous small muscles in the hand.

HIP JOINT (Fig. 17:7)

This *ball and socket* joint is formed by the cup-shaped acetabulum of the innominate bone and the almost spherical head of the femur. The capsular ligament includes most of the neck of the femur. The cavity is deepened by the *acetabular labrum*, a ring of fibrocartilage attached to the rim of the acetabulum. This adds stability to the joint without limiting its range of movement. The ligament of the head of the femur extends from the shallow depression in the middle of the head of the femur to the acetabulum. It conveys a blood vessel to the head of the femur. Synovial membrane covers both sides of the acetabular labrum and forms a sleeve around the ligament of the head of the femur. There are three important ligaments that surround and strengthen the capsule. They are the iliofemoral, ischiofemoral and pubofemoral ligaments.

MUSCLES AND MOVEMENTS (Figs 17:8 and 17:9)

Flexion: psoas, iliacus, rectus femoris and sartorius.
Extension: gluteus maximus and the hamstrings.
Abduction: gluteus medius and minimus, sartorius and others.
Adduction: adductor group.
Lateral rotation: mainly gluteal muscles and adductor group.
Medial rotation: gluteus medius and minimus and others.

Psoas muscle arises from the transverse processes and bodies of the lumbar vertebrae. It passes across the bones of the greater pelvis and behind the inguinal ligament to be inserted into the femur. Together with the iliacus it flexes the hip joint.

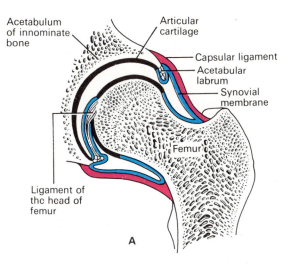

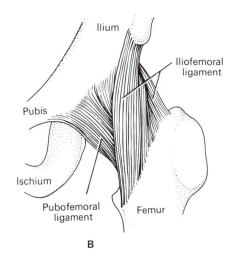

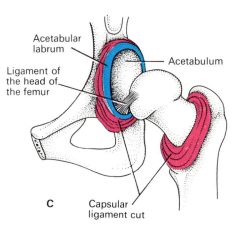

Figure 17:7 The hip joint. Anterior view. A. Section. B Supporting ligaments. C. Head of femur and acetabulum separated to show acetabular labrum and ligament of head of femur.

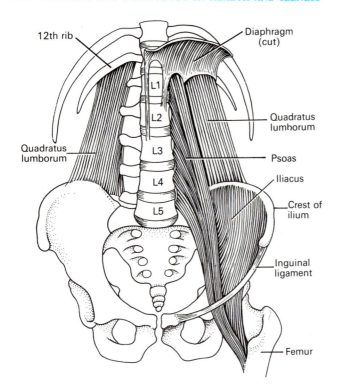

Figure 17:8   The muscles of the posterior abdominal wall and pelvis which flex the hip joint.

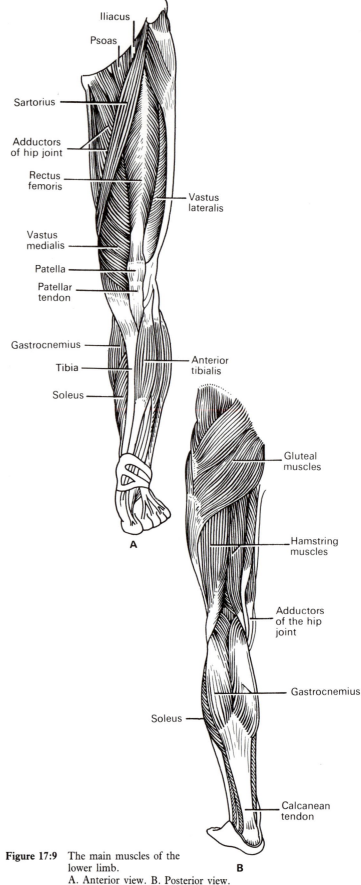

Figure 17:9   The main muscles of the lower limb.
A. Anterior view. B. Posterior view.

Iliacus muscle lies in the iliac fossa of the innominate bone. It originates from the iliac crest, passes over the iliac fossa and joins the tendon of the psoas muscle to be inserted into the femur. The combined action of iliacus and psoas flexes the hip joint.

Quadriceps femoris is a group of four muscles lying on the front of the thigh. They are *rectus femoris* and *three vasti*. The rectus femoris originates from the ilium and the three vasti from the upper end of the femur. Together they pass over the front of the knee joint to be inserted into the tibia by the patellar tendon. Only the rectus femoris flexes the hip joint. Together the group acts as a very strong extensor of the knee joint.

Gluteal muscles consist of the gluteus maximus, medius and minimus which together form the fleshy part of the buttock. They originate from the ilium and sacrum and are inserted into the femur. They cause extension, abduction and medial rotation at the hip joint.

Sartorius is the longest muscle in the body. It originates from the anterior superior iliac spine and passes obliquely across the hip joint, thigh and knee joint to be inserted into the medial surface of the upper part of the tibia. It is associated with flexion and abduction at the hip joint and flexion at the knee.

Adductor group lies on the medial aspect of the thigh. They originate from the pubic bone and are inserted into the linea aspera of the femur. They adduct the thigh.

KNEE JOINT (Fig. 17:10)

This *hinge* joint is formed by the condyles of the femur, the condyles of the tibia and the posterior surface of the patella. The anterior part of the capsule consists of the tendon of the quadriceps femoris muscle which also supports the patella. Intracapsular structures include two cruciate ligaments which cross each other, extending from the intercondylar notch of the femur to the intercondylar eminence of the tibia. They help to stabilise the joint.

Semilunar cartilages or menisci are incomplete discs of white fibrocartilage lying on top of the articular condyles of the tibia. They are wedge-shaped, being thicker at their outer edges. They help to stabilise the joint by preventing lateral displacement of the bones.

Bursae and *pads of fat* are numerous. They prevent friction between a bone and a ligament or tendon and between the skin and the patella. Synovial membrane covers the cruciate ligaments and the pads of fat. The menisci are not covered with synovial membrane because they are weight-bearing. The most important strengthening ligaments are the medial and lateral ligaments.

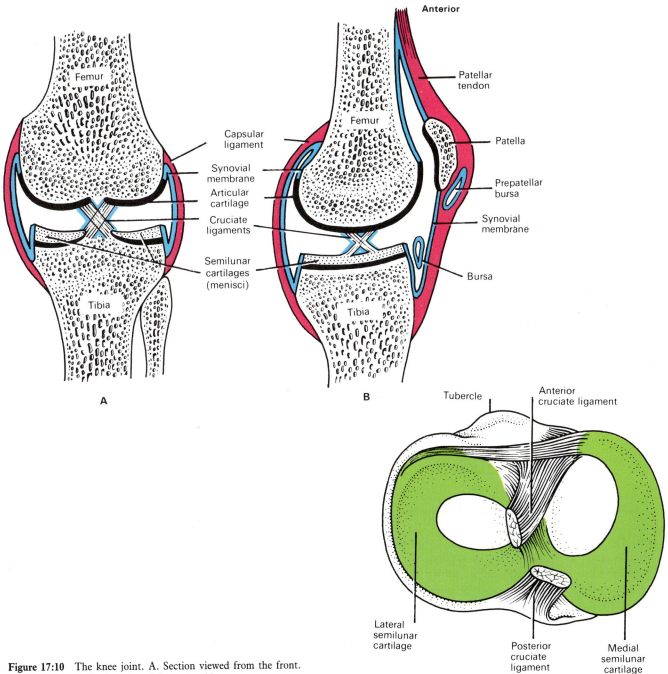

Figure 17:10   The knee joint. A. Section viewed from the front.
B. Section viewed from the side.
C. The superior surface of the tibia showing the semilunar cartilages and the cruciate ligaments.

MUSCLES AND MOVEMENTS (Fig. 17:9)

The movements are flexion, extension and a rotatory movement which 'locks' the joint when it is fully extended. When the joint is locked, balance is maintained with less muscular effort than when it is flexed.

> *Flexion* (bending backwards): gastrocnemius and hamstrings.
> *Extension* (straightening): quadriceps femoris muscle.

Hamstring muscles lie on the posterior aspect of the thigh. They originate from the ischium and are inserted into the upper end of the tibia. They are *biceps femoris*, *semimembranosus* and *semitendinosus muscles*. They flex the knee joint.

Gastrocnemius forms the bulk of the calf of the leg. It arises by two heads, one from each condyle of the femur, and passes down behind the tibia to be inserted into the calcaneus by the *calcanean tendon* (*Achilles* tendon). It crosses both knee and ankle joints, causing flexion at the knee and *plantarflexor* at the ankle.

Quadriceps femoris (described above) extends the knee joint.

ANKLE JOINT (Fig. 17:11)

This *hinge* joint is formed by the distal end of the tibia and its malleolus (medial malleolus), the distal end of the fibula (lateral malleolus) and the talus. There are four important ligaments strengthening this joint. They are the anterior, posterior and lateral ligaments and the deltoid, a very strong medial ligament.

MUSCLES AND MOVEMENTS (Fig. 17:9)

> *Flexion* (dorsiflexion): anterior tibialis assisted by the muscles which extend the toes.
> *Extension* (plantarflexion): gastrocnemius and soleus assisted by the muscles which flex the toes.

The movements of *inversion* and *eversion* occur between the tarsal bones and not at the ankle joint.

Anterior tibialis muscle originates from the upper end of the tibia, lies on the anterior surface of the leg and is inserted into the middle cuneiform bone by a long tendon. It is associated with dorsiflexion of the foot.

Soleus is one of the main muscles of the calf of the leg, lying immediately deep to the gastrocnemius. It originates from the heads and upper parts of the fibula and the tibia. Its tendon joins that of the gastrocnemius so that they have a common insertion into the calcaneus by the calcanean tendon. It causes plantarflexion at the ankle and helps to stabilise the joint when the individual is standing up.

Gastrocnemius (described above) is a powerful plantarflexor.

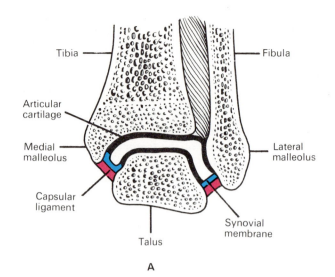

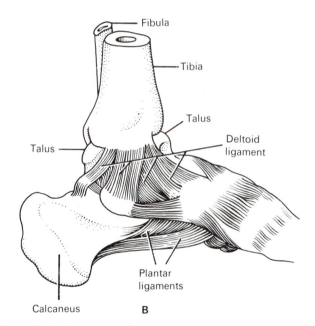

Figure 17:11   The left ankle joint. A. Section viewed from the front. B. Supporting ligaments. Medial view.

JOINTS OF THE FOOT AND TOES

There are a number of synovial joints between the tarsal bones, between the tarsal and metatarsal bones, between the metatarsals and proximal phalanges and between the phalanges. Movements are produced by muscles in the leg with long tendons which cross the ankle joint, and by muscles of the foot. The tendons crossing the ankle joint are encased in synovial sheaths and are held close to the bones by strong transverse ligaments. They move smoothly within their sheaths as the joints move. In addition to moving the joints of the foot these muscles support the arches of the foot and help to maintain body balance.

DISORDERS OF JOINTS

The tissues involved in diseases of the synovial joints are synovial membrane, hyaline cartilage and bone.

INFLAMMATORY DISEASES OF JOINTS (Arthritis)

RHEUMATOID ARTHRITIS (Sero-positive, non-infective, RA)

This is an autoimmune disease affecting synovial membrane. The cause is not clearly understood but it may be that development of the autoimmunity is initiated by microbial infection, possibly by viruses, in genetically susceptible people. Antigen/antibody complexes (*rheumatoid factors*) are formed and are present in the blood. They appear early in severe cases of sudden onset, and later when the disease develops gradually. RA is an acute febrile condition, usually with periods of remission of varying lengths of time. The joints most commonly affected are those of the hands and feet, but in severe cases most of the synovial joints may be involved. With each febrile exacerbation of the disease there is additional and cumulative damage to the joints, leading to increasing deformity, pain and loss of function. The primary changes that may be *reversible* include hypertrophy and hyperplasia of synovial cells and fibrinous inflammatory effusion into the joint. If the disease progresses there are further secondary changes which may be *irreversible*, including:

1. Erosion of articular cartilage and the growth of granulation tissue (*pannus*) that separates the bones and distorts the shape of the joint.
2. Fibrosis of pannus which causes adhesions between the bones, limiting movement
3. Ossification of the fibrosed pannus, further restricting joint movement
4. Spread of granulation tissue to tendons
5. Weakening and atrophy of muscles possibly due to limited exercise
6. Development of collagen rheumatoid nodules outside the joints, e.g., in pressure areas such as the elbows, over the knuckles and in the lungs, pleura, heart and eyes
7. Enlargement of lymph nodes and spleen

In the later stages of the disease the inflammation and fever are less marked and movement is limited by deformity of the joint, muscle weakness and pain.

OTHER TYPES OF POLYARTHRITIS

This group of autoimmune inflammatory arthritic diseases has many characteristics similar to rheumatoid arthritis but the rheumatoid factor is absent. The causes are not known but genetic features may be involved. The joints affected are mainly those of the axial skeleton. The diseases include:

1. Ankylosing spondylitis in which the sacroiliac and vertebral joints become ossified
2. Psoriatic arthritis that occurs in a high proportion of people who suffer from skin psoriasis
3. Reiter's syndrome (polyarthritis with urethritis and conjunctivitis) which, it is believed, may be precipitated by infection with *Chlamydia trachomatis*
4. Arthritis that complicates acute rheumatic fever

ACUTE INFECTIVE ARTHRITIS

Many commonly occurring microbes may be carried in the blood to the joints from foci of infection elsewhere in the body. In most cases the joint has been damaged by previous injury or arthritic disease. The outcome may be:

1. Resolution of the infection with no residual ill-effects
2. Suppuration followed by healing with the formation of fibrous tissue that may become ossified
3. Development of chronic infection, especially in brucellosis, gonorrhoea and tuberculosis

TRAUMATIC INJURY TO JOINTS

SPRAINS, STRAINS AND DISLOCATIONS

These damage the soft tissues, tendons and ligaments round the joint without penetrating the joint capsule. In dislocations there may be additional damage to intracapsular structures by stretching, e.g., long head of biceps muscle in the shoulder joint, cruciate ligaments in the knee joint, ligament of head of femur in the hip joint. If repair is incomplete there may be some loss of stability which increases the risk of repeated injury.

PENETRATING INJURIES

These may be caused by a compound fracture of one of the articulating bones, or trauma caused by, e.g., gun shot. Healing may be uneventful or it may be delayed by:

1. The presence in the joint of tissue fragments or sequestra too large to be removed by phagocytes
2. Incomplete healing of torn ligaments inside the capsule
3. Infection that may be blood-borne or enter through broken skin

When healing is incomplete there is a tendency for irreversible degenerative changes to occur.

OSTEOARTHRITIS (Osteoarthrosis, OA)

This is a degenerative non-inflammatory disease. Osteoarthrosis is the more appropriate name but is less commonly used. Articular cartilage gradually becomes thinner because its replacement does not keep pace with its removal. Eventually the bony articular surfaces come in contact and the bones begin to degenerate. There is abnormal bone repair and the articular surfaces become mis-shapen. Chronic inflammation develops with effusion into the joint, possibly due to irritation caused by tissue debris not removed by phagocytes. Sometimes there is abnormal outgrowth of cartilage at the edges of bones which becomes ossified, forming *osteophytes*.

PRIMARY OSTEOARTHRITIS

In this type the cause is unknown. Changes may be due to acceleration of the normal ageing process in joints that have had excessive use.

CERVICAL SPONDYLOSIS

Osteophytes may develop round the margins of joints of the vertebral column, commonly in the cervical region. They may cause damage to the nervous system, varying from compression of individual spinal nerves to spinal cord injury.

SECONDARY OSTEOARTHRITIS

This occurs in joints in which cartilage has already been damaged due to:
1. Congenital deformity of bones, e.g., in congenital dislocation of the hip there are unusual stresses in the joint
2. Trauma, e.g., intracapsular fracture of a bone, injury to intracapsular structures
3. Disease, e.g., inflammatory diseases, incomplete repair following injury, haemophilia with repeated haemorrhages into the joints, peripheral nerve lesions, diabetes mellitus

GOUT

This is caused by the deposition of sodium urate crystals in joints and tendons. Acute inflammation is due to chemotaxic substances released by phagocytes that have ingested the crystals. It occurs in some people whose blood uric acid is abnormally high due to either overproduction or defective excretion by the kidneys. Uric acid is a waste product of the breakdown of cell nuclei and is produced in excess when there is large-scale cell destruction, e.g., following trauma or treatment with cytotoxic drugs and in anaemia, starvation and malignancy. Defective excretion occurs in some kidney conditions. Episodes of arthritis lasting days or weeks are interspersed with periods of remission. After repeated acute attacks permanent damage may occur. The joints most commonly affected are the metatarsophalangeal joint of the big toe, the ankle, knee, wrist and elbow.

HAEMOPHILIC ARTHROPATHY

In haemophilia, slight injury may lead to haemorrhage into a joint. Repeated bleeding episodes may damage articular cartilage and lead to osteoarthritis.

The Muscular System

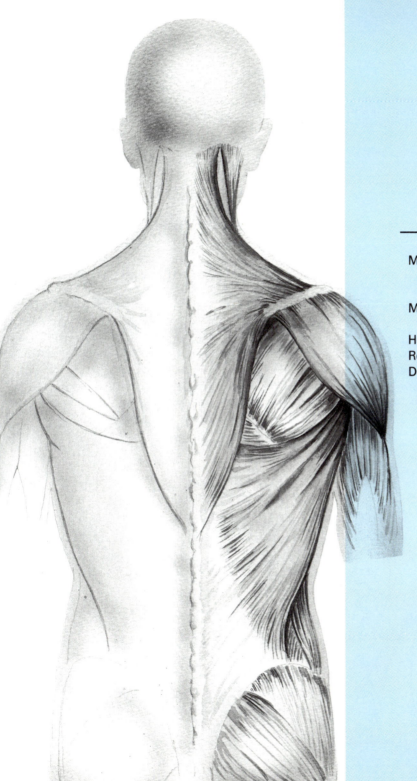

18. The Muscular System

The skeletal muscles described in this chapter are those not directly involved in the movements of the joints of the limbs.

MUSCLES OF THE FACE AND THE NECK
(Fig. 18:1)

MUSCLES OF THE FACE

There are a large number of muscles involved in changing facial expression and with movement of the lower jaw during chewing and speaking. Only the main muscles are described here. Except where indicated the muscles are present in pairs, one on each side.

Occipitofrontalis (unpaired) consists of a posterior muscular part over the occipital bone, an anterior part over the frontal bone and an extensive flat tendon or *aponeurosis* that stretches over the dome of the skull and joins the two muscular parts. It raises the eyebrows.

Levator palpebrae superioris extends from the posterior part of the orbital cavity to the upper eyelid. It raises the eyelid.

Orbicularis oculi surrounds the eye, eyelid and orbital cavity. It closes the eye and when strongly contracted 'screws up' the eyes.

Buccinator. This flat muscle of the cheek draws the cheeks in towards the teeth in chewing and in forcible expulsion of air from the mouth ('the trumpeter's muscle').

Orbicularis oris (unpaired) surrounds the mouth and blends with the muscles of the cheeks. It closes the lips and, when strongly contracted, shapes the mouth for whistling.

Masseter is a broad muscle, extending from the zygomatic arch to the angle of the jaw. In chewing it draws the mandible up to the maxilla and exerts considerable pressure on the food.

Temporalis covers the squamous part of the temporal bone. It passes behind the zygomatic arch to be inserted into the coronoid process of the mandible. It closes the mouth and assists with chewing.

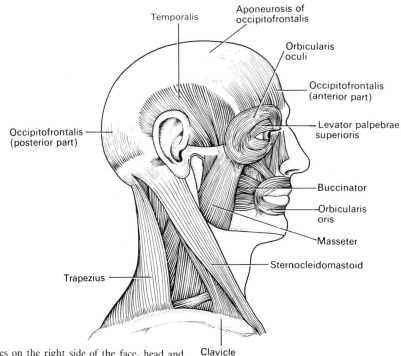

Figure 18:1   The main muscles on the right side of the face, head and neck.

338

Pterygoid muscle extends from the sphenoid bone to the mandible. It closes the mouth and pulls the lower jaw forward.

MUSCLES OF THE NECK

There are a great many muscles situated in the neck but only the two largest are considered here.

Sternocleidomastoid muscle arises from the manubrium of the sternum and the clavicle and extends upwards to the mastoid process of the temporal bone. It assists in turning the head from side to side. When the muscle on one side contracts it draws the head towards the shoulder. When both contract together they flex the cervical vertebrae and draw the sternum and clavicle upwards when the head is maintained in a fixed position, e.g., in forced respiration.

Trapezius muscle covers the shoulder and the back of the neck. The upper attachment is to the occipital protuberance; the medial attachment is to the transverse processes of the cervical and thoracic vertebrae; and the lateral attachment is to the clavicle and to the spinous and acromion processes of the scapula. It pulls the head backwards, squares the shoulders and controls the movements of the scapula when the shoulder joint is in use.

MUSCLES OF THE BACK (Fig. 18:2)

There are six pairs of large muscles in the back in addition to those that form the posterior abdominal wall. The arrangement of these muscles is the same on each side of the vertebral column. They are:

Trapezius
Teres major — described in Chapter 17 in association with joints
Psoas
Quadratus lumborum
Sacrospinalis
Latissimus dorsi

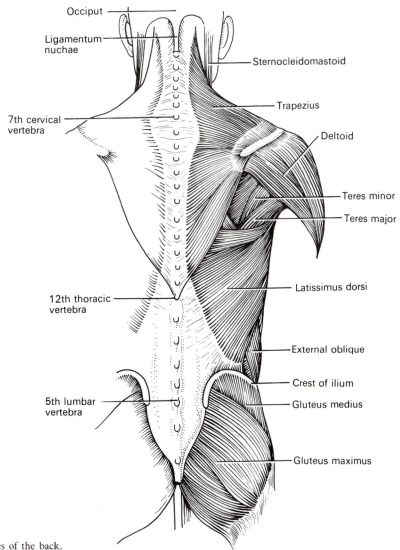

Figure 18:2 The main muscles of the back.

Quadratus lumborum originates from the crest of the ilium then it passes upwards, parallel and close to the vertebral column and is inserted into the 12th rib. Together the two muscles fix the lower rib during respiration and cause extension of the vertebral column (bending backwards). If one muscle contracts it causes lateral flexion of the lumbar region of the vertebral column.

Sacrospinalis is a group of muscles, lying between the spinous and transverse processes of the vertebrae. They originate from the sacrum and are finally inserted into the occipital bone. Their contraction causes extension of the vertebral column.

MUSCLES OF THE ABDOMINAL WALL
(Figs 18:3 and 18:4)

ANTERIOR AND LATERAL PARTS

There are four pairs of muscles, arranged in four layers, that make up the anterior and lateral parts of the abdominal wall. From the surface inwards they are:

Rectus abdominis
External oblique
Internal oblique
Transversus abdominis

The anterior abdominal wall is divided longitudinally by a very strong midline tendinous cord, the *linea alba*, which extends from the xiphoid process of the sternum to the symphysis pubis. The structure of the abdominal wall on each side of the linea alba is identical.

Rectus abdominis is the most superficial muscle. It is broad and flat, originating from the transverse part of the pubic bone then passing upwards to be inserted into the lower ribs and the xiphoid process of the sternum. Medially the two muscles are attached to the linea alba.

External oblique extends from the lower ribs *downwards and forward* to be inserted into the iliac crest and, by an aponeurosis, to the linea alba.

Internal oblique lies deep to the external oblique. Its fibres arise from the crest of the ilium and by a broad band of fascia from the spinous processes of the lumbar vertebrae. The fibres pass *upwards towards the midline* to be inserted into the lower ribs and, by an aponeurosis, into

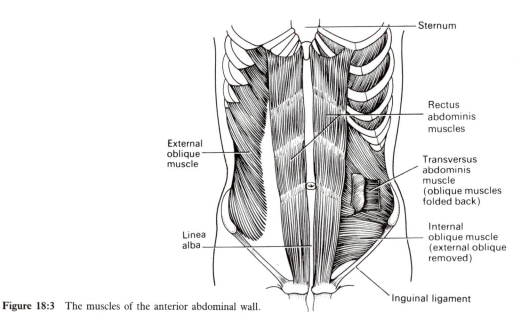

Figure 18:3 The muscles of the anterior abdominal wall.

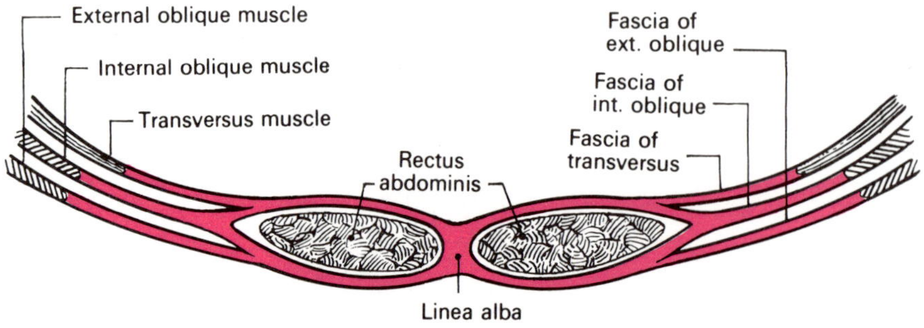

Figure 18:4 Diagram of the arrangement of the fascia of the muscles of the anterior abdominal wall. Cross section.

the linea alba. The fibres are at right angles to those of the external oblique.

Transversus abdominis is the deepest muscle of the abdominal wall. The fibres arise from the iliac crest and the lumbar vertebrae and pass across the abdominal wall to be inserted into the linea alba by an aponeurosis. The fibres are at right angles to those of the rectus abdominis.

Functions

The main function of the four pairs of muscles is to form the strong muscular anterior wall of the abdominal cavity. When the muscles contract together they:

1. Compress the abdominal organs
2. Flex the vertebral column in the lumbar region

Contraction of the muscles on one side only bends the trunk towards that side. Contraction of the oblique muscles on one side rotates the trunk.

INGUINAL CANAL

This canal is 2.5 to 4 cm long and passes obliquely through the abdominal wall. It runs parallel to and immediately in front of the transversalis fascia and part of the inguinal ligament. In the male it contains the *spermatic cord* and in the female, the *round ligament*. It constitutes a weak point in the otherwise strong abdominal wall through which herniation may occur (see p. 175).

POSTERIOR ABDOMINAL WALL

The muscles of the posterior part of the abdominal wall have been described already (see Figs 18:5 and 18:6). They are:

Quadratus lumborum
Psoas
Internal oblique
Transversus abdominis

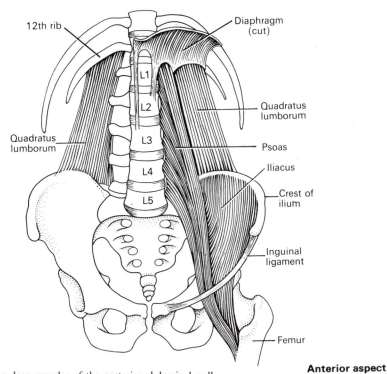

Figure 18:5   The deep muscles of the posterior abdominal wall.

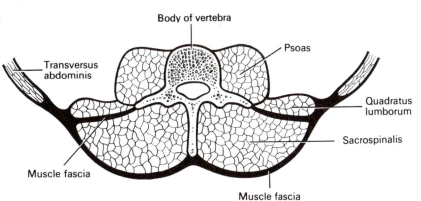

Figure 18:6   Transverse section of a lumbar vertebra and its associated muscles.

MUSCLES OF THE PELVIC FLOOR
(Fig. 18:7)

The pelvic floor is divided into two identical parts at the midline. Each half is made up of muscles and fascia which unite in the midline. The muscles are:

Levator ani

Coccygeus

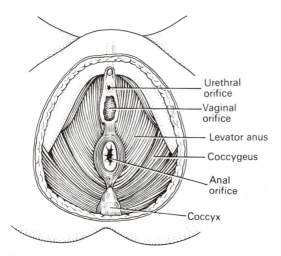

Figure 18:7   The muscles of the pelvic floor.

Levator ani are broad flat muscles, forming the anterior part of the pelvic floor. They originate from the inner surface of the true pelvis and unite in the midline. Together they form a sling which supports the pelvic organs.

Coccygeus are triangular sheets of muscle and tendinous fibres situated behind the levator ani. They originate from the medial surface of the ischium and are inserted into the sacrum and coccyx. They complete the formation of the pelvic floor which is perforated in the male by the urethra and anus, and in the female by the urethra, vagina and anus.

MUSCLES OF RESPIRATION

The muscles of respiration have been described previously (Ch. 7). They are:

External intercostal	11 pairs
Internal intercostal	11 pairs
Diaphragm	1

HEALING OF MUSCLE FIBRES

Muscle fibres may be damaged accidentally or be cut during surgery. The extent of the damage determines the mode and effectiveness of healing. In all cases, damaged tissue is removed by phagocytosis and replaced by granulation tissue.

In *slight injury* the small gap in the muscle fibre is bridged by outgrowths from the surviving ends of the fibre, completely restoring its integrity.

In *more extensive injury* the muscle fibre outgrowths may not be able to extend far enough into the granulation tissue to restore it completely. When this happens the remaining granulation tissue becomes fibrosed and scar tissue forms. In time this contracts and may restrict joint movement.

In *very extensive injury* repair is by fibrosis. In some cases, such as following the crushing of a limb, there may be systemic effects (*crush syndrome*). Sustained pressure, e.g., on a limb, causes ischaemia resulting in massive muscle necrosis. When pressure is relieved and the circulation restored, myoglobin and other necrotic products enter the blood and pass to the kidneys. If there is too much of this material for the kidneys to cope with, death may result from acute kidney failure. A common complication of this type of injury is infection, especially by anaerobic microbes e.g., *Clostridium perfringens* (*Cl. welchii*) and other clostridia, causing *gas gangrene*.

REPAIR OF NERVES SUPPLYING MUSCLES

A *motor unit* consists of a lower motor neurone (LMN) and the muscle fibres it supplies. When the nerve supply is cut the muscle ceases to contract and gradually atrophies but, if the nerve regenerates, muscle fibre function is restored. The axon of the LMN in a peripheral nerve divides into numerous terminal branches each of which supplies a muscle fibre. In a bundle of muscle fibres the nerve supply is derived from several LMNs. Nerve supply to muscle fibres may be restored by:

1. Regeneration of the nerve if cut near the parent cell and if the cut ends are in close apposition (Fig. 18:8).
2. By the outgrowth of new terminal nerve fibres from the other axons supplying adjacent muscle fibres in the same bundle (Fig. 18:9)

DISEASES OF MUSCLES

MYASTHENIA GRAVIS

This is an autoimmune disease of unknown origin in which there is defective muscle stimulation. Antibodies develop which damage the acetylcholine receptors of neuromuscular junctions, blocking the transmission of the nerve impulses to muscle fibres. This causes progressive and

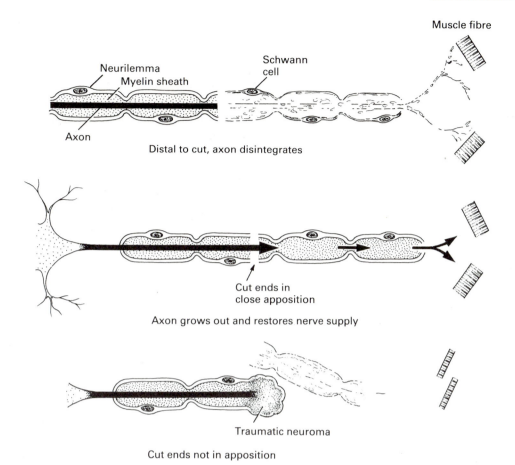

Figure 18.8 Regrowth of a peripheral nerve.

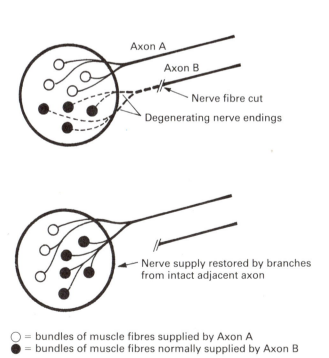

○ = bundles of muscle fibres supplied by Axon A
● = bundles of muscle fibres normally supplied by Axon B

Figure 18:9   Restoration of nerve supply to muscle.

extensive muscle weakness. Extraocular and eyelid muscles are affected first, followed by those of the neck and limbs. There are periods of remission, relapses being precipitated by, e.g., severe muscular exercise, infections, emotional disturbances, pregnancy.

MYOPATHIES

PROGRESSIVE MUSCULAR DYSTROPHIES

In this group of inherited diseases there is progressive degeneration of groups of muscles. The main differences in the types are the:
1. Age of onset
2. Rate of progression
3. Groups of muscles involved

DUCHENNE TYPE

The muscle abnormality is present before birth but may not be evident until the child is about 5 years of age. Wasting and weakness begin in muscles of the lower limbs then spread to the upper limbs, progressing without remission. Death usually occurs in adolescence, often from infection.

FACIO-SCAPULO-HUMERAL DYSTROPHY

This disease usually begins in adolescence and the younger the age of onset the more rapidly it progresses. Muscles of the face and upper limbs are affected first. This is a chronic condition that usually progresses slowly and may not cause complete disability.

MYOTONIC DYSTROPHY

This disease usually begins in adult life. Muscles contract and relax slowly, often seen as difficulty in releasing an object held in the hand. Muscles of the tongue and the face are first affected then muscles of the limbs. Other conditions associated with myotonic dystrophy include:

1. Premature cataract
2. Atrophy of the gonads
3. Defects in the conducting system of the heart
4. Endocrine disturbances, e.g., in insulin utilisation

The disease progresses without remission and with increasing disability. Death usually occurs in middle age.

Bibliography

Anderson J E 1983 Grant's atlas of anatomy, 8th edn. Williams and Wilkins, Baltimore

Anderson J R (ed) 1980 Muir's textbook of pathology, 11th edn. Edward Arnold, London

Anderson W A D, Scotti T M 1980 Synopsis of pathology, 10th edn. Mosby, St Louis

Bell G H, Emslie-Smith D, Paterson C R 1980 Textbook of physiology, 10th edn. Churchill Livingstone, Edinburgh

Brobeck J R (ed) 1979 Best and Taylor's physiological basis of medical practice, 10th edn. Williams and Wilkins, Baltimore

Cantarow A, Schepartz B 1967 Biochemistry, 4th edn. Saunders, Philadelphia

Emery A E H 1983 Elements of medical genetics, 6th edn. Churchill Livingstone, Edinburgh

Govan A D T, Macfarlane P S, Callander R 1981 Pathology illustrated. Churchill Livingstone, Edinburgh

Guyton A C 1977 Basic human physiology, 2nd edn. Saunders, Philadelphia

Ham A W 1979 Histology, 8th edn. Pitman, London; Lippincott, Philadelphia

Hare R, Cooke E M 1984 Bacteriology and immunity for nurses, 6th edn. Churchill Livingstone, Edinburgh

Lippold O C J, Winton F R (eds) 1979 Human physiology, 7th edn. Churchill Livingstone, Edinburgh

McGilvery R W, Goldstein G W 1983 Biochemistry: a functional approach, 3rd edn. Saunders, Philadelphia

Macleod J (ed) 1984 Davidson's principles and practice of medicine, 14th edn. Churchill Livingstone, Edinburgh

McNaught A B, Callander R 1983 Illustrated physiology, 4th edn. Churchill Livingstone, Edinburgh

Merck manual of diagnosis and therapy 1982 14th edn

Robbins S L, Cotran R S 1984 Pathologic basis of disease, 3 rd edn. Saunders, Philadelphia

Sharp D W A 1983 The Penguin dictionary of chemistry. Penguin, Harmondsworth

Turk D C et al 1983 A short textbook of medical microbiology, 5th edn. Hodder & Stoughton, London

Uvarov et al 1986 The Penguin dictionary of science. Penguin, Harmondsworth

Walmsley R, Murphy T R 1972 Jamieson's illustrations of regional anatomy (series), 9th edn. Churchill Livingstone, Edinburgh

Walter J B, Israel M S 1979 General pathology, 5th edn. Churchill Livingstone, Edinburgh

Warwick R, Williams P L (eds) 1980 Gray's anatomy, 36th edn. Longman

Weir D M 1983 Immunology, 5th edn. Churchill Livingstone, Edinburgh

Index

Page numbers referring to anatomy and physiology are in ordinary type. Those referring to diseases are in italics.